On the Origin (and Evolution) of Baryonic Galaxy Halos

Special Issue Editors

Duncan A. Forbes
Ericson D. Lopez

MDPI • Basel • Beijing • Wuhan • Barcelona • Belgrade

MDPI

Special Issue Editors
Duncan A. Forbes
Swinburne University
Australia

Ericson D. Lopez
Quito Astronomical Observatory of National Polytechnic School
Ecuador

Editorial Office
MDPI AG
St. Alban-Anlage 66
Basel, Switzerland

This edition is a reprint of the Special Issue published online in the open access journal *Galaxies* (ISSN 2075-4434) in 2017 (available at: http://www.mdpi.com/journal/galaxies/special_issues/baryonic_galaxy_halos).

For citation purposes, cite each article independently as indicated on the article page online and as indicated below:

Lastname, F. M.; Lastname, F. M. Article title. *Journal Name* **Year**, *Article number*, page range.

First Edition 2018

ISBN 978-3-03842-722-3 (Pbk)
ISBN 978-3-03842-721-6 (PDF)

Cover photo courtesy of Duncan A. Forbes and Ericson D. Lopez

Table of Contents

About the Special Issue Editors

Duncan A. Forbes is a Professor at Swinburne University, Australia, where he has been a faculty member since 2000. Professor Forbes completed his PhD. at the University of Cambridge, UK, and was a Postdoctoral Researcher at the University of California, Santa Cruz. He was a senior lecturer at the University of Birmingham. His research interests are in galaxy formation and globular cluster systems.

Ericson D. Lopez obtained his PhD. at Main Astronomical Observatory of Russian Academy of Science, Saint Petersburg, Russia. Afterwards, he obtained a postdoctoral researcher position at the Department of Astronomy of Sao Paulo University, Sao Paulo, Brazil, and another postdoctoral re-searcher position at the Space Telescope Science Institute, Baltimore, USA. Since 1996, he has been the Director of the Quito Astronomical Observatory and Professor at National Polytechnic School, Quito, Ecuador. His research interests are in high-energy astrophysics, AGNs, blazars, relativistic jets, relativistic astrophysics, and cosmology.

galaxies

MDPI

Editorial

A Conference on the Origin (and Evolution) of Baryonic Galaxy Halos

Duncan Forbes [1,*] and Ericson Lopez [2]

1   Centre for Astrophysics and Supercomputing, Swinburne University, Hawthorn VIC 3122, Australia
2   Observatorio Astronomico de Quito, Escuela Politecnica Nacional, Quito 17-01-165, Ecuador;
    ericsson.lopez@epn.edu.ec
*   Correspondence: dforbes@swin.edu.au

Academic Editor: Emilio Elizalde
Received: 10 May 2017; Accepted: 10 May 2017; Published: 17 May 2017

Abstract: A conference was held in March 2017 in the Galapagos Islands on the topic of The Origin (and Evolution) of Baryonic Galaxy Halos. It attracted some 120 researchers from around the world. They presented 68 talks (nine of which were invited) and 30 posters over five days. A novel element of the talk schedule was that participants were asked which talks they wanted to hear and the schedule was made up based on their votes and those of the Scientific Organizing Committee SOC . The final talk schedule had 34% of the talks given by women. An emphasis was given to discussion time directly after each talk. Combined with limited/no access to the internet, this resulted in high level of engagement and lively discussions. A prize was given to the poster voted the best by participants. A free afternoon included organized excursions to see the local scenery and wildlife of the Galapagos (e.g., the giant tortoises). Four public talks were given, in Spanish, for the local residents of the town. A post-conference survey was conducted, with most participants agreeing that the conference met their scientific needs and helped to initiate new research directions. Although it was challenging to organize such a large international meeting in such an isolated location as the Galapagos Islands (and much credit goes to the Local Organizing Committee LOC and staff of Quito Astronomical Observatory for their logistical efforts, organizing the meeting for over a year), it was very much a successful conference. We hope it will play a small part in further developing astronomy in Ecuador.

Keywords: galaxies; formation; evolution; halos

1. Conference Introduction

Galaxy halos provide important clues to the origin and evolution of galaxies. We decided that the time was ripe to bring together the latest simulations and deep observations of galaxy halos in a conference focusing on the baryonic (star and gas) component of halos. In particular, deep, wide, and detailed observations of galaxy halos beyond the Local Group are becoming more ubiquitous. Simulations that incorporate realistic baryonic physics in a cosmological context have also made significant progress in recent years in modelling galaxy halos. These simulations predict outer halo regions that differ strongly in their formation processes and properties from the well-studied inner regions of galaxies. Halos have long dynamical times, and as such preserve the unique signatures of galaxy assembly.

Where better to focus on the Origin (and Evolution) of Baryonic Galaxy Halos than the Galapagos Islands? The islands were of course the inspiration for Charles Darwin's discovery on the origin and evolution of species. The islands are relatively cheap compared to some USA and European locations, and have the necessary infrastructure to guarantee a successful meeting. The key to a successful conference is the people and a conference structure that facilities presentations and discussion. In order to encourage the attendance of Ph.D. students, we completely waived the registration fee for students.

We were very pleased to attract 120 researchers from all over the globe (i.e., Africa, Asia, Europe, North America, South America (including a dozen from Ecuador), and Australia). A photograph of the conference participants is shown in Figure 1. The conference included an official dinner and wildlife excursions for participants and their guests.

Another key element of choosing the Galapagos was to help promote Ecuadorian astronomy and the wider region. The conference received considerable media coverage, including television in Quito. We hope that the meeting provides a long-lasting legacy for astronomy in Ecuador. In an effort to bring astronomy to the Islands, two nights of public talks on astronomy were given in Spanish by Ricardo Salinas, Ericson Lopez, Wladimir Banda, and Carlos Frenk.

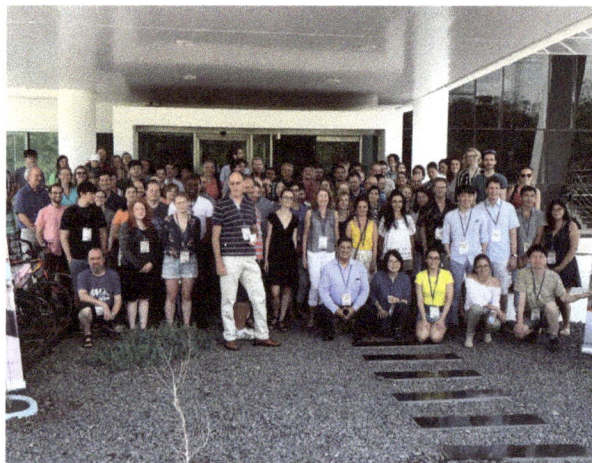

Figure 1. Conference photograph showing participants outside of the conference venue, the CIER (Centro de Informacion de Energia Renovable) building.

2. Voting for Talks

At the suggestion of SOC member Michael Merrifield, we tried an experiment to get participants more involved. Using Survey Monkey, we asked everyone registered to vote for their top 50 talks (including their own!). Some 2/3 of participants voted. The most voted for talk (the "people's choice") was by Evan Skillman. Combined with the votes from the SOC members, a draft schedule was made up. This schedule required only minor tweaking to ensure a balanced program. The final schedule consisted of 68 talks, with 23 by women (34%, which also reflects the overall attendance fraction). Several talk slots were allocated to Ph.D. students. Overall, voting for the talk schedule received positive feedback from participants.

3. Conference Structure

The conference was held at the new Centro de Informacion de Energia Renovable (CIER)—the renewable energy center, part of the electricity company of Galapagos (Elecgalapagos)—in Puerto Ayora on the island of Santa Cruz in the Galapagos. A reception was held on the Sunday night, with the conference starting at 9 a.m. on Monday 13 March 2017. The conference was opened by Dr. Alberto Celi, the Vice-Rector for research and innovation at the Escuela Politecnica Nacional (EPN) University in Quito (host of the conference through its Quito Astronomical Observatory). A short welcome speech was also given by a representative from the local Ministry of Tourism. The basic format was invited talks of 25 min plus 10 min discussion, long talks 20 min plus 5 min, and short talks of 10 min plus 5 min. Thus, a key emphasis was placed on having plenty of time for discussion immediately after each talk. Indeed a notable aspect of the conference was the lively discussion throughout the meeting, which extended

Galaxies **2017**, *5*, 23

to the coffee breaks. Internet was generally unavailable during the conference (satellite service is very expensive on the islands), but many participants felt this contributed to a more engaged audience and lively discussion. Some 30 posters were also presented. A couple of coffee breaks were dedicated to poster time and a chance to chat with the author. Voting forms were given out to everyone, and the best poster was determined. A small prize was given to Gwendolyn Eadie at the end of the conference for her poster. Thus, dedicated coffee breaks and poster voting ensured engagement with the posters.

4. Post-Conference

Many participants chose to stay on after the conference and see the sights in the islands and/or in Quito. We followed up the conference by requesting PDF files of each talk so that participants and those unable to attend the conference can see the talks in detail at the conference homepage. In the longer-term, a conference proceedings book will be published. We also conducted a short survey to gauge the success of the conference with participants. When asked about the statement "The Galapagos galaxy halos conference meet my scientific needs", over 90% said they Agreed or Strongly Agreed. Similarly, to the statement "The galaxy halos conference helped me to initiate new research project or direction", over 90% Agreed or Strongly Agreed.

The articles that follow in this Special Issue include some of the talks and posters given at the conference. We hope that this forms a useful collection of works that reflects the current state-of-the-art on the Origin and Evolution of Baryonic Halos and promote the development of astronomical research in Ecuador and its neighboring countries.

Acknowledgments: We thank Escuela Politecnica Nacional, Centro de Educacion Continua (CEC-EPN), Parque Nacional Galapagos, and Centro de Informacion de Energia Renovable for their support. We would like to thank the Scientific Organising Committee: Jean Brodie, Carlos Frenk, Lars Hernquist, Claudia Mendes de Oliveira, Michael Merrifield, Marina Rejkuba and Aaron Romanowsky. We also thank the Local Organising Committee: Jairo Armijos, Wladimir Banda and Crispin Logan. Finally, a special thanks to the staff from Quito Astronomical Observatory for their logistical support that made the conference a great success.

Conflicts of Interest: The authors declare no conflict of interest.

![galaxies logo] *galaxies*

MDPI

Article

A Photometric Study of Giant Ellipticals and Their Stellar Halos With VST

Marilena Spavone [1,*], Massimo Capaccioli [1,2], Nicola R. Napolitano [1], Enrichetta Iodice [1], Aniello Grado [1], Luca Limatola [1], Andrew P. Cooper [3], Michele Cantiello [4], Duncan A. Forbes [5], Maurizio Paolillo [2] and Pietro Schipani [1]

[1] INAF-Astronomical Observatory of Capodimonte, Salita Moiariello 16, I80131 Naples, Italy; capaccioli@na.infn.it (M.C.); napolita@na.astro.it (N.R.N.); iodice@na.astro.it (E.I.); agrado@na.astro.it (A.G.); limatola@na.astro.it (L.L.); schipani@na.astro.it (P.S.)

[2] Department of physics, University of Naples Federico II, C.U. Monte Sant'Angelo, Via Cinthia, 80126 Naples, Italy; paolillo@na.infn.it

[3] Institute for Computational Cosmology, Durham DH1 3, UK; a.p.cooper@durham.ac.uk

[4] INAF-Astronomical Observatory of Teramo, Via Maggini, 64100 Teramo, Italy; cantiell@oa-teramo.inaf.it

[5] Centre for Astrophysics & Supercomputing, Swinburne University, Hawthorn 3122, Australia; dforbes@swin.edu.au

* Correspondence: spavone@na.astro.it

Academic Editor: Emilio Elizalde
Received: 15 June 2017 ; Accepted: 20 July 2017 ; Published: 26 July 2017

Abstract: Observations of diffuse starlight in the outskirts of galaxies are thought to be a fundamental source of constraints on the cosmological context of galaxy assembly in the ΛCDM model. Such observations are not trivial because of the extreme faintness of such regions. In this work, we investigated the photometric properties of six massive early-type galaxies (ETGs) in the VST Elliptical GAlaxies Survey (VEGAS) sample (NGC 1399, NGC 3923, NGC 4365, NGC 4472, NGC 5044, and NGC 5846) out to extremely low surface brightness levels with the goal of characterizing the global structure of their light profiles for comparison to state-of-the-art galaxy formation models. We carried out deep and detailed photometric mapping of our ETG sample taking advantage of deep imaging with VST/OmegaCAM in the g and i bands. By fitting the light profiles, and comparing the results to simulations of elliptical galaxy assembly, we have identified signatures of a transition between relaxed and unrelaxed accreted components and can constrain the balance between in situ and accreted stars. The very good agreement of our results with predictions from theoretical simulations demonstrates that the full VEGAS sample of $\sim$ 100 ETGs will allow us to use the distribution of diffuse light as a robust statistical probe of the hierarchical assembly of massive galaxies.

Keywords: techniques: image processing; galaxies: elliptical and lenticular, cD; galaxies: fundamental parameters; galaxies: formation; galaxies: halos

---

1. Introduction

Theories of galaxy formation within the currently accepted Λ Cold Dark Matter cosmological paradigm predict that galaxies grow through a combination of in situ star formation and accretion of stars from other galaxies [1]. The ratio of stellar mass contributed by these two modes of growth is expected to change systematically over the lifetime of a galaxy as its dark matter halo and star formation efficiency evolve (e.g., [2]). Accreted stars are expected to dominate in the outer parts of galaxies because they have much lower binding energies in the host galaxy than stars formed by dissipative collapse. Since dynamical timescales are long in these outer regions, phase-space substructures related to accretion, such as streams and caustics, can persist over many gigayears.

The structural properties of the outer parts of galaxies and their correlations with stellar mass and other observables might therefore provide ways of testing theoretical predictions of growth by accretion. In this paper we use extremely deep images of six massive early-type galaxies (ETGs) from the VEGAS survey (described below) to constrain the properties of their accreted stellar components.

In ETGs the connections between different mechanisms of mass growth and the 'structural components' inferred from images are not straightforward. If the bulk of the stars are really accreted, then the accreted component (or 'spheroid' or 'classical bulge') should be identified with at least the structural component that dominates the observed stellar mass. However, other empirical 'components' might also be accreted. In situ stars in ETGs are extremely difficult to distinguish if the also follow a spheroidal, dispersion supported spatial distribution and have old, metal-rich stellar populations resembling those of the dominant accreted component(s) with which they have been thoroughly mixed by violent relaxation.

Cosmological dynamical simulations can help by suggesting plausible interpretations for features in the surface brightness profiles of ETGs in the context of specific galaxy formation theories. In particular, simulated galaxies show evidence of substructure in the form of inflections ('breaks'), at which the surface brightness profile either becomes steeper or shallower (e.g., [3,4]). These inflections also correspond to variations in the ratio between individual accreted components as a function of radius [5–7].

Using different techniques with observations of different depths, several authors have concluded that the profiles of massive ETGs are not well described by a single Sérsic $r^{1/n}$ law component, once thought to be near universal for spheroidal galaxies. Taking advantage of the wide field of view and high spatial resolution of the VLT Survey Telescope (VST; [8]) at the ESO Cerro Paranal Observatory (Chile), we carried out deep and detailed photometric mapping of six massive early-type galaxies (ETGs) in the VEGAS sample.

2. The VEGAS Survey

The VST Elliptical GAlaxies Survey (VEGAS, [9]) is a deep multi-band (g, r, i) imaging survey of early-type galaxies in the southern hemisphere carried out with VST at the ESO Cerro Paranal Observatory (Chile). The large field of view (FOV) of the OmegaCAM mounted on VST (one square degree matched by pixels 0.21 arcsec wide), together with its high efficiency and spatial resolution allows us to map with a reasonable integration time the surface brightness of a galaxy out to isophotes encircling about 95% of the total light.

The expected depths at a signal-to-noise ratio (S/N) of >3 in the g, r, and i bands are 31, 28, and 27 mag/arcsec^2, respectively. The main science goals of the VEGAS survey are: (1) to study the 2D light distribution out to at least ~10 effective radii, R_e, focusing on the galaxy structural parameters and the diffuse light component, highlighting the presence of inner substructures as a signature of recent cannibalism events and/or inner discs and bars fuelling the active nucleus that is present in almost all objects of our sample; (2) to map the surface brightness profile and isophote geometry out to 10 R_e or more; (3) to analyse the colour gradients and their connection with galaxy formation theories, also taking advantage of stellar population synthesis techniques; (4) to study the external low surface brightness structures of the galaxies and the connection with the environment; (5) to make a census of small stellar systems (GCs, ultra-compact dwarfs and galaxy satellites) out to ~20 Re from the main galaxy center, and their photometric properties (e.g., GC luminosity function and colors, and their radial changes out to several Re), allowing us to study the properties of GCs in the outermost "fossil" regions of the host galaxy.

The data used in this work consist of exposures in g and i SDSS bands obtained with VST + OmegaCAM, both in service and visitor mode, for six giant ETGs: NGC 3923, NGC 4365, NGC 5044 and NGC 5846, and those of NGC 4472 and NGC 1399 (published by Capaccioli et al. [9] and Iodice et al. [10], respectively). More details about the observing strategy can be found in Spavone et al. [11] and in Section 5.

3. Fitting the Light Distribution

Since there is considerable evidence in the literature that the light profiles of many of the most massive ETGs are not well fitted by a single Sérsic law and at least one additional component is needed [12,13], our analysis focusses on the fit of projected one-dimensional (ellipsoidally averaged) surface brightness profiles of our sample galaxies (see Section 5 for details).

We adopt an empirically motivated, two-component approach most common in the literature, as well as an alternative approach, which is motivated by the predictions of numerical simulations, in which we fitted the surface brightness profiles of our galaxies with three components: two dominant Sérsic components and an outer exponential component.

Theoretical models suggest that massive ETGs accumulate the bulk of their stellar mass by accretion. For this reason, the accreted component in these galaxies should be identified with the component dominating the stellar mass. From an observational point of view, it is not straightforward to separate the in situ and the accreted component in ETGs, since they have similar physical properties and are well mixed together. The overall profile is comprised of different contributions and for this reason theory suggests that the surface brightness profile of ETGs should be described by the superposition of different components.

3.1. Two Components Fits

We first present models of the surface brightness profiles of galaxies in our sample with a double Sérsic law [14,15],

$$\mu(R) = \mu_e + k(n) \left[\left(\frac{R}{r_e} \right)^{1/n} - 1 \right], \tag{1}$$

where $k(n) = 2.17n - 0.355$, R is the galactocentric radius, and r_e and μ_e are the effective radius and surface brightness. We found that this model converges to a best-fit solution with a physically meaningful value for only two galaxies, NGC5044 and NGC 5846. For the cases in which our double-Sérsic fit did not converge, we imposed an exponential profile ($n = 1$) on the outer component, given by the equation

$$\mu(R) = \mu_0 + 1.086 \times R/r_h, \tag{2}$$

where μ_0 and r_h are the central surface brightness and exponential scale length, respectively. The result of these fits and their residuals are shown in Figure 1.

We found that the inner components of each fit have effective radii $r_e \sim$5–25 kpc (45–202 arcsec), with an average value of $r_e \sim$12 kpc, and Sérsic indices $n \sim$3–6, with an average value of $n \sim$4.3. These values are consistent with those reported by Gonzalez et al. [16] and Donzelli et al. [12], who for their samples of BCGs found $r_e \sim$5–15 kpc and $n \sim$4.4.

The relative contribution of the outer halo with respect to the total galaxy light (f_h) estimated from these our two component fits, ranges between 27% to 64%. Since there is no clear reason to believe that in massive elliptical galaxies the outer component in a fit such as this accounts for most of the accreted mass, these halo mass fractions should be considered a lower limit for the total accreted mass.

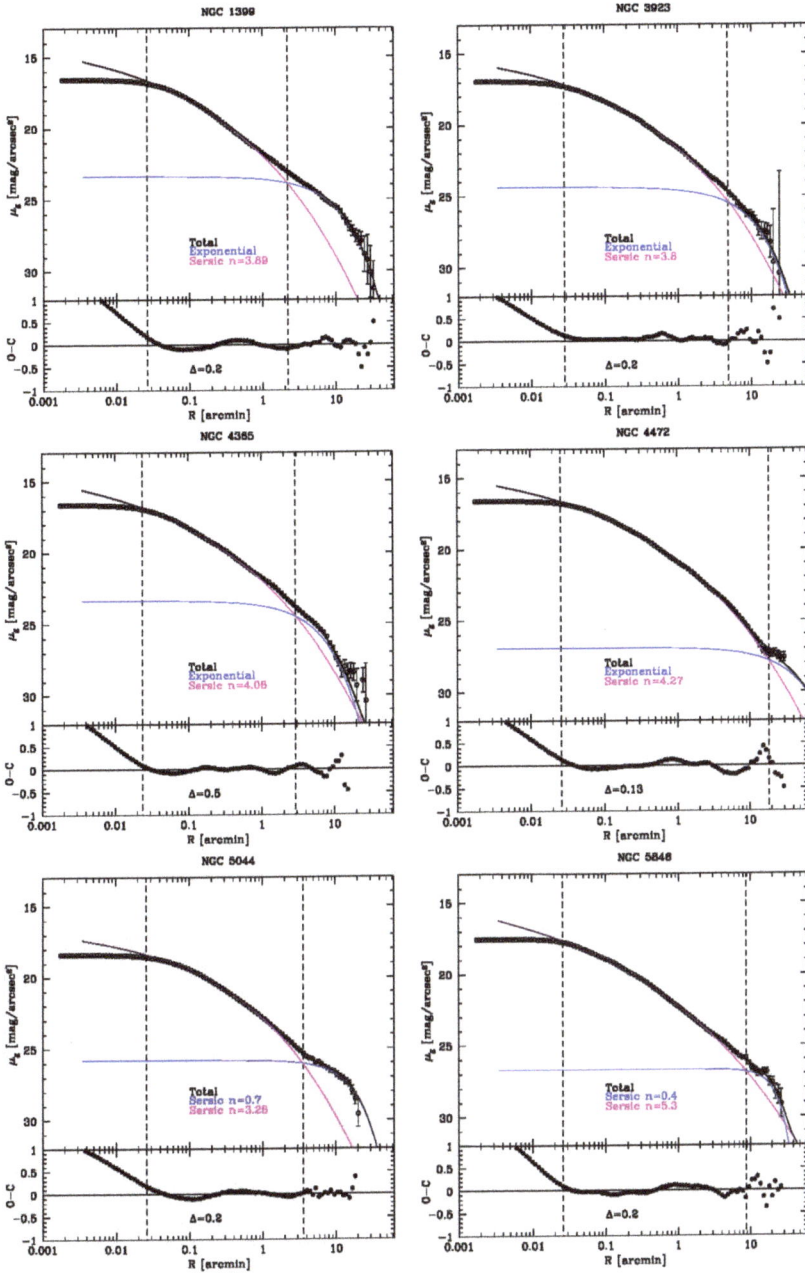

Figure 1. VST g-band profiles of NGC 1399, NGC 3923, NGC 4365, NGC 4472, NGC 5044, and NGC 5846 plotted on a logarithmic scale. The blue line is a fit to the outer regions with an exponential component, for NGC 1399, NGC 3923, NGC 4365, and NGC 4472, and with a Sérsic component for NGC 5044 and NGC 5846. The magenta line is a fit to the inner regions with a Sérsic profile, and the black line is the sum of the components in each fit. The dashed lines indicate the core of the galaxy ($1.5 \times FWHM$), which was excluded in the fit, and the transition point between the two components, respectively.

3.2. Three Components Fits

Numerical simulations predict that stars accreted by BCGs account for most of the total galaxy stellar mass (~90% on average), while in situ stars significantly contribute to the surface brightness profile only out to R ~10 kpc [3,4,17]. The overall accreted profile is built up by contributions from several significant progenitors. For this reason, theory suggests that the surface brightness profile of an ETG should be well described by the superposition of an inner Sérsic profile representing the (sub-dominant) in situ component in the central regions, another Sérsic profile representing the (dominant) superposition of the relaxed, phase-mixed accreted components, and an outer diffuse component representing unrelaxed accreted material (streams and other coherent concentrations of debris), which does not contribute any significant surface density to the brighter regions of the galaxy.

Following these theoretical predictions, we described the surface brightness profiles of our six galaxies with a three-component model: a Sérsic profile for the centrally concentrated in situ stars, a second Sérsic for the relaxed accreted component, and an exponential component for the diffuse and unrelaxed outer envelope. To mitigate the degeneracy in parameters and provide estimates of accreted components that are closely comparable to the results of numerical simulations, we fixed n ~2 for the in situ component of our three-component fits [3]. This value has been chosen because it is a representative value from the simulations. The results of these fits are shown in Figure 2. Looking at the rms scatter Δ, of each fit, we can clearly see that by adding the third component we achieve an improvement of at least 10% for each galaxy.

From this plot it appears that, as argued by Cooper et al. [17], the radius R_{tr} identified in Figure 1 marks the transition between different accreted components in different states of dynamical relaxation, rather than that between in situ and accreted stars.

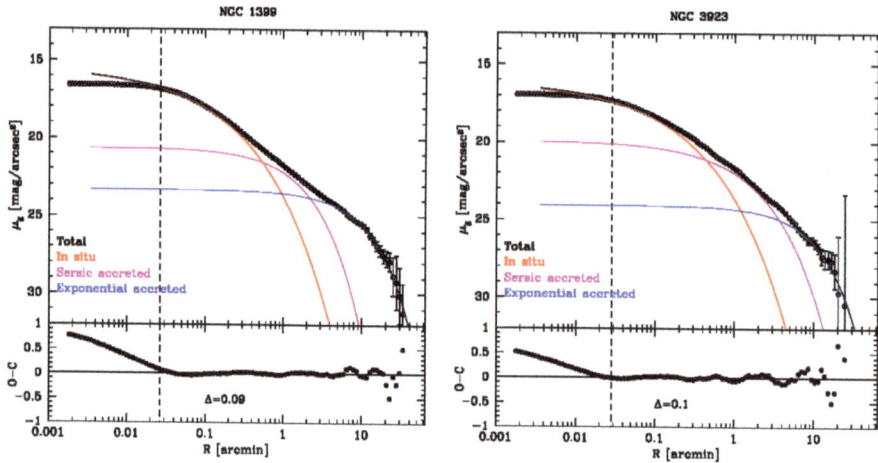

Figure 2. *Cont.*

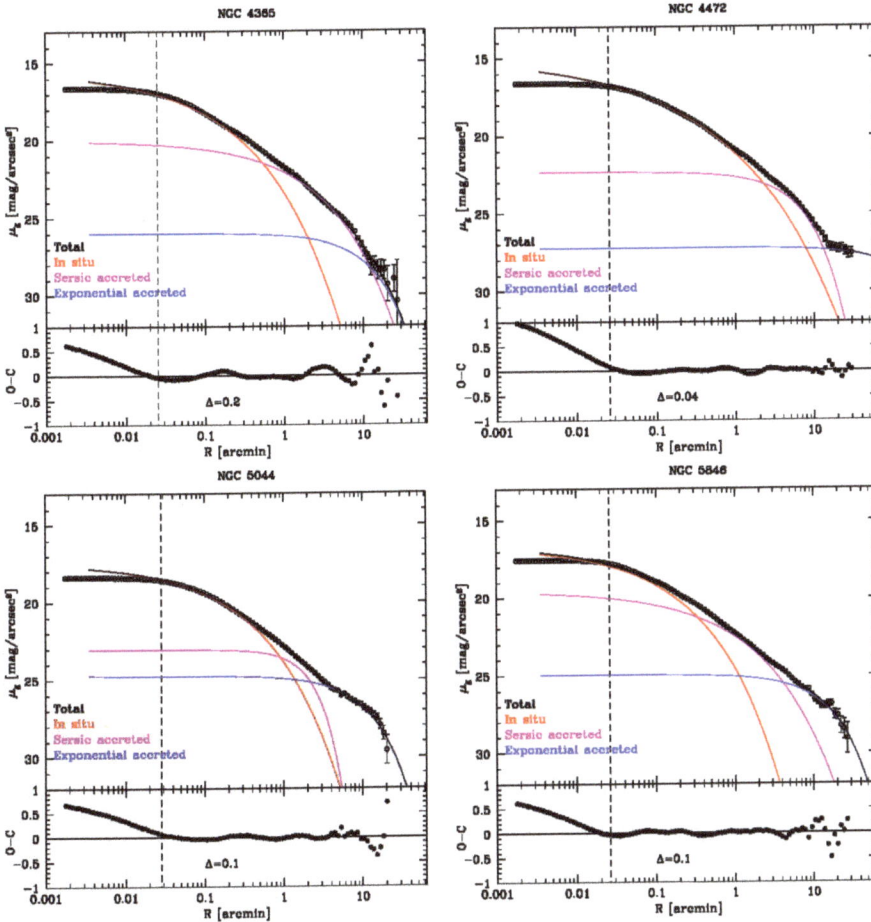

Figure 2. VST *g* band profiles of NGC 1399, NGC 3923, NGC 4365, NGC 4472, NGC 5044, and NGC 5846, fitted with a three-component model motivated by the predictions of theoretical simulations.

4. Comparison With Theoretical Predictions for Accreted Mass Fractions

In the previous section, we identified inflections in the surface brightness profiles of galaxies in our sample that may correspond to transitions between regions dominated by debris from different accreted progenitors (or ensembles of progenitors) in different dynamical states. From our fitting procedure, we estimated the contributions of outer exponential 'envelopes' to the total galaxy stellar mass (derived by using colours), which range from 28% to 60% for the galaxies in our sample, and the fraction of total accreted mass, which range from 83% to 95%.

In Figure 3 we compare the accreted mass ratios we infer from our observations (filled red triangles) with other observational estimates for BCGs by Seigar et al. [13], Bender et al. [18] and Iodice et al. [10], theoretical predictions from semi-analytic particle-tagging simulations by [3,17], and the Illustris cosmological hydrodynamical simulations [4]. We find that the stellar mass fraction of the accreted component derived for galaxies in our sample is fully consistent both with published data for other BCGs (despite considerable differences in the techniques and assumptions involved) and with the theoretical models by [3,17].

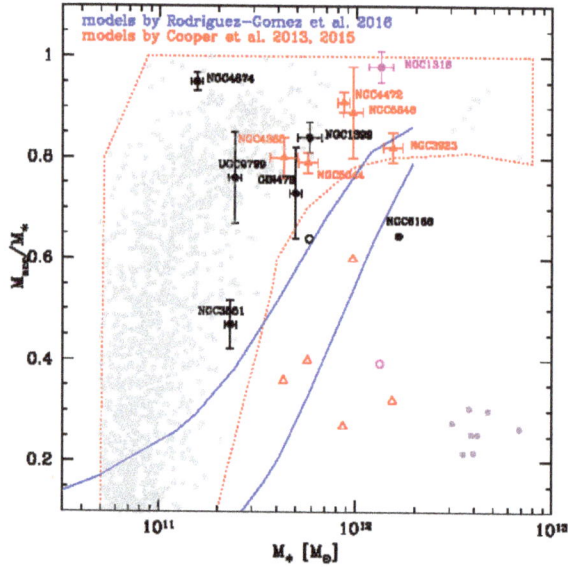

Figure 3. Accreted mass fraction vs. total stellar mass for early-type galaxies (ETGs). Our VST Elliptical GAlaxies Survey (VEGAS) measurements are given as red filled and open triangles (see text for details). Black circles correspond to other BCGs from the literature [10,13,18]. Pink points are for NGC1316 [19]. Red and blue regions indicate the predictions of cosmological galaxy formation simulations by [3,17] and Rodriguez-Gomez et al. [4], respectively, while grey points are from Cooper et al. [3]. Purple-grey points (between 3×10^{12} and 7×10^{12} solar masses) show the mass fraction associated with the streams from Table 1 in Cooper et al. [17] for comparison to the observations shown by open symbols.

In Figure 3 we also compare the stellar mass fractions obtained for the outermost exponential component of our multicomponent fit (open red triangles) with the mass fraction associated with unbound debris streams from surviving cluster galaxies in the simulations of Cooper et al. [17]. We found that the mass fraction in this component of our fits is consistent with these values from the simulations, suggesting that such components may give a crude estimate of the mass distribution associated with dynamically unrelaxed components originating from disrupting or recently disrupted galaxies, as argued by Cooper et al. [17].

5. Materials and Methods

The data used were collected with the VST/OmegaCAM in March and April of 2015 within the Italian Guaranteed Time Observation (GTO). This work is based on visitor mode observations taken at the ESO La Silla Paranal Observatory within the VST Guaranteed Time Observations, Programme IDs 090.B-0414(D), 091.B-0614(A), 094.B-0496(A), 094.B-0496(B), 094.B-0496(D) and 095.B-0779(A).

The VEGAS survey and the data reduction procedure adopted in this work are described in details by Capaccioli et al. [9] and Spavone et al. [11]. The background estimate and subtraction is the most critical operation in deep photometric analysis because it affects the ability of detecting and measuring the faint outskirts of galaxies. For this reason we decided to adopt a *step-dither* observing strategy for galaxies with large angular extent, consisting of a cycle of short exposures centred on the target and on offset fields ($\Delta = \pm 1$ degree). With such a technique the background can be estimated from exposures taken as close as possible, in space and time, to the scientific images. This ensures better accuracy, reducing the uncertainties at very faint surface brightness levels, as found by Ferrarese et al. [20].

This observing strategy allowed us to build an average sky background of the night, which was subtracted from each science image.

The isophotal analysis has been performed on the sky-subtracted mosaics in both the *g* and *i* bands, where all the bright sources have been masked. We used the IRAF[1] task ELLIPSE to extract azimuthally averaged intensity profiles in elliptical annuli out to the edges of the frame.

To fit the light distribution of our galaxies, we adopt the same approach described by Seigar et al. [13] and performed least-square fits using a Levenberg–Marquardt algorithm, in which the function to be minimized is the rms scatter, defined as $\Delta = \sqrt{\frac{\sum_{i=1}^{m} \delta_i^2}{m}}$, where *m* is the number of data points and δ_i is the *i*th residual. In all the fit presented above, the innermost seeing-dominated regions ($\sim 1.5 \times FWHM$), indicated with dashed lines, were excluded.

6. Conclusions

We have presented new deep photometry in the *g* and *i* bands for six giant ETGs in the VST Early-type Galaxy Survey (VEGAS): NGC 1399, NGC 3923, NGC 4365, NGC 4472, NGC 5044, and NGC 5846. In particular, we studied the shapes of their surface brightness profiles to obtain evidence of structural variations that may constrain their assembly history.

Our analysis suggests that the surface brightness profiles of the galaxies in our study are best reproduced by multicomponent models. We took two approaches to constructing such models. We Adopt an empirically motivated, two-component approach most common in the literature, as well as an alternative approach, which is motivated by the predictions of numerical simulations, in which we fitted the surface brightness profiles of our galaxies with three components: two dominant Sérsic components and an outer exponential component. To mitigate some of the degeneracy in this approach, we fixed the Sérsic index of the inner component to a representative value from simulations. Compared to the traditional empirical fit, this approach allows us to make a more meaningful estimate of the total contribution of accreted stars.

The mass fractions in the exponential components of our two-component profile decompositions are in good agreement with the mass fractions associated with streams from surviving galaxies in the simulations of Cooper et al. [17]. This suggests these outer exponential components may give a crude estimate of the stellar mass fraction associated with recently disrupted galaxies. We find values for this fraction ranging from 28% to 64%.

For all the galaxies in our study we can identify at least one inflection in the surface brightness profile. These inflections occur at very faint surface brightness levels ($24.0 \leq \mu_g \leq 27.8$ mag/arcsec^2). They appear to correlate with changes in the trend of ellipticity, position angle, and colour with radius, where the isophotes become flatter and misaligned and the colours become bluer beyond the inflections (see Spavone et al. [11] for details). This suggests that these inflections mark transitions between physically distinct components (or ensembles of similar components) in different states of dynamical relaxation.

It is encouraging that we see a variety of profile inflections in our photometric investigation of this small subset of the VEGAS sample and that these are broadly consistent with the expectations of state-of-the-art theoretical models. Our results suggests that with the complete sample of extremely deep surface brightness profiles from the full survey, we will be able to investigate the late stages of massive galaxy assembly statistically, thereby distinguishing dynamically evolved systems from those that are still reaching dynamical equilibrium and probing the balance between in situ star formation and accretion across a wide range of stellar mass. This is a promising route to constraining cosmological models of galaxy formation such as those we have compared with here, which predict

---

[1]   IRAF (Image Reduction and Analysis Facility) is distributed by the National Optical Astronomy Observatories, which is operated by the Associated Universities for Research in Astronomy, Inc. under cooperative agreement with the National Science Foundation.

fundamental, relatively tight correlations between the present-day structure of massive galaxies and the growth histories of their host dark matter halos.

Acknowledgments: M. Spavone wishes to thank the ESO staff of the Paranal Observatory for their support during the observations at VST. APC is supported by a COFUND/Durham Junior Research Fellowship under EU grant [267209] and acknowledges support from STFC (ST/L00075X/1). The data reduction for this work was carried out with the computational infrastructure of the INAF-VST Center at Naples (VSTceN). This research made use of the NASA/IPAC Extragalactic Database (NED), which is operated by the Jet Propulsion Laboratory, California Institute of Technology, under contract with the National Aeronautics and Space Administration, and has been partly supported by the PRIN-INAF "Galaxy evolution with the VLT Survey Telescope (VST)" (PI A. Grado). NRN, EI, and MP have been supported by the PRIN-INAF 2014 "Fornax Cluster Imaging and Spectroscopic Deep Survey" (PI. N.R. Napolitano). MS, EI, and M. Cantiello acknowledge finacial support from the VST project (P.I. M. Capaccioli).

Author Contributions: M.S. wrote this paper. M.S., M.C., E.I., N.R.N. and A.P.C. conceived and designed the experiments. M.S. performed the experiments. A.G. and L.L. reduced the data. D.A.F., M.Ca., M.P. and P.S. helped to write and revise the paper.

Conflicts of Interest: The authors declare no conflict of interest.

References

1. White, S.D.M.; Frenk, C.S. Galaxy formation through hierarchical clustering. *Astrophys. J.* **1991**, *379*, 52.

2. Guo, Q.; White, S.D.M. Galaxy growth in the concordance *Lambda*CDM cosmology. *Mon. Not. R. Astron. Soc.* **2008**, *384*, 2.

3. Cooper, A. P.; D'Souza, R.; Kauffmann, G.; Wang, J.; Boylan-Kolchin, M.; Guo, Q.; Frenk, C.S.; White, S.D.M. Galactic accretion and the outer structure of galaxies in the CDM model. *Mon. Not. Roy. Astro. Soc.* **2013**, *434*, 3348.

4. Rodriguez-Gomez, V.; Pillepich, A.; Sales, L.V.; Genel, S.; Vogelsberger, M.; Zhu, Q.; Wellons, S.; Nelson, D.; Torrey, P.; Springel, V.; et al. The stellar mass assembly of galaxies in the Illustris simulation: growth by mergers and the spatial distribution of accreted stars. *Mon. Not. R. Astron. Soc.* **2016**, *458*, 2371.

5. Amorisco, N.C. Contributions to the accreted stellar halo: an atlas of stellar deposition. *arXiv* **2015**, arXiv:1511.08806.

6. Cooper, A.P.; Cole, S.; Frenk, C.S.; White, S.D.M.; Helly, J.; Benson, A.J.; De Lucia, G.; Helmi, A.; Jenkins, A.; Navarro, J.F.; et al. Galactic stellar haloes in the CDM model. *Mon. Not. R. Astron. Soc.* **2010**, *406*, 744.

7. Deason, A.J.; Belokurov, V.; Evans, N.W.; Johnston, K.V. Broken and Unbroken: The Milky Way and M31 Stellar Halos. *Astrophys. J.* **2013**, *763*, 113.

8. Capaccioli, M.; Schipani, P. The VLT Survey Telescope Opens to the Sky: History of a Commissioning. *Messenger* **2011**, *146*, 2.

9. Capaccioli, M.; Spavone, M.; Grado, A.; Iodice, E.; Limatola, L.; Napolitano, N.R.; Cantiello, M.; Paolillo, M.; Romanowsky, A.J.; Forbes, D.A.; et al. VEGAS: A VST Early-type GAlaxy Survey. I. Presentation, wide-field surface photometry, and substructures in NGC 4472. *Astron. Astrophys.* **2015**, *581*, A10.

10. Iodice, E.; Capaccioli, M.; Grado, A.; Limatola, L.; Spavone, M.; Napolitano, N.R.; Paolillo, M.; Peletier, R.F.; Cantiello, M.; Lisker, T.; et al. The Fornax Deep Survey with VST. I. The Extended and Diffuse Stellar Halo of NGC 1399 out to 192 kpc. *Astrophys. J.* **2016**, *820*, 42.

11. Spavone, M.; Capaccioli, M.; Napolitano, N.R.; Iodice, E.; Grado, A.; Limatola, L.; Cooper, A. P.; Cantiello, M.; Forbes, D.A.; Paolillo, M.; Schipani, P. VEGAS: A VST Early-type GAlaxy Survey. II. Photometric study of giant ellipticals and their stellar halos. *Astron. Astrophys.* **2017**, *603*, 38.

12. Donzelli, C.J.; Muriel, H.; Madrid, J.P. The Luminosity Profiles of Brightest Cluster Galaxies. *Astrophys. J.* **2011**, *195*, 15.

13. Seigar, M.S.; Graham, A.W.; Jerjen, H. Intracluster light and the extended stellar envelopes of cD galaxies: an analytical description. *Mon. Not. R. Astron. Soc.* **2007**, *378*, 1575.

14. Caon, N.; Capaccioli, M.; D'Onofrio, M. On the Shape of the Light Profiles of Early Type Galaxies. *Mon. Not. R. Astron. Soc.* **1993**, *265*, 1013.

15. Sérsic, J.L. Influence of the atmospheric and instrumental dispersion on the brightness distribution in a galaxy. *Boletin de la Asociacion Argentina de Astronomia La Plata Argentina* **1963**, *6*, 41.

16. Gonzalez, A.H.; Zabludoff, A.I.; Zaritsky, D. Structural Properties of Brightest Cluster Galaxies. *Astron. Astrophys.* **2003**, *285*, 67.

17. Cooper, A.P.; Gao, L.; Guo, Q.; Frenk, C.S.; Jenkins, A.; Springel, V.; White, S.D.M. Surface photometry of brightest cluster galaxies and intracluster stars in *Lambda*CDM. *Mon. Not. R. Astron. Soc.* **2015**, *451*, 2703.

18. Bender, R.; Kormendy, J.; Cornell, M.E.; Fisher, D.B. Structure and Formation of cD Galaxies: NGC 6166 in ABELL 2199. *Astrophys. J.* **2015**, *807*, 56.

19. Iodice, E.; Spavone, M.; Capaccioli, M.; Peletier, R.F.; Richtler, T.; Hilker, M.; Mieske, S.; Limatola, L.; Grado, A.; Napolitano, N.R.; et al. The Fornax Deep Survey with VST. II. Fornax A: A Two-phase Assembly Caught in the Act. *Astrophys. J.* **2017**, *839*, 21.

20. Ferrarese, L.; Côté, P.; Cuillandre, J.C.; Gwyn, S.D.J.; Peng, E.W.; MacArthur, L.A.; Duc, P.; Boselli, A.; Mei, S.; Erben, T.; et al. The Next Generation Virgo Cluster Survey (NGVS). I. Introduction to the Survey. *Astrophys. J.* **2012**, *200*, 4.

galaxies

MDPI

Article

Assembly Pathways and the Growth of Massive Early-Type Galaxies

Duncan Forbes

Centre for Astrophysics & Supercomputing, Swinburne University, Hawthorn VIC 3122, Australia;
dforbes@swin.edu.au

Academic Editor: Emilio Elizalde
Received: 20 April 2017; Accepted: 1 June 2017; Published: 7 June 2017

Abstract: Based on data from the SAGES Legacy Unifying Globulars and GalaxieS (SLUGGS) survey, I present results on the assembly pathways, dark matter content and halo growth of massive early-type galaxies. Using galaxy starlight information we find that such galaxies had an early dissipative phase followed by a second phase of halo growth from largely minor mergers (and in rare cases major mergers). Thus our result fits in well with the two-phase scenario of galaxy formation. We also used globular cluster radial velocities to measure the enclosed mass within 5 effective radii. The resulting dark matter fractions reveal a few galaxies with very low dark matter fractions that are not captured in the latest cosmological models. Multiple solutions are possible, but none yet is convincing. Translating dark matter fractions into epochs of halo assembly, we show that low mass galaxies tend to grow via gas-rich accretion, while high mass galaxies grow via gas-poor mergers.

Keywords: galaxies; formation; evolution; halos

1. Assembly Pathways

The SAGES Legacy Unifying Globulars and GalaxieS (SLUGGS) survey (Brodie et al. 2014) [1] is studying 25 nearby, massive (log stellar mass $\sim$11), early-type galaxies. The survey uses the DEIMOS multi-slit instrument on the Keck II telescope to obtain spectra of both the underlying starlight and surrounding globular clusters (GCs). The data have the advantage of reaching out to $\sim$3 R_e (effective radii) for starlight and $\sim$10 R_e for GCs. From this data, we can probe the dark matter content, assembly pathways and halo growth of these galaxies.

A key aim of near-field cosmology is to determine the assembly history of an individual galaxy. The hydro-zoom cosmological simulations of Naab et al. (2014) [2] showed that the assembly histories of massive galaxies are preserved in the 2D kinematics of present day galaxies. In the Naab et al. simulations massive galaxies form in two phases—the first in-situ phase at high redshift results in a compact, massive object (a red nugget); the second phase after redshift 2 is dominated by accretion of ex-situ stars i.e., formed in external galaxies. In this picture low mass galaxies have a high in-situ formed fraction, whereas high mass galaxies are largely built by accretion. A schematic of this two-phase galaxy formation is given in Figure 1.

Naab et al. used 3 kinematic diagnostics to classify galaxies into one of six assembly pathways. The SLUGGS starlight data provides similar diagnostics out to 3 R_e. They are radial lambda (spin) profiles, 2D kinematic maps and higher order velocity moments h3 and h4 vs V/σ. Using these 3 diagnostics (see Figure 2 for examples) we have classified each galaxy into a Nabb et al. assembly class. We find the most common pathway (14/24) to be class A—these reveal disk-like kinematics with slowly rotating stellar halos whose mass growth is due to minor mergers. Three galaxies are classified as class E which show disturbed kinematics, including rolling, double sigma profile and a decoupled core. Galaxies with high accretion fractions tend to be old with shallow metallicity gradients. For further details see Forbes et al. (2016) [3].

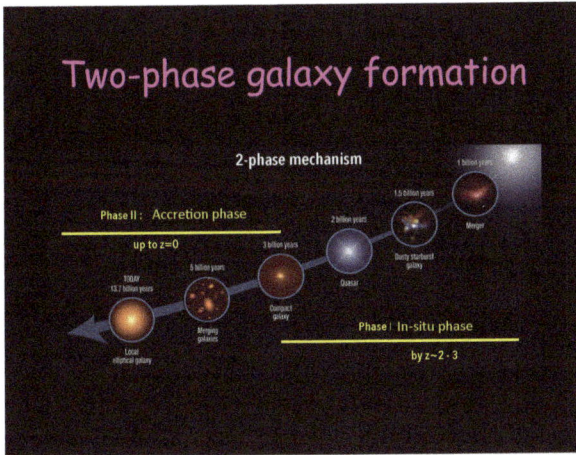

Figure 1. Two-phase galaxy formation. This schematic illustrates the two-phase galaxy formation scenario as described by the cosmological simulations of Naab et al. (2014) [2]. The first phase, at high redshift, is a dissipative one that results in the in-situ formation of a compact, massive core (red nugget) and perhaps an AGN. In he second phase, after redshift z ~2, growth is dominated by accretion from minor or major mergers which leads to the formation of the galaxy halo. Credit: NASA, ESA, S. Toft, A. Feild.

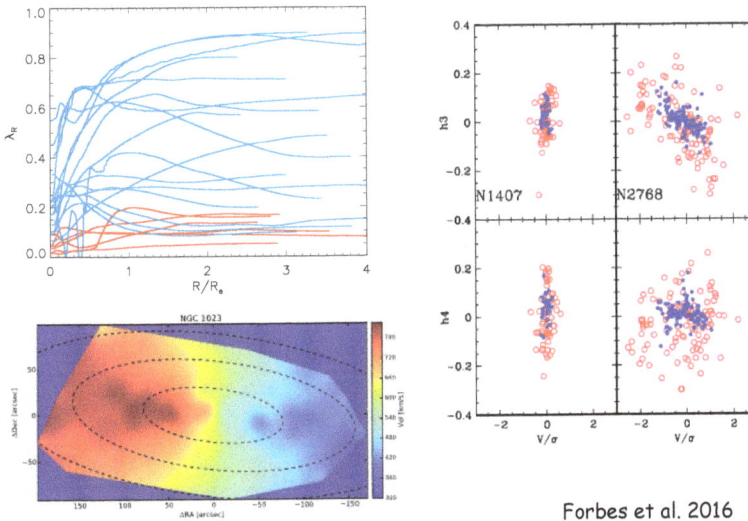

Figure 2. SLUGGS diagnostics. The kinematic diagnostics used to place each SLUGGS galaxy in one of six Naab et al. (2014) [2] assembly history classes are: radial lambda (spin) profiles (top left), 2D kinematic maps (lower left) and higher order velocity moments vs. V/σ (right). In the spin profiles, blue lines are centrally fast rotators and red lines are for centrally slow rotators. In the higher order moments, blue symbols are data within 1 R_e, and red symbols for data outside 1 R_e.

2. Dark Matter Content

The SLUGGS survey has now produced over 4000 high quality GC radial velocities out to ~10 R_e (Forbes et al. 2017) [4]. Using these as discrete tracers of the galaxy halo, we employ the Tracer Mass Estimator (TME) of Watkins et al. (2010) [5] to estimate the enclosed mass within 5 R_e for the SLUGGS galaxies and a few early-type galaxies using literature GC velocities. We correct the mass estimates for bulk rotation of the GC system, and make assumptions for the galaxy potential, tracer density slope and anisotropy (here we assume isotropic orbits). Further details can be found in Alabi et al. (2016, 2017) [6,7]. We find good agreement with other mass tracers such as planetary nebulae and X-ray emission (Alabi et al. 2016) [6]. From the enclosed total mass within 5 R_e we subtract the stellar mass (assuming a Kroupa IMF) to calculate the dark matter fraction within 5 R_e. In Figure 3 we show the derived dark matter fractions within 5 R_e vs galaxy stellar mass from our analysis (Alabi et al. 2017) [7] and two sets of cosmological simulations. The simulations are with (Remus et al. 2017) [8] and without (Wu et al. 2014) [9] AGN feedback. The plot also shows a simple galaxy model based on galaxy scaling relations and an NFW dark matter halo. This model lies between the two simulations. Although some galaxies are consistent with the model predictions, we find a large range to dark matter fractions that is not captured by the models. A similar trend for low dark matter fractions in galaxies with stellar masses just below log M = 11 was seen in the ATLAS3D data within ~1 R_e by Cappellari et al. (2013) [10].

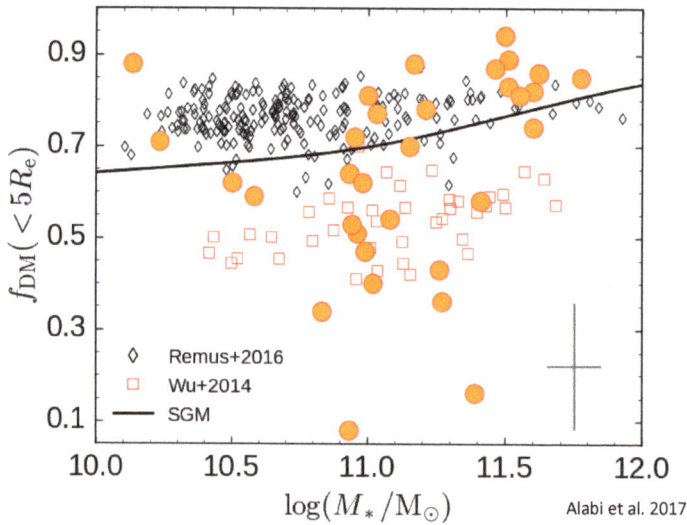

Figure 3. Dark matter fraction vs simulations. The dark matter fraction within 5 effective radii is shown against stellar mass. The SLUGGS data from Alabi et al. (2017) [7] are shown by filled orange circles, with a typical error bar at the lower right. The black and red open symbols are the cosmological simulations of Remus et al. (2017) [8] and Wu et al. (2014) [9] respectively. The solid black line shows a simple galaxy model based on scaling relations and a NFW dark matter halo. The data show some agreement with the model galaxies, but reveal a considerable range in dark matter fractions for log stellar masses of ~11 that is not captured in the models.

A number of possible solutions to explain the very low dark matter fractions we observe include:

- Mass-anisotropy degeneracy: There is a well-known degeneracy between the mass and orbits of the tracers. We have rederived the mass for both radial and tangential orbits, finding that although it makes a small 20% difference at high and low galaxy masses, the difference at log M ~ 11 is negligible.
- Stellar IMF: We assumed a standard Kroupa IMF. Although there is evidence that the IMF gets steeper in higher mass galaxies (van Dokkum & Conroy 2012) [11], this would be expected to increase the baryonic mass within 5 R_e and hence drive down the DM fractions of the most massive galaxies—which is not seen in Figure 3.
- Peculiar effective radii: If the very low DM fraction galaxies have effective radii that deviate strongly from the standard size-mass relation this may explain their low DM fractions. This may be a partial answer to explaining the properties of some of the galaxies.
- Baryonic feedback/halo expansion: Feedback from an AGN may lead to adiabatic expansion pushing the DM from central regions to the outer halo regions. In principle, this would lower DM fractions. This effect, like that of a steeper IMF, should be most strong for the highest mass galaxies and is not seen.
- Environment: Halos that collapse early in a dense environment may be expected to have higher DM fractions. Of our very low DM galaxies, 2 are in the field and 3 are in groups—none are in clusters. More data are warranted to explore this possible environmental trend.
- Morphology: Lenticular galaxies may behave more like spiral galaxies than ellipticals, with different DM fractions. Of our very low DM galaxies, 2 are S0, 2 are clear elliptical and the other has an intermediate classification of E/S0. Like environment, more galaxies are needed to decide if morphology is a factor in low DM fractions.
- Self-interacting DM: Self-interacting DM (SIDM) is thought to lead to galaxies that are stellar dominated in their central regions. Unfortunately few models exist. Recently, Di Cintio et al. (2017) [12] modelled two log M ≤ 11 galaxies involving SIDM (see Figure 3). These galaxies have DM fractions of ~55%, so lower than the Remus et al. cold DM models (and similar to the Wu et al. models) but still somewhat higher than our very low DM fraction galaxies. SIDM remains a possible/partial answer which requires further investigation.

To summarise, a non-standard IMF and AGN driven feedback are unlikely causes of low DM fractions in log M ≤ 11 galaxies as they should have a stronger influence on more massive galaxies. Variations in the galaxy size-mass relation may be a partial solution. A larger sample will however allow us to better test for trends with environment and morphology, which are unclear currently due to small sample size in the key mass range. A larger sample will also allow us to better define the stellar mass associated with the minimum DM fraction and to constrain future SIDM models (which remain as a possible solution to the problem).

3. Halo Growth

Our measurements also allow us to estimate the dark matter volume density within 5 R_e. The density of galaxy halos is in turn related to the epoch of halo assembly. In Figure 4 we show the inferred epoch of halo assembly vs the luminosity-weighted mean age of the stars in each SLUGGS galaxy. The plot is coded by stellar mass. The upper left of the diagram shows those galaxies for which the halo assembled, on average, before most star formation occurred, i.e., halo growth continued via gas-rich accretion that led to star formation. Galaxies in this part of the diagram tend to be of low mass. In the lower right part of the diagram, halos continue to be assembled after the main epoch of star formation, i.e., growth is dominated by gas-poor accretion. High mass galaxies tend to be located in this part of the diagram.

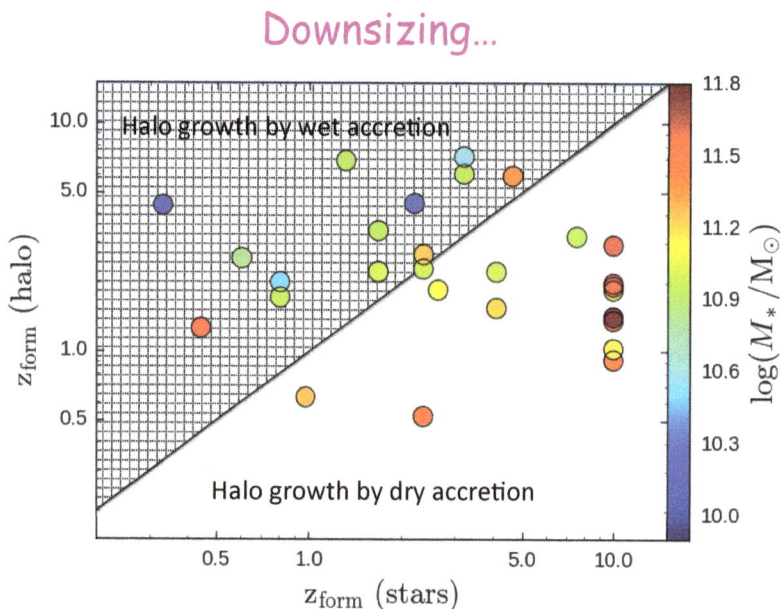

Figure 4. Downsizing—*the early formation and completion of star formation in massive galaxies.* The assembly epoch of galaxy halos, derived from the dark matter density within 5 effective radii, is shown vs the luminosity-weighted mean age of the stars in each SLUGGS galaxy. Symbols are colour-coded by their stellar mass. Galaxies in the upper left tend to be of low stellar mass for which halo growth continues via gas-rich accretion. Whereas, for high mass galaxies growth is dominated by gas-poor mergers.

4. Conclusions

Here we have used starlight and globular cluster data from the SLUGGS survey to probe the dark matter content, assembly history and halo growth of massive early-type galaxies. By comparing kinematic diagnostics of SLUGGS galaxies with the cosmological simulations of Naab et al. (2014) [2], we classified galaxies into six different assembly pathways. We find that massive galaxies had an early dissipative phase and a second phase of halo growth from largely minor mergers (and in rare cases major mergers). See Forbes et al. (2016) [3] for further details. This result fits in well with the two-phase scenario of galaxy formation.

We also used globular cluster radial velocities to measure the enclosed mass within 5 effective radii. Calculating dark matter fractions within this radius revealed that a few galaxies with log stellar mass ~11 have very low dark matter fractions indicating diffuse halos, and that this behaviour is not captured in the latest cosmological models. Multiple solutions are possible, but none yet is convincing. See Alabi et al. (2017) [7] for further details. Converting dark matter fractions into halo densities and hence epochs of halo assembly, we show that low mass galaxies tend to grow via gas-rich accretion, while high mass galaxies grow via gas-poor mergers.

Acknowledgments: I thank the SLUGGS survey team for their contribution to the research presented here, especially Busola Alabi. This work was supported by ARC grant DP130100388.

Conflicts of Interest: The author declares no conflict of interest.

References

1. Brodie, J.P.; Romanowsky, A.J.; Strader, J.; Forbes, D.A.; Foster, C.; Jennings, Z.G.; Pastorello, N.; Pota, V.; Usher, C.; Blom, C.; et al. The SAGES Legacy Unifying Globulars and GalaxieS Survey (SLUGGS): Sample Definition, Methods, and Initial Results. *Astrophys. J.* **2014**, *796*, doi:10.1088/0004-637X/796/1/52.

2. Naab, T.; Oser, L.; Emsellem, E.; Cappellari, M.; Krajnović, D.; McDermid, R.M.; Alatalo, K.; Bayet, E.; Blitz, L.; Bois, M.; et al. The ATLAS^{3D} project—XXV. Two-dimensional kinematic analysis of simulated galaxies and the cosmological origin of fast and slow rotators. *Mon. Not. R. Astron. Soc.* **2014**, *444*, 3357–3387.

3. Forbes, D.A.; Romanowsky, A.J.; Pastorello, N.; Foster, C.; Brodie, J.P.; Strader, J.; Usher, C.; Pota, V. The SLUGGS survey: The assembly histories of individual early-type galaxies. *Mon. Not. R. Astron. Soc.* **2016**, *457*, 1242–1256.

4. Forbes, D.A.; Alabi, A.; Brodie, J.P.; Romanowsky, A.J.; Strader, J.; Foster, C.; Usher, C.; Spitler, L.; Bellstedt, S.; Pastorello, N.; et al. The SLUGGS Survey: A Catalog of Over 4000 Globular Cluster Radial Velocities in 27 Nearby Early-type Galaxies. *Astron. J.* **2017**, *153*, 114.

5. Watkins, L.L.; Evans, N.W.; An, J.H. The masses of the Milky Way and Andromeda galaxies. *Mon. Not. R. Astron. Soc.* **2010**, *406*, 264–278.

6. Alabi, A.; Forbes, D.A.; Romanowsky, A.J.; Brodie, J.P.; Strader, J.; Janz, J.; Pota, V.; Pastorello, N.; Usher, C.; Spitler, L.R.; et al. The SLUGGS survey: The mass distribution in early-type galaxies within five effective radii and beyond. *Mon. Not. R. Astron. Soc.* **2016**, *460*, 3838–3860.

7. Alabi, A.; Forbes, D.A.; Romanowsky, A.J.; Brodie, J.P.; Strader, J.; Janz, J.; Usher, C.; Spitler, L.R.;Bellstedt, S.; Ferré-Mateu, A. The SLUGGS Survey: Dark matter fractions at large radii and assembly epochs of early-type galaxies from globular cluster kinematics. *Mon. Not. R. Astron. Soc.* **2017**, in press.

8. Remus, R.-S.; Dolag, K.; Naab, T.; Burkert, A.; Hirschmann, M.; Hoffmann, T.L.; Johansson, P.H. The co-evolution of total density profiles and central dark matter fractions in simulated early-type galaxies. *Mon. Not. R. Astron. Soc.* **2017**, *464*, 3742–3756.

9. Wu, X.; Gerhard, O.; Naab, T.; Oser, L.; Martinez-Valpuesta, I.; Hilz, M.; Churazov, E.; Lyskova, N. The mass and angular momentum distribution of simulated massive early-type galaxies to large radii. *Mon. Not. R. Astron. Soc.* **2014**, *438*, 2701–2715.

10. Cappellari, M.; Scott, N.; Alatalo, K.; Blitz, L.; Bois, M.; Bournaud, F.; Bureau, M.; Crocker, A.F.; Davies, R.L.; Davis, T.A.; et al. The ATLAS^{3D} projec—XV. Benchmark for early-type galaxies scaling relations from 260 dynamical models: mass-to-light ratio, dark matter, Fundamental Plane and Mass Plane. *Mon. Not. R. Astron. Soc.* **2013**, *432*, 1709–1741.

11. Dokkum, P.G.; Conroy, C. The Stellar Initial Mass Function in Early-type Galaxies from Absorption Line Spectroscopy. I. Data and Empirical Trends. *Astrophys. J.* **2012**, *760*, 70.

12. Di Cintio, A.; Tremmel, M.; Governato, F.; Pontzen, A.; Zavala, J.; Bastidas Fry, A.; Brooks, A.; Vogelsberger, M. A rumble in the dark: Signatures of self-interacting dark matter in Super-Massive Black Hole dynamics and galaxy density profiles. *Mon. Not. R. Astron. Soc.* **2017**, submitted.

galaxies

MDPI

Conference Report

Better Galactic Mass Models through Chemistry

Robyn E. Sanderson [1,2,*,†], **Andrew R. Wetzel** [1,3,4], **Sanjib Sharma** [5] **and Philip F. Hopkins** [1]

[1] TAPIR, California Institute of Technology, Pasadena, CA 91125, USA; awetzel@ucdavis.edu (A.R.W.); phopkins@caltech.edu (P.F.H.)

[2] Department of Astronomy, Columbia University, New York, NY 10025, USA

[3] Department of Physics, University of California, Davis, CA 95616, USA

[4] The Observatories of the Carnegie Institution for Science, Pasadena, CA 91101, USA

[5] Sydney Institute for Astronomy, School of Physics, University of Sydney, Sydney NSW 2006, Australia; sanjib.sharma@gmail.com

* Correspondence: robyn@caltech.edu; Tel.: +1-626-395-6426

† NSF Astronomy & Astrophysics Postdoctoral Fellow.

Academic Editors: Duncan A. Forbes and Ericson D. Lopez

Received: 29 July 2017; Accepted: 11 August 2017; Published: 21 August 2017

Abstract: With the upcoming release of the Gaia catalog and the many multiplexed spectroscopic surveys on the horizon, we are rapidly moving into a new data-driven era in the study of the Milky Way's stellar halo. When combined, these data sets will give us a many-dimensional view of stars in accreted structures in the halo that includes both dynamical information about their orbits and chemical information about their formation histories. Using simulated data from the state-of-the-art Latte simulations of Milky-Way-like galaxies, which include hydrodynamics, feedback, and chemical evolution in a cosmological setting, we demonstrate that while dynamical information alone can be used to constrain models of the Galactic mass distribution in the halo, including the extra dimensions provided by chemical abundances can improve these constraints as well as assist in untangling different accreted components.

Keywords: galaxy: kinematics and dynamics; galaxy: halo; galaxy: abundances; galaxy: structure; galaxy: formation; cosmology: dark matter; methods: statistical; methods: numerical

1. Introduction

Our knowledge of the Galaxy in which we live, the Milky Way (MW), is poised to undergo a revolution in the next ten years thanks to a new generation of state-of-the-art stellar surveys. These include photometric surveys like PanSTARRS [1] and LSST [2], spectroscopic surveys like 4MOST [3], DESI [4], WEAVE [5], and Subaru-PFS [6], and astrometric surveys starting with Gaia [7] and extending to LSST and WFIRST [8]. The Galactic renaissance will also include far greater insight into the chemical abundances of stars thanks to efforts like the Cannon [9], which can translate the abundance patterns generated from smaller high-resolution spectroscopic surveys like APOGEE [10], GALAH [11], and the Gaia-ESO survey [12] into the larger medium- and low-resolution spectroscopic surveys listed above. It is not an exaggeration to say that ten years from now we can expect to have complete phase-space information for stars in the Galaxy nearly to its virial radius, complemented by 10–20 dimensions of chemical abundance information. This new high-dimensional view of the Galaxy demands new approaches to understanding its contents. Here we discuss how this combined phase and abundance space will be uniquely powerful for setting constraints on the MW's dark matter (DM) distribution and untangling the building blocks of its accreted stellar halo, allowing us to use our Galaxy as a time machine to study ancient dwarf galaxies.

Currently our knowledge of the distribution of stars in the MW is limited to the region within about 120 kpc [13], less than half of what we imagine to be the full extent of its dark halo. Because of this limited information, our knowledge of the shape and mass of the MW's DM halo is equally limited: measurements of its total mass disagree by a factor of ~4 [14] and our understanding of its shape and radial profile is still poorly constrained enough to be controversial [15,16] especially in light of new estimates of the mass of its largest companion, the Large Magellanic Cloud [17,18]. This makes it difficult to place our Galaxy in a cosmological context in order to use it as a test of DM theories, since many predictions from cosmological simulations scale with mass and concentration [19].

2. Methods

Stars in the accreted stellar halo of the Milky Way provide one of the few sources of information about the outer extents of the Galactic dark halo. Most of these stars are part of structures called tidal streams (e.g., [20]) created by the disruption of ancient dwarf satellite galaxies of the MW, and are expected to reach to the MW's virial radius and beyond [21,22]. Since the stars in tidal streams all have similar orbits, tidal streams are, in principle, more sensitive to the radial profile and shape of the halo than the orbits of individual bound dwarf galaxies or globular clusters, and so are valuable for constraining the MW's shape and mass. Furthermore the approximately isotropic nature of accretion (although see [23]) samples orbits well out of the plane of the Galactic disk and at much larger distances, giving us a more complete map of the gravitational forces exerted by our Galaxy than we could otherwise achieve.

2.1. The Action Space of Stellar Orbits

One approach to mapping the Milky Way's dark halo with tidal streams is to work in the space of their actions $\mathbf{J}$, which are defined for stars on bound orbits in a system described by a Hamiltonian $\mathcal{H}$ as a canonical transformation of their positions $\mathbf{x}$ and momenta $\mathbf{p}$, such that for each component x_i,

$$J_i = \oint p_i dx_i, \tag{1}$$

where the integral is over one cycle of the coordinate x_i. When defined in this way, a star's actions $\mathbf{J}$ are adiabatic invariants (i.e., integrals of the motion) and the angles θ evolve linearly in time with frequencies Ω:

$$\theta_i = \Omega_i t + \theta_{i,0}, \tag{2}$$

$$\Omega_i = \frac{\partial \mathcal{H}}{\partial J_i}. \tag{3}$$

For more background on action-angle variables we refer the reader to the discussions in Chapter 3 of [24] or Chapter 10 of [25]; for more detailed information on the techniques described in this section see [26,27].

For a tidal stream, in which all the stars began life within the small (compared to the MW) phase-space volume of a dwarf galaxy, their actions are persistently similar unless the host galaxy (in this case the MW) undergoes a major merger that non-adiabatically transforms the gravitational potential. The actions thus conveniently label stars in each tidal stream by their orbit family, and the small cluster in $\mathbf{J}$ representing the stream can be treated using linear transformations [28]. Furthermore, a stellar halo made of accreted small galaxies, as predicted by theories of cold DM, will exhibit a high degree of clustering in the space of its actions even for structures that are spread across the Galaxy (Figure 1, panels A and B). This space is thus ideal for separating different accreted structures from one another [29–31].

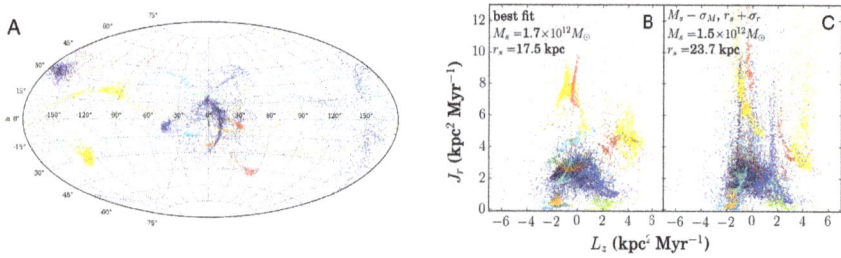

Figure 1. Panel A: tidal streams in an accreted mock stellar halo from [32] shown distributed on the sky in Galactic coordinates. Each different color represents stars from a different progenitor dwarf galaxy. **Panel B:** The same mock stellar halo shown in a projection of action space, calculated using the best-fit spherical NFW approximation for the gravitational potential of the dark halo. The stars in each progenitor are now clustered together around the parent orbits of the accreted galaxies they formed from. **Panel C:** The same mock stellar halo with actions calculated using a potential that is roughly 1σ away from the best-fit approximation. The clustering is notably less pronounced in this incorrect action space. Figure adapted from [27].

Studying tidal streams in action space requires assuming a global gravitational potential, but the structure of action space itself gives insight into the correct potential to use. As shown by comparing panels B and C of Figure 1, only when the assumed potential is a close representation of the true mass distribution will the action space of the stellar demonstrate the expected degree of clustering [26,33,34]. A statistical measure of the degree of clustering can thus be used as the figure of merit to determine the best-fit potential [26,27,35]. We will describe the quantification of clustering in Section 2.2. The advantage of this method of determining the Galactic gravitational potential, especially in the context of the many upcoming stellar surveys, is twofold. First, it does not require that individual streams are identified or stars assigned to membership in a particular stream, merely that the stars in the fitting sample come mainly from the accreted stellar halo so that the assumption of an intrinsically clustered distribution is satisfied. Second, the method is agnostic as to which labels are used as dimensions, so long as the stars in each stream demonstrate some correlation with each other in those labels and at least some of them depend on the gravitational potential. This means that if chemical abundance information is also available for the stars in the fitting sample, these labels can be added as extra dimensions. The chemical abundances of stars from a given progenitor galaxy should be correlated through their shared star formation history, so this extra information should improve the ability to constrain the gravitational potential by, for example, differentiating stars from progenitors with different masses that accreted on similar orbits.

2.2. Measuring Clustering in Actions

To quantify the degree of clustering in the distribution of actions $\mathbf{J}$ and abundances $\mathbf{Z}$, given a guess for the gravitational potential Φ, we use the *mutual information* (MI) statistic [36], defined as the relative entropy between the target distribution $p_\Phi(\mathbf{J}, \mathbf{Z})$ and a "shuffled" version $\tilde{p}_\Phi(\mathbf{J}, \mathbf{Z})$:

$$MI(\Phi) \equiv \int p_\Phi(\mathbf{J}, \mathbf{Z}) \ln \frac{p_\Phi(\mathbf{J}, \mathbf{Z})}{\tilde{p}_\Phi(\mathbf{J}, \mathbf{Z})} d\mathbf{J} d\mathbf{Z} \qquad (4)$$

The distribution $\tilde{p}_\Phi$ is constructed by randomizing the $\mathbf{J}$ and $\mathbf{Z}$ coordinates of each star relative to one another. This transformation preserves the one-dimensional marginal distributions along each coordinate, and is equivalent to the product of the marginals:

$$\tilde{p}_\Phi(\mathbf{J}, \mathbf{Z}) = \prod_i p_\Phi(J_i | J_{\neq i}, \mathbf{Z}) \prod_j p_\Phi(Z_j | Z_{\neq j}, \mathbf{J}). \qquad (5)$$

To calculate the MI for a given set of stars, we first calculate the stellar actions assuming the potential Φ and use the density estimation code EnLink [37] to calculate the densities p_Φ and $\tilde{p}_\Phi$ in $(\mathbf{J}, \mathbf{Z})$ space at the location of each star. We then use Monte Carlo integration to estimate the MI by summation over all stars in the sample:

$$MI(\Phi) = \sum_{i=1}^{N_*} \ln \frac{p_{i,\Phi}(\mathbf{J}, \mathbf{Z})}{\tilde{p}_{i,\Phi}(\mathbf{J}, \mathbf{Z})} \qquad (6)$$

As shown in Figure 2, the more clustered the distribution is, the greater the difference with the product of its marginals, and hence the larger the value of $MI(\Phi)$. The maximum value of $MI(\Phi)$ identifies the most clustered distribution and hence the best-fit gravitational potential.

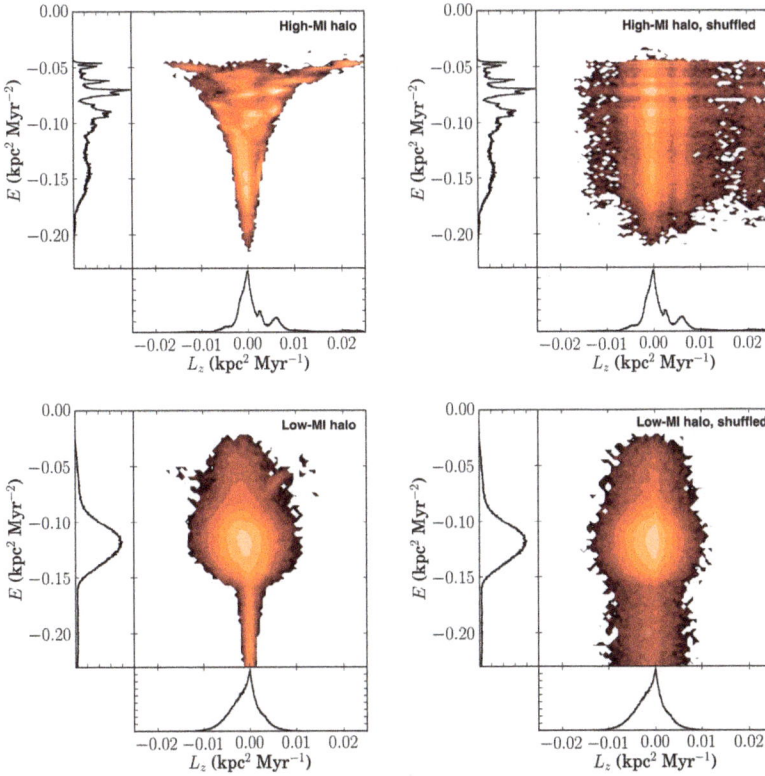

Figure 2. Measuring the degree of clustering using the mutual information. All panels show the distribution of energies and angular momenta for a region of a simulated stellar halo from one of the Latte galaxies [38]. The colors denote log projected density, with dark red denoting low density regions and light yellow regions of high density. The side-plots show the one-dimensional projected density along each dimension. The top row shows a case with a high degree of mutual information (MI, Equation (4)), where the distribution obtained by randomizing the coordinates of the stars while retaining the one-dimensional projected distributions (**right**) differs greatly from the original correlated distribution (**left**). The bottom row shows a case with a low degree of mutual information, where the distribution is similar whether or not the coordinates of the stars are shuffled. The mutual information can thus be used to measure the degree of clustering in the space of actions or constants of motion for a set of stars.

2.3. Simulating the Accreted Stellar Halo

To test the hypothesis that adding chemical abundance information can improve constraints on the Galactic gravitational potential, we used the simulated galaxy m12i from the Latte simulation suite, described in [38]. This MW-like simulated galaxy is the product of a high-resolution, cosmological zoom-in simulation that uses the FIRE-2 feedback recipe [39] to follow star formation and the evolution of stellar populations. The simulation tracks the evolution of ten chemical elements produced by supernovae (types I and II) and winds from evolved stars. For the examples shown here we used the second-highest resolution version of the simulation, in which star particles have $m_p \sim$56,500 M$_\odot$. For this preliminary test, we selected star particles in a region between 50 and 100 kpc from the main galaxy's center to avoid the galactic stellar and gas disk.

Using this selection as the fitting sample, we modeled the galactic mass distribution with a spherical Navarro-Frenk-White form [40] with scale radius r_s and scale density ρ_s, parameterized in terms of the enclosed mass M_s at the scale radius,

$$M_s \equiv M_{encl}(r_s) = 4\pi\rho_s r_s^3 \left(\ln 2 - \frac{1}{2} \right),$$ (7)

In terms of M_s and r_s, the potential is

$$\Phi(r) = -G \frac{M_s}{\ln 2 - 1/2} \frac{\ln(1 + r/r_s)}{r}$$ (8)

where G is Newton's constant. The choice of M_s over ρ_s as the mass parameter is motivated by the sensitivity of stellar orbits to the enclosed mass rather than the local density, and avoids severe degeneracies between parameters during the fit. This functional form ignores the triaxiality of the halo but can adequately approximate the radial profile of the gravitational potential over a limited radial range. To check the results of the fit using the halo sample, we also fit the same functional form directly to the gravitational potentials of the stars as calculated directly by the simulation (Figure 3).

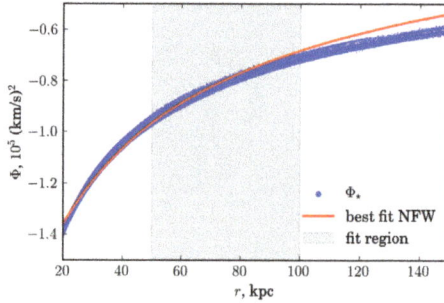

Figure 3. Gravitational potential Φ_* of star particles in the simulated galaxy calculated directly from the simulation (blue) as a function of radius r, with the best direct fit of the NFW functional form (red) over the range shaded in gray. The density of stars is higher at smaller r within the range being fit, so agreement at smaller r is prioritized (consistent with the behavior of the action-space fitting scheme).

3. Results

Calculating the MI for a distribution involves estimating the density from a finite number of points, so adding extra dimensions is costly and can introduce noise in the MI value and hence in the best fit. So before adding all ten chemical abundances tracked by the simulation, we first looked at the change in the MI when adding each abundance separately to the actions, while assuming the potential fit directly to the values calculated from the simulation and shown in Figure 3. A larger increase in MI when a given abundance is added signals a larger degree of clustering in that dimension and should correspond to more additional constraining power on the potential when added to the actions. The left panel of Figure 4 shows that indeed some abundances appear to contain more extra information than others, probably as a result of the assumptions about how different elements are produced in the simulation. Guided by this result we selected nitrogen, calcium, and iron to use as additional labels in fitting the gravitational potential. We caution that the result that these three abundances were most informative should not be over-interpreted, given the simplified yield tables used in the simulations; this is more a proof of concept and a demonstration of a technique that could be used on future data sets.

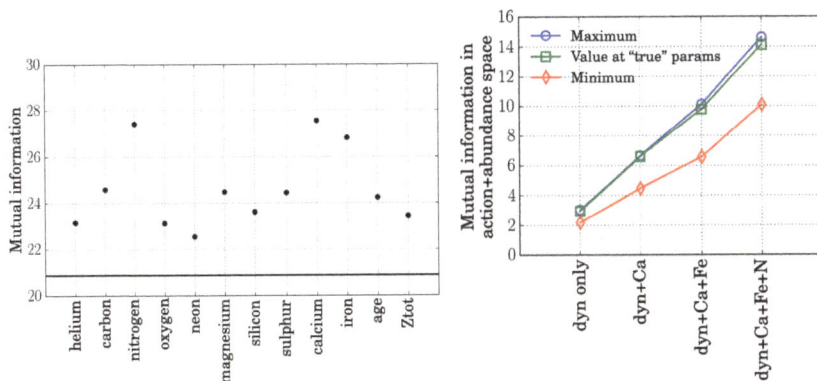

Figure 4. Left: Mutual information of stars in the Latte halo sample, in the space of constants of motion plus one additional label as shown on the horizontal axis. The solid black line shows the value of the mutual information (MI) calculated for constants of motion alone. The elements nitrogen, calcium, and iron appear to be most strongly correlated with the orbits of stars in the tidal streams. **Right**: MI values at the best-fit (blue), direct-fit (green), and far from the best fit (red) parameters for the gravitational potential, for the Latte stellar halo sample. When additional abundances (Ca, Fe, N) are included as extra dimensions, the contrast between the peak and the background increases relative to the fit when only actions are used ("dyn only").

Next we carried out a fit of the gravitational potential by varying the parameters M_s and r_s of the model over an iteratively refined grid in parameter space, first using only the actions and then adding the extra chemical abundance dimensions one at a time. We then looked at the contrast between the maximum value of the MI (at the best-fit potential) and the value far from the best-fit peak, and likewise at the value of the MI for the parameters obtained when the NFW function was fit to the potential values calculated directly from the simulation (i.e., Figure 3). We compare to this baseline MI far from the best fit because the MI is an entropy and therefore an extensive quantity, meaning that its absolute value will increase somewhat when a dimension is added even if that dimension is not informative. However, as is evident from the right panel of Figure 4, the *difference* between the peak and baseline MI values increases as new abundance dimensions are introduced, suggesting that the height of the best-fit peak is increasing (and therefore the confidence interval around the best fit should shrink).

Once the best-fit potential is obtained, the different building blocks of the accreted stellar halo can be separated from one another by running a group-finding algorithm on the same space of actions and abundances. Here again one expects that adding abundance information to constants-of-motion space can help distinguish the different components. Figure 5 illustrates how groups found in constants of motion space alone using EnLink (left) indeed represent different tidal streams (center) but also have distinct, though overlapping, abundance distributions (right). This bodes well for future attempts to reconstruct the component dwarf galaxies of the Milky Way's stellar halo.

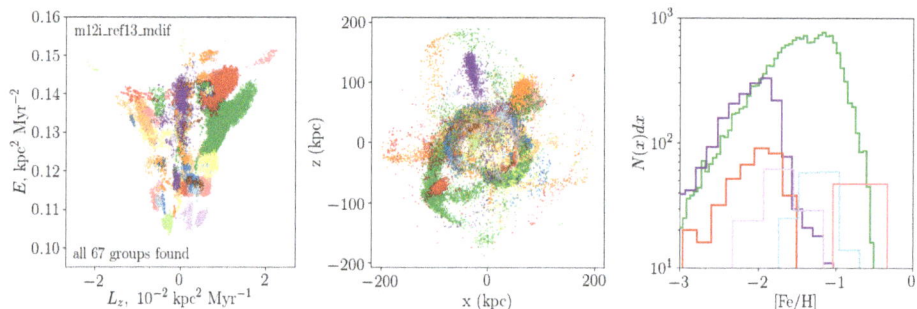

Figure 5. **Left:** groups found by EnLink group finder [37] in the space of constants of motion for star particles with $50 < d < 300$ kpc in a high-resolution simulated Milky Way ($m_p = 7070\,M_\odot$) from the Latte suite, including a more realistic model for the turbulent diffusion of chemical elements [39] that results in more plausible abundance distributions. Different groups are plotted in different colors. **Center:** the same groups shown in Cartesian coordinates (with origin at the center of the main simulated galaxy) show the features of tidal streams. **Right:** a selection of 5 groups shows the variety of different distributions in iron abundance among the different building blocks of the stellar halo.

4. Discussion

These results, obtained using the star particle data from a realistic simulation of a MW-like galaxy, indicate that studying tidal streams in the combined space of orbits and abundances is a promising direction both for constraining the Galactic gravitational potential and reconstructing the destroyed galaxies of the halo. Most interestingly, these results were obtained with a simulation in which the diffusion of chemical elements into star-forming gas was *not* implemented, which produces broader abundance distributions compared to those shown in Figure 5, and broader than the known abundance distributions of bound MW dwarf galaxies (Escala et al., in prep). Thus these results can be understood as a conservative estimate for the extra information contributed by elemental abundances.

On the other hand, these results assumed perfect data without observational uncertainties; furthermore, those uncertainties depend strongly on the type of stellar tracer for both phase-space coordinates and abundances. For example, variable-star standard candles like RR Lyra stars will have extremely precise distances [13,41] but pose difficulties when obtaining precise radial velocities or elemental abundances thanks to the same physics that makes them standard candles. Conversely, bright, cool giant stars will be visible to great distances and have well-constrained abundances and RVs, but less precise distances. Even more fundamentally, different stellar tracers have different specific frequencies depending on the underlying metal content of the population; for example, RR Lyr are more common in metal-poor populations and M giants are more common in metal-rich populations. As a result, they probe different regions of the Galaxy, and different regions in the stellar halo in particular, as illustrated in the left panel of Figure 6.

To better account for both the variety of stellar tracers and the range of expected observational errors, we are now updating the synthetic survey code Galaxia [42] to self-consistently resample the high-resolution Latte simulations with synthetic stars drawn from PARSEC isochrones [43]. A preliminary example is shown in the right panel of Figure 6. When complete, this machinery will allow us to make robust, detailed comparisons between the Milky Way and simulated galaxies, and accurately forecast the performance of various methods to constrain the Galactic gravitational potential and untangle the accretion history of the stellar halo.

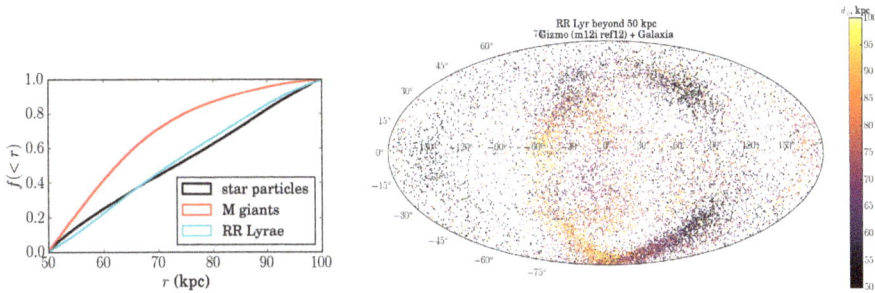

Figure 6. **Left**: Cumulative distance distributions for two stellar tracers, M giants (red) and RR Lyr (cyan), generated from the simulated Latte halo used in this work using a modification of the code Galaxia [42]. While the RR Lyr closely trace the distribution of star particles from the simulation (black), the M giants have significantly different, far more centrally concentrated distribution. **Right**: Sky distribution of the synthetic RR Lyr survey generated from the Latte halo. The color scale indicates Galactocentric distance.

Acknowledgments: R.S. is supported by an NSF Astronomy & Astrophysics Postdoctoral Fellowship under grant NSF-1400989. A.W. was supported by a Caltech-Carnegie Fellowship, in part through the Moore Center for Theoretical Cosmology and Physics at Caltech, and by NASA through grant HST-GO-14734 from STScI. Numerical calculations were run on allocations TG-AST130039 & TG-AST150080 granted by the Extreme Science and Engineering Discovery Environment (XSEDE) supported by the NSF, and the NASA High-End Computing (HEC) Program through the NASA Advanced Supercomputing (NAS) Division at Ames Research Center on allocation SMD-16-7592.

Author Contributions: R.S. developed the mass-modeling technique used for this work and generated the mock halos on which tests were performed, presented the work at the conference and wrote this conference proceedings. The high-resolution simulations of the Milky Way used for this work were run by A.W. using code (GIZMO/FIRE) developed by A.W. and P.H. S.S. kindly provided the group finder used in this work and was the original author of the Galaxia resampling code modified for this work by R.S.

Conflicts of Interest: The authors declare no conflict of interest.

References

1. Chambers, K.C.; Magnier, E.A.; Metcalfe, N.; Flewelling, H.A.; Huber, M.E.; Waters, C.Z.; Denneau, L.; Draper, P.W.; Farrow, D.; Finkbeiner, D.P.; et al. The Pan-STARRS1 Surveys. *arXiv* **2016**, arXiv:astro-ph.IM/1612.05560.

2. LSST Science Collaboration; Abell, P.A.; Allison, J.; Anderson, S.F.; Andrew, J.R.; Angel, J.R.P.; Armus, L.; Arnett, D.; Asztalos, S.J.; Axelrod, T.S.; et al. LSST Science Book, Version 2.0. *arXiv* **2009**, arXiv:astro-ph.IM/0912.0201.

3. De Jong, R.S.; Bellido-Tirado, O.; Chiappini, C.; Depagne, É.; Haynes, R.; Johl, D.; Schnurr, O.; Schwope, A.; Walcher, J.; Dionies, F.; et al. 4MOST: 4-Metre Multi-Object Spectroscopic Telescope. In Proceedings of the SPIE Astronomical Instrumentation and Telescopes Conference, Amsterdam, The Netherlands, 1–6 July 2012; McLean, I.S., Ramsay, S.K., Takami, H., Eds.; SPIE: Bellingham, DC, USA, 2012; Volume 8446.

4. Flaugher, B.; Bebek, C. The Dark Energy Spectroscopic Instrument (DESI). In Proceedings of the Ground-based and Airborne Instrumentation for Astronomy V, Montréal, QC, Canada, 22 June 2014; Ramsay, S.K., McLean, I.S., Takami, H., Eds.; SPIE: Bellingham, DC, USA, 2014; Volume 9147.

5. Dalton, G.; Trager, S.C.; Abrams, D.C.; Carter, D.; Bonifacio, P.; Aguerri, J.A.L.; MacIntosh, M.; Evans, C.; Lewis, I.; Navarro, R.; et al. WEAVE: The next generation wide-field spectroscopy facility for the William Herschel Telescope. In Proceedings of the Ground-based and Airborne Instrumentation for Astronomy IV, Amsterdam, The Netherlands, 1 July 2012; Ramsay, S.K., Takami, H., Eds.; SPIE: Bellingham, DC, USA, 2012; Volume 8446.

6. Chiba, M.; Cohen, J.; Wyse, R.F.G. Galactic Archaeology with the Subaru Prime Focus Spectrograph. In Proceedings of the International Astronomical Union, Honolulu, HI, USA, 3 August 2015; Bragaglia, A., Arnaboldi, M., Rejkuba, M., Romano, D., Eds.; IAU: Paris, France, 2016; Volume 317, pp. 280–281.

7. Gaia Collaboration; Prusti, T.; de Bruijne, J.H.J.; Brown, A.G.A.; Vallenari, A.; Babusiaux, C.; Bailer-Jones, C.A.L.; Bastian, U.; Biermann, M.; Evans, D.W.; et al. The Gaia mission. *Astron. Astrophys.* **2016**, *595*, A1.

8. Spergel, D.; Gehrels, D.; Baltay, C.; Bennett, D.; Breckinridge, J.; Donahue, M.; Dressler, A.; Gaudi, B.S.; Greene, T.; Guyon, O.; et al. Wide-Field InfrarRed Survey Telescope-Astrophysics Focused Telescope Assets WFIRST-AFTA 2015 Report. *arXiv* **2015**, arXiv:1503.03757.

9. Ness, M.; Hogg, D.W.; Rix, H.W.; Ho, A.Y.Q.; Zasowski, G. The Cannon: A data-driven approach to Stellar Label Determination. *Astrophys. J. Suppl. Ser.* **2015**, *808*, 21.

10. Alam, S.; Albareti, F.D.; Prieto, C.A.; Anders, F.; Anderson, S.F.; Andrews, B.H.; Armengaud, E.; Aubourg, E.; Bailey, S.; Bautista, J.E.; et al. The Eleventh and Twelfth Data Releases of the Sloan Digital Sky Survey: Final Data from SDSS-III. *Astrophys. J. Suppl. Ser.* **2015**, *219*, 27.

11. Martell, S.L.; Sharma, S.; Buder, S.; Duong, L.; Schlesinger, K.J.; Simpson, J.; Lind, K.; Ness, M.; Marshall, J.P.; Asplund, M.; et al. The GALAH survey: Observational overview and Gaia DR1 companion. *Mon. Not. R. Astron. Soc.* **2017**, *465*, 3203–3219.

12. Gilmore, G.; Randich, S.; Asplund, M.; Binney, J.; Bonifacio, P.; Drew, J.; Feltzing, S.; Ferguson, A.; Jeffries, R.; Micela, G.; et al. The Gaia-ESO Public Spectroscopic Survey. *Messenger* **2012**, *147*, 25–31.

13. Sesar, B.; Hernitschek, N.; Dierickx, M.I.P.; Fardal, M.A.; Rix, H.W. The > 100 kpc Distant Spur of the Sagittarius Stream and the Outer Virgo Overdensity, as Seen in PS1 RR Lyrae Stars. *arXiv* **2017**, arXiv:1706.10187.

14. Eadie, G.M.; Springford, A.; Harris, W.E. Bayesian Mass Estimates of the Milky Way: Including Measurement Uncertainties with Hierarchical Bayes. *Astrophys. J. Suppl. Ser.* **2017**, *835*, 167.

15. Law, D.R.; Majewski, S.R.; Johnston, K.V. Evidence for a Triaxial Milky Way Dark Matter Halo from the Sagittarius Stellar Tidal Stream. *Astrophys. J. Suppl. Ser.* **2009**, *703*, L67–L71.

16. Vera-Ciro, C.; Helmi, A. Constraints on the Shape of the Milky Way Dark Matter Halo from the Sagittarius Stream. *Astrophys. J. Suppl. Ser.* **2013**, *773*, L4.

17. Peñarrubia, J.; Gómez, F.A.; Besla, G.; Erkal, D.; Ma, Y.Z. A timing constraint on the (total) mass of the Large Magellanic Cloud. *Mon. Not. R. Astron. Soc.* **2016**, *456*, L54–L58.

18. Laporte, C.F.P.; Gómez, F.A.; Besla, G.; Johnston, K.V.; Garavito-Camargo, N. Response of the Milky Way's disc to the Large Magellanic Cloud in a first infall scenario. *arXiv* **2016**, arXiv:1608.04743.

19. Mao, Y.Y.; Williamson, M.; Wechsler, R.H. The Dependence of Subhalo Abundance on Halo Concentration. *Astrophys. J. Suppl. Ser.* **2015**, *810*, 21.

20. Newberg, H.J.; Yanny, B.; Rockosi, C.; Grebel, E.K.; Rix, H.W.; Brinkmann, J.; Csabai, I.; Hennessy, G.; Hindsley, R.B.; Ibata, R.; et al. The Ghost of Sagittarius and Lumps in the Halo of the Milky Way. *Astrophys. J. Suppl. Ser.* **2002**, *569*, 245–274.

21. Bochanski, J.J.; Willman, B.; Caldwell, N.; Sanderson, R.; West, A.A.; Strader, J.; Brown, W. The Most Distant Stars in the Milky Way. *Astrophys. J. Suppl. Ser.* **2014**, *790*, L5.

22. Sanderson, R.E.; Secunda, A.; Johnston, K.V.; Bochanski, J.J. New views of the distant stellar halo. *Mon. Not. R. Astron. Soc.* **2017**, in press.

23. Arnold, V.I.; Shandarin, S.F.; Zeldovich, I.B. The large scale structure of the universe. I - General properties One- and two-dimensional models. *Geophys. Astrophys. Fluid Dyn.* **1982**, *20*, 111–130.

24. Binney, J.; Tremaine, S. *Galactic Dynamics*, 2nd ed.; Princeton University Press: Princeton, NJ, USA, 2008.

25. Goldstein, H.; Poole, C.; Safko, J. *Classical Mechanics*, 3rd ed.; Addison Wesley: Boston, MA, USA, 2002.

26. Sanderson, R.E.; Helmi, A.; Hogg, D.W. Action-space Clustering of Tidal Streams to Infer the Galactic Potential. *Astrophys. J. Suppl. Ser.* **2015**, *801*, 98.

27. Sanderson, R.E.; Hartke, J.; Helmi, A. Modeling the Gravitational Potential of a Cosmological Dark Matter Halo with Stellar Streams. *Astrophys. J. Suppl. Ser.* **2017**, *836*, 234.

28. Helmi, A.; White, S.D.M. Building up the stellar halo of the Galaxy. *Mon. Not. R. Astron. Soc.* **1999**, *307*, 495–517.

29. Helmi, A.; de Zeeuw, P.T. Mapping the substructure in the Galactic halo with the next generation of astrometric satellites. *Mon. Not. R. Astron. Soc.* **2000**, *319*, 657–665.

30. Sharma, S.; Johnston, K.V.; Majewski, S.R.; Muñoz, R.R.; Carlberg, J.K.; Bullock, J. Group Finding in the Stellar Halo Using M-giants in the Two Micron All Sky Survey: An Extended View of the Pisces Overdensity? *Astrophys. J. Suppl. Ser.* **2010**, *722*, 750–759.

31. Gómez, F.A.; Helmi, A.; Brown, A.G.A.; Li, Y.S. On the identification of merger debris in the Gaia era. *Mon. Not. R. Astron. Soc.* **2010**, *408*, 935–946.

32. Cooper, A.P.; Cole, S.; Frenk, C.S.; White, S.D.M.; Helly, J.; Benson, A.J.; De Lucia, G.; Helmi, A.; Jenkins, A.; Navarro, J.F.; et al. Galactic stellar haloes in the CDM model. *Mon. Not. R. Astron. Soc.* **2010**, *406*, 744–766.

33. Peñarrubia, J.; Koposov, S.E.; Walker, M.G. A Statistical Method for Measuring the Galactic Potential and Testing Gravity with Cold Tidal Streams. *Astrophys. J. Suppl. Ser.* **2012**, *760*, 2.

34. Magorrian, J. Bayes versus the virial theorem: Inferring the potential of a galaxy from a kinematical snapshot. *Mon. Not. R. Astron. Soc.* **2014**, *437*, 2230–2248.

35. Sanderson, R.E. Inferring the Galactic Potential with GAIA and Friends: Synergies with Other Surveys. *Astrophys. J. Suppl. Ser.* **2016**, *818*, 41.

36. Kullback, S.; Leibler, R.A. On Information and Sufficiency. *Ann. Math. Stat.* **1951**, *22*, 79–86.

37. Sharma, S.; Johnston, K.V. A Group Finding Algorithm for Multidimensional Data Sets. *Astrophys. J. Suppl. Ser.* **2009**, *703*, 1061–1077.

38. Wetzel, A.R.; Hopkins, P.F.; Kim, J.H.; Faucher-Giguère, C.A.; Kereš, D.; Quataert, E. Reconciling Dwarf Galaxies with ΛCDM Cosmology: Simulating a Realistic Population of Satellites around a Milky Way-mass Galaxy. *Astrophys. J. Lett.* **2016**, *827*, L23.

39. Hopkins, P.F.; Wetzel, A.; Keres, D.; Faucher-Giguere, C.A.; Quataert, E.; Boylan-Kolchin, M.; Murray, N.; Hayward, C.C.; Garrison-Kimmel, S.; Hummels, C.; et al. FIRE-2 Simulations: Physics versus Numerics in Galaxy Formation. *arXiv* **2017**, arXiv:1702.06148.

40. Navarro, J.F.; Frenk, C.S.; White, S.D.M. The Structure of Cold Dark Matter Halos. *Astrophys. J. Suppl. Ser.* **1996**, *462*, 563.

41. Beaton, R.L.; Freedman, W.L.; Madore, B.F.; Bono, G.; Carlson, E.K.; Clementini, G.; Durbin, M.J.; Garofalo, A.; Hatt, D.; Jang, I.S.; et al. The Carnegie-Chicago Hubble Program. I. An Independent Approach to the Extragalactic Distance Scale Using Only Population II Distance Indicators. *Astrophys. J. Suppl. Ser.* **2016**, *832*, 210.

42. Sharma, S.; Bland-Hawthorn, J.; Johnston, K.V.; Binney, J. Galaxia: A Code to Generate a Synthetic Survey of the Milky Way. *Astrophys. J. Suppl. Ser.* **2011**, *730*, 3.

43. Bressan, A.; Marigo, P.; Girardi, L.; Salasnich, B.; Dal Cero, C.; Rubele, S.; Nanni, A. PARSEC: Stellar tracks and isochrones with the PAdova and TRieste Stellar Evolution Code. *Mon. Not. R. Astron. Soc.* **2012**, *427*, 127–145.

galaxies

MDPI

conference Report

Constraints on the Formation of M31's Stellar Halo from the SPLASH Survey

Karoline Gilbert [1,2]

[1]   Space Telescope Science Institute, 3700 San Martin Dr, Baltimore, MD 21218, USA; kgilbert@stsci.edu;
       Tel.: +1-410-338-2475
[2]   Department of Physics & Astronomy, Johns Hopkins University, Baltimore, MD 21218, USA

Academic Editors: Duncan A. Forbes and Ericson D. Lopez
Received: 16 August 2017; Accepted: 13 September 2017; Published: 30 September 2017

Abstract: The SPLASH (Spectroscopic and Photometric Landscape of Andromeda's Stellar Halo) Survey has observed fields throughout M31's stellar halo, dwarf satellites, and stellar disk. The observations and derived measurements have either been compared to predictions from simulations of stellar halo formation or modeled directly in order to derive inferences about the formation and evolution of M31's stellar halo. We summarize some of the major results from the SPLASH survey and the resulting implications for our understanding of the build-up of M31's stellar halo.

Keywords: galaxies: evolution; galaxies: halos; galaxies: individual (M31)

1. Introduction

Observations of stellar halos provide the potential to decipher both the early and recent merger histories of nearby galaxies (Figure 1). Recent accretion events are visible as strong enhancements in stellar density, and typically as kinematically cold features in stellar velocity distributions. These features can be modeled in detail to determine the properties of the progenitor system and its orbit (e.g., [1]). Simulations of stellar halo formation in a cosmological context make predictions for the global properties (e.g., stellar density profiles, metallicity gradients, and fraction of stars formed in situ and accreted) of stellar halos for hosts spanning a range of masses and merger histories (e.g., [2–5]). Suites of simulated stellar halos enable comparisons of the observed global properties of stellar halos to simulations to make inferences about the early accretion history of the halos.

The proximity of the Andromeda galaxy (M31) provides a unique opportunity to study the global properties of a stellar halo in great detail. The ability to make measurements of individually resolved stars is powerful, enabling measurements of the stellar density to very low surface brightness, stellar line of sight velocities, chemical abundances, and star formation histories. This has led to a significant investment of time spent observing M31, including by the SPLASH (Spectroscopic and Photometric Landscape of Andromeda's Stellar Halo), PAndAS (Pan-Andromeda Archaeological Survey; e.g., [6,7]), and PHAT (Panchromatic Hubble Andromeda Treasury; [8]) teams. Here we summarize the contributions to understanding the formation and evolution of M31's stellar halo made by the SPLASH survey.

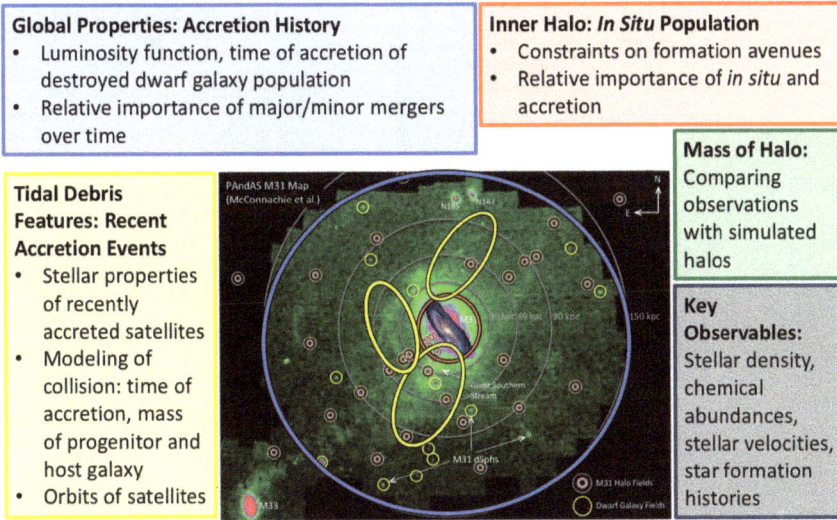

Figure 1. A summary of what can be learned from observations of stellar halos. The underlying figure (Figure 2 of [9]) portrays the stellar surface density of M31's stellar halo, and was generated from resolved star counts derived from images obtained with the MegaCam instrument on the Canada-France-Hawaii Telescope by the PAndAS survey [6].

2. The SPLASH Survey

The SPLASH collaboration has obtained photometry and spectroscopy of fields throughout M31's stellar halo and disk.

Photometry was primarily obtained with the Mosaic camera on the Kitt Peak 4 m Mayall Telescope, imaging 78 fields in the broad-band M and T_2 filters, and the narrow-band DDO51 filter (PIs S. Majewski and R. Beaton). The DDO51 filter overlaps the surface-gravity-sensitive Mg b and MgH stellar ab-sorption features, which are strong in dwarf stars but weak in red giant branch (RGB) stars. This filter combination allows for the photometric pre-selection of stars likely to be M31 red giant branch stars (e.g., [10]), greatly increasing our spectroscopic efficiency over a large region of M31's stellar halo. We have also utilized broad-band photometry obtained with SuprimeCam on Subaru, and Hubble Space Telescope imaging of M31's disk.

Spectroscopy of individual stars was obtained with the DEIMOS spectrograph on the Keck II 10 m telescope, primarily at $R \sim 6000$. The SPLASH collaboration has obtained more than 20,000 M31 stellar spectra in ∼170 spectroscopic masks targeting Andromeda's disk, dwarf galaxies, and halo, in fields ranging from 2 to 230 kpc in projected distance from Andromeda's center (Figure 2).

The SPLASH dataset has led to the discovery and characterization of Andromeda's extended metal-poor stellar halo [11–14]. In addition to the studies discussed below (global properties, identification and characterization of tidal debris features), it has also been used to study Andromeda's dwarf satellites [15–21] and to characterize M31's stellar disk [22].

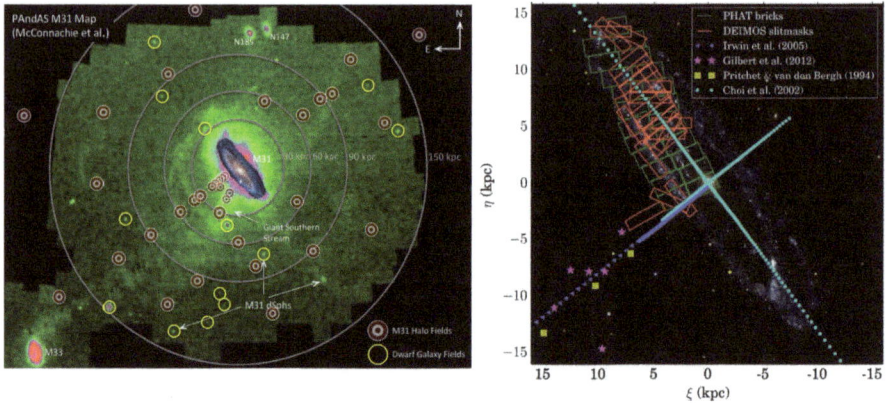

Figure 2. The locations of M31 halo and disk spectroscopic fields in the SPLASH survey. (**Left**) Fields targeting M31's halo, tidal debris features, and dwarf galaxies, superimposed on a stellar density map created from the Pan-Andromeda Archaeological Survey (PAndAS) observations [23] (Figure 2 of Gilbert et al. [9]). (**Right**) Fields targeting M31's disk; many of these spectroscopic masks were designed using photometry from the Hubble Space Telescope Multi Cycle Treasury Program PHAT (Figure 3 of Dorman et al. [24]).

3. The Properties of M31's Halo Measured by SPLASH

Below, we briefly summarize a selection of measurements made with the SPLASH dataset and their implications for the formation and evolution of Andromeda and its stellar halo.

In each study, spectroscopic and photometric measurements were used to identify secure samples of M31 RGB stars, enabling us to remove Milky Way dwarf star contaminants (e.g., [12]). The stellar velocity distributions in each field were used to identify stars that were likely to be associated with kinematically cold tidal debris features (e.g., [25–27]).

3.1. Global Properties of Andromeda's Stellar Halo

3.1.1. Surface Brightness and Metallicity Profiles

SPLASH observations in 38 halo fields, spanning all quadrants of the halo and ranging from 9 to 230 kpc in projected distance from the center of M31 (Figure 2), have been used to measure the surface brightness and metallicity profiles of M31's stellar halo.

The surface brightness profile of Andromeda's stellar halo is consistent with a single power-law with a power law index of -2.2 ± 0.2, extending to a projected distance of more than 175 kpc from M31's center (Figure 3; [27]). This is true regardless of whether tidal debris features are included in the profile, although the inclusion of tidal debris features does affect the normalization and index of the power-law fit. A similar surface brightness profile was found using the PAndAS star count data [7], and an extension of a power-law profile into the inner regions of M31 (to 3 kpc in projected distance from M31's center) was observed in blue horizontal branch stars with PHAT data [28,29].

There is no sign of a downward break in the surface brightness profile, out to ~2/3 of the estimated virial radius (~260 kpc; [30]). This was found to be atypical of simulated stellar halos, most of which do display a break in the surface brightness profile well within the range of projected radii probed by the SPLASH dataset. In the Bullock and Johnston [2] simulated halos, the only halo without a downward break in the stellar density profile is the one with many recent low-mass accretions. Future observations of the outer halo of M31—including measurements of the [Fe/H] and [α/Fe] abundances [31,32]—could confirm whether the stars in the outer halo are consistent with such a scenario: the mean [Fe/H] of a dwarf galaxy is strongly correlated with its luminosity (e.g., [33]), while the [α/Fe] abundance is sensitive to the timescale over which stars formed (e.g., [34]). Together, the [Fe/H] and [α/Fe] abundances of halo stars can thus be used to infer the luminosity and time since accretion of the dwarf satellite progenitors (e.g., [1,2,35]).

The SPLASH data also reveal that Andromeda's stellar halo has a significant gradient in metallicity out to projected distances of ~100 kpc [36], with a total decrease of ~1 dex. A significant gradient is observed whether or not tidal debris features are included in the measurement. The [Fe/H] measurements shown in Figure 3 are for a spectroscopically selected set of M31 RGB stars, but are based on photometry. Spectroscopic estimates of [Fe/H] for a subset of the highest S/N spectra, based on the equivalent width of the calcium triplet, result in a consistent gradient of metallicity with radius. A decrease in metallicity with radius was also observed in the PAndAS data [7].

Global Properties of M31's Halo
Surface Brightness and Metallicity Profile to 180 kpc

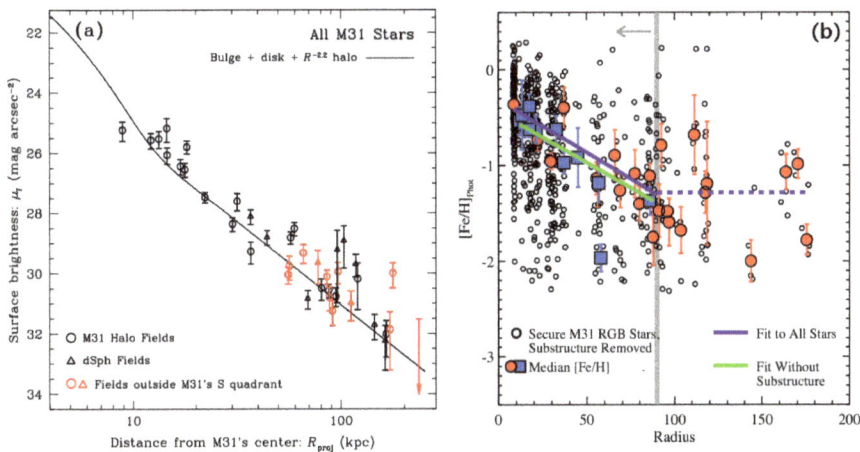

Figure 3. (**a**) The surface brightness profile of M31's stellar halo from the SPLASH survey (Figure 6 of Gilbert et al. [27]). The number of spectroscopically confirmed M31 RGB and Milky Way (MW) dwarf stars is used to determine the surface brightness in each SPLASH spectroscopic field. The surface brightness profile of halo stars after tidal debris features have been removed follows an $r^{-2.2}$ power law to large distances from M31's center. (**b**) Metallicity profile of M31's stellar halo from the SPLASH survey (Figure 10 of Gilbert et al. [36]). Metallicity estimates are based on comparing the position of the star in the color magnitude diagram to a grid of isochrones at the distance of M31. All spectroscopically confirmed M31 RGB stars that are not associated with tidal debris features are shown as small, open black points; the median metallicity and estimated uncertainty in the median value are shown as large, solid red circles (fields without tidal debris features) and blue squares (fields with tidal debris features). No tidal debris features are identified in fields greater than 90 kpc from the center of M31, in which the samples of M31 stars are small. M31's halo shows a significant gradient with metallicity to at least 100 kpc. The solid lines show fits to the median metallicity in fields out to 90 kpc, while the dashed line shows the median metallicity of stars in fields at projected distances >90 kpc from M31's center.

Based on comparisons with simulations of stellar halo formation [3,5], the observed large-scale metallicity gradient extending over ~100 kpc may indicate that the majority of the stars in the halo were contributed by one to a few early and relatively massive accretion events (see discussion in [36]). This hypothesis can be tested with future observations of [Fe/H] and [α/Fe] abundances of stars in M31's halo.

3.1.2. Stellar Velocity Dispersion

Gilbert et al. [9] have modeled the velocity distribution of stars in M31's stellar halo, using a set of 50 spectroscopic fields. We used Markov Chain Monte Carlo techniques to sample the parameter space of a Gaussian mixture model that includes the M31 halo, all known M31 tidal debris features in our fields, as well as three components for the MW contamination (disk, thick disk, and halo). The model also includes the probability that a star is an M31 RGB star or a MW dwarf star as a prior. This procedure was performed in a series of radial bins, in order to trace the change in velocity dispersion of M31's stellar halo with projected distance from M31's center.

M31's stellar halo appears to have a fairly flat velocity dispersion profile with radius (Figure 4). The velocity dispersion of M31's stellar halo was found to be significantly flatter than that observed in M31's globular cluster system by the PAndAS team [37]. It should be noted that in addition to probing two distinct halo tracers, these studies have two significant differences, each of which could affect the measured slope of the power-law: [37] included rotation in the model, but did not account for tidal debris features, while [9] fit for tidal debris features, but (as shown here) did not include rotation in the model.

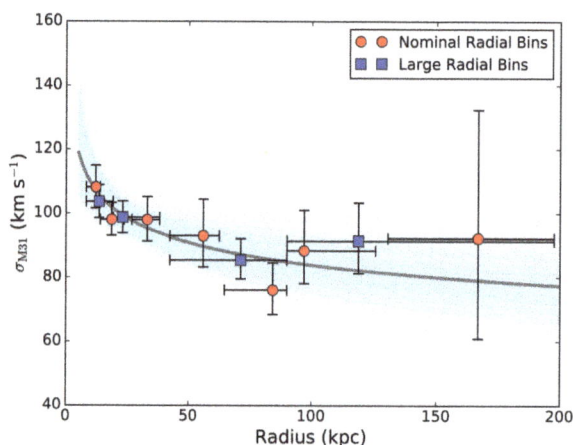

Figure 4. The stellar velocity dispersion of M31's halo as a function of projected distance from M31's center (Figure 10 of Gilbert et al. [9]). Points, placed at the median radius of all stars in the bin (with error bars denoting the full range of radii), show the 50th percentile of the marginalized one-dimensional posterior probability distribution for the velocity dispersion, with the error bar denoting the 16th to 84th percentile. The light blue curves show a subset of power-law fits to random draws from the posteriors; the dark gray curve shows a power-law defined by the 50th percentile values for the normalization and slope distributions. The velocity dispersion of stars in M31's halo remains relatively flat, with only a weak gradient, out to large radii.

This analysis provides a third global profile for comparing to simulations of stellar halo formation. Furthermore, since all known M31 tidal debris features in the spectroscopic fields were included in the model, we now have more precise constraints on the kinematical properties of these features, along with a significantly improved understanding of the uncertainties on the relevant parameters.

In the future, this analysis can be leveraged to provide a constraint on M31's total mass, using a separate set of halo tracers than the globular cluster and dwarf satellite systems. This can be done either through the use of the profile itself in conjunction with mass estimators such as that of Wolf et al. [38], or by using the velocities of the stars themselves along with tracer mass estimator techniques (e.g., [37,39]).

3.2. Tidal Debris Features

The velocity distributions of spectroscopically confirmed M31 stars in individual fields have been analyzed to identify and characterize tidal debris features, including measurements along the giant southern stream (GSS)—the most prominent tidal stream in M31's stellar halo [25–27,35,40,41].

These measurements also include the discovery of the continuation of the GSS [25,42,43], which forms a shell system in the inner regions of M31's halo. In conjunction with the giant southern stream itself, observations of the shells created by the progenitor of the GSS provide sensitive constraints for modeling the interaction of the progenitor with M31, as shown in Figure 5. Modeling of these features yields detailed inferences for the properties of the progenitor and its orbit (a dwarf galaxy with stellar mass comparable to that of the Large Magellanic Cloud (log $M = 9.5 \pm 0.1$), with the disruptive pericentric passage occurring 760 ± 50 Myr ago), as well as constraints on the total mass of M31 ($M_{200} \sim 2 \times 10^{12} \, M_\odot$) [44].

Detailed Dissection of Past Collision Events

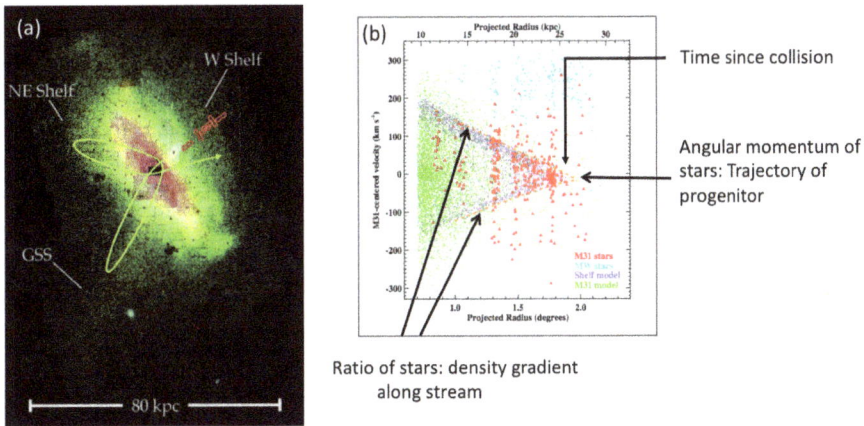

Figure 5. Shell systems, such as that created by the progenitor of the giant southern stream (GSS), provide exquisite observational constraints for modeling the orbit and properties of the dwarf satellite progenitor. Since confirmed as a shell system through the prediction and discovery of the southeast shelf [25,42], observations of the western shelf (shown here), in conjunction with previous observations of the GSS and southeast shelf, were used to determine the mass and disruption time of the progenitor, and to constrain the mass of M31 [44]. (**a**) Star count map of M31's inner halo, derived from the INT survey of M31 (image created by M. Irwin [45]), with the location of the SPLASH spectroscopic slitmasks covering the western shelf region overlaid in red (Figure 1 of Fardal et al. [43]). The green curve shows the progenitor's path in the model of Fardal et al. [43]. (**b**) M31-centric velocity ersus projected distance from M31's center for observed M31 stars (large red points) and simulated particles (Figure 8 of Fardal et al. [43]). A quantitative comparison of data with the simulations can constrain the orbit and properties of the progenitor system: the ratio of stars in the upper and lower caustic of the shell feature constrains the density gradient along the stream, while the location of the tip of the feature in projected distance and velocity is sensitive to the time since the disruptive pericentric passage and the angular momentum of the stars, respectively.

3.3. Andromeda's Inner Halo

Dorman et al. [24] modeled the relative contributions of M31's bulge, disk, and halo with radius, using a combination of stellar kinematics in M31's disk from SPLASH (Figure 2), the stellar luminosity function from resolved HST imaging (PHAT; [8] also shown in Figure 2), and unresolved surface photometry. The best fitting models had a disk fraction that was in significant tension with the fraction of stars with disk-like (dynamically cold) kinematics as measured from the velocity distribution of the stars. The disk fraction favored by analyzing all datasets in conjunction is $5.2 \pm 2.1\%$ higher than the fraction of stars with disk-like kinematics. Earlier work by Dorman et al. [46] also measured significant rotation for stars with halo-like kinematics in the inner halo of M31.

This is the first observational evidence for a population of halo stars that formed in the disk and were subsequently dynamically heated into halo-like orbits. While cosmological hydrodynamical simulations of stellar halo formation predict this as a mechanism for forming a portion of the inner regions of stellar halos (e.g., [47–49]), simulations differ in their predictions of the relative importance of this mechanism in terms of the fraction of stellar halo stars originally formed in a disk (see Section 6.1 of Dorman et al. [46] for a detailed discussion). Further observations, such as measurements of the [Fe/H] and [α/Fe] abundances of stars with halo-like kinematics, could lead to stronger constraints on the distribution of stars heated from M31's disk, since the different star formation histories of disk stars and halo stars will result in significant differences in the distribution of [α/Fe] as a function of [Fe/H].

4. Conclusions

The Andromeda system provides a unique testbed for furthering our physical understanding of galaxy formation and evolution. By observing individual stars throughout M31's stellar halo, we can probe the early accretion history of M31, constrain the fraction of stars in the inner halo that were once part of M31's disk, model the orbits and properties of recently disrupted dwarf satellites, and constrain the total mass of M31. Figure 6 summarizes the inferences that have been made regarding the formation of M31's stellar halo through the use of the SPLASH dataset. Further observations—in particular chemical abundance measurements of individual stars—hold much promise for furthering our understanding of the formation of M31's stellar halo.

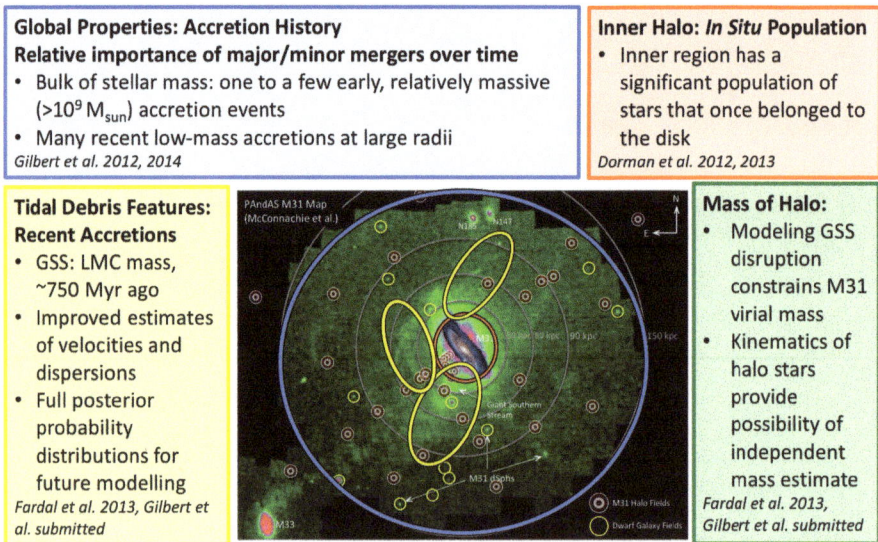

Figure 6. A summary of the major inferences that can be made about the formation of M31's stellar halo from measurements made using the SPLASH dataset. The underlying figure is the same as in Figure 1.

Acknowledgments: I thank the SPLASH team for their contributions to the work summarized here.

Conflicts of Interest: The author declares no conflict of interest.

Abbreviations

The following abbreviations are used in this manuscript:

M31   Andromeda
MW   Milky Way
PAndAS   Pan-Andromeda Archaeological Survey
PHAT   Panchromatic Hubble Andromeda Treasury
RGB   red giant branch
SPLASH   Spectroscopic and Photometric Landscape of Andromeda's Stellar Halo

References

1.   Johnston, K.V.; Bullock, J.S.; Sharma, S.; Font, A.; Robertson, B.E.; Leitner, S.N. Tracing Galaxy Formation with Stellar Halos. II. Relating Substructure in Phase and Abundance Space to Accretion Histories. *Astrophys. J.* **2008**, *689*, 936–957.

2.   Bullock, J.S.; Johnston, K.V. Tracing Galaxy Formation with Stellar Halos. I. Methods. *Astrophys. J.* **2005**, *635*, 931–949.

3.   Cooper, A.P.; Cole, S.; Frenk, C.S.; White, S.D.M.; Helly, J.; Benson, A.J.; De Lucia, G.; Helmi, A.; Jenkins, A.; Navarro, J.F.; et al. Galactic stellar haloes in the CDM model. *Mon. Not. R. Astron. Soc.* **2010**, *406*, 744–766.

4.   Font, A.S.; McCarthy, I.G.; Crain, R.A.; Theuns, T.; Schaye, J.; Wiersma, R.P.C.; Vecchia, C.D. Cosmological simulations of the formation of the stellar haloes around disc galaxies. *Mon. Not. R. Astron. Soc.* **2011**, *416*, 2802–2820.

5.   Tissera, P.B.; Beers, T.C.; Carollo, D.; Scannapieco, C. Stellar haloes in Milky Way mass galaxies: From the inner to the outer haloes. *Mon. Not. R. Astron. Soc.* **2014**, *439*, 3128–3138.

6.   McConnachie, A.W.; Irwin, M.J.; Ibata, R.A.; Dubinski, J.; Widrow, L.M.; Martin, N.F.; Côté, P.; Dotter, A.L.; Navarro, J.F.; Ferguson, A.M.N.; et al. The remnants of galaxy formation from a panoramic survey of the region around M31. *Nature* **2009**, *461*, 66–69.

7.   Ibata, R.A.; Lewis, G.F.; McConnachie, A.W.; Martin, N.F.; Irwin, M.J.; Ferguson, A.M.N.; Babul, A.; Bernard, E.J.; Chapman, S.C.; Collins, M.; et al. The Large-scale Structure of the Halo of the Andromeda Galaxy. I. Global Stellar Density, Morphology and Metallicity Properties. *Astrophys. J.* **2014**, *780*, 128.

8.   Dalcanton, J.J.; Williams, B.F.; Lang, D.; Lauer, T.R.; Kalirai, J.S.; Seth, A.C.; Dolphin, A.; Rosenfield, P.; Weisz, D.R.; Bell, E.F.; et al. The Panchromatic Hubble Andromeda Treasury. *Astrophys. J. Suppl.* **2012**, *200*, 18.

9.   Gilbert, K.M.; Tollerud, E.; Beaton, R.L.; Guhathakurta, P.; Bullock, J.S.; Chiba, M.; Kalirai, J.S.; Kirby, E.N.; Majewski, S.R.; Tanaka, M. Global Properties of M31's Stellar Halo From the SPLASH Survey: III. Measuring the Stellar Velocity Dispersion Profile. *Astrophys. J.* **2017**, submitted.

10. Palma, C.; Majewski, S.R.; Siegel, M.H.; Patterson, R.J.; Ostheimer, J.C.; Link, R. Exploring Halo Substructure with Giant Stars. IV. The Extended Structure of the Ursa Minor Dwarf Spheroidal Galaxy. *Astron. J.* **2003**, *125*, 1352–1372.

11. Guhathakurta, P.; Ostheimer, J.C.; Gilbert, K.M.; Rich, R.M.; Majewski, S.R.; Kalirai, J.S.; Reitzel, D.B.; Patterson, R.J. Discovery of an extended halo of metal-poor stars in the Andromeda spiral galaxy. *ArXiv* **2005**, arXiv:astro-ph/0502366.

12. Gilbert, K.M.; Guhathakurta, P.; Kalirai, J.S.; Rich, R.M.; Majewski, S.R.; Ostheimer, J.C.; Reitzel, D.B.; Cenarro, A.J.; Cooper, M.C.; Luine, C.; et al. A New Method for Isolating M31 Red Giant Stars: The Discovery of Stars out to a Radial Distance of 165 kpc. *Astrophys. J.* **2006**, *652*, 1188–1212.

13. Kalirai, J.S.; Gilbert, K.M.; Guhathakurta, P.; Majewski, S.R.; Ostheimer, J.C.; Rich, R.M.; Cooper, M.C.; Reitzel, D.B.; Patterson, R.J. The Metal-poor Halo of the Andromeda Spiral Galaxy (M31). *Astrophys. J.* **2006**, *648*, 389–404.

14. Courteau, S.; Widrow, L.M.; McDonald, M.; Guhathakurta, P.; Gilbert, K.M.; Zhu, Y.; Beaton, R.L.; Majewski, S.R. The Luminosity Profile and Structural Parameters of the Andromeda Galaxy. *Astrophys. J.* **2011**, *739*, 20–36.

15. Majewski, S.R.; Beaton, R.L.; Patterson, R.J.; Kalirai, J.S.; Geha, M.C.; Muñoz, R.R.; Seigar, M.S.; Guhathakurta, P.; Gilbert, K.M.; Rich, R.M.; et al. Discovery of Andromeda XIV: A Dwarf Spheroidal Dynamical Rogue in the Local Group? *Astrophys. J. Lett.* **2007**, *670*, L9–L12.
16. Kalirai, J.S.; Zucker, D.B.; Guhathakurta, P.; Geha, M.; Kniazev, A.Y.; Martínez-Delgado, D.; Bell, E.F.; Grebel, E.K.; Gilbert, K.M. The SPLASH Survey: A Spectroscopic Analysis of the Metal-Poor, Low-Luminosity M31 dSph Satellite Andromeda X. *Astrophys. J.* **2009**, *705*, 1043–1055.
17. Howley, K.M.; Geha, M.; Guhathakurta, P.; Montgomery, R.M.; Laughlin, G.; Johnston, K.V. Darwin Tames an Andromeda Dwarf: Unraveling the Orbit of NGC 205 Using a Genetic Algorithm. *Astrophys. J.* **2008**, *683*, 722–749.
18. Kalirai, J.S.; Beaton, R.L.; Geha, M.C.; Gilbert, K.M.; Guhathakurta, P.; Kirby, E.N.; Majewski, S.R.; Ostheimer, J.C.; Patterson, R.J.; Wolf, J. The SPLASH Survey: Internal Kinematics, Chemical Abundances, and Masses of the Andromeda I, II, III, VII, X, and XIV Dwarf Spheroidal Galaxies. *Astrophys. J.* **2010**, *711*, 671–692.
19. Tollerud, E.J.; Beaton, R.L.; Geha, M.C.; Bullock, J.S.; Guhathakurta, P.; Kalirai, J.S.; Majewski, S.R.; Kirby, E.N.; Gilbert, K.M.; Yniguez, B.; et al. The SPLASH Survey: Spectroscopy of 15 M31 Dwarf Spheroidal Satellite Galaxies. *Astrophys. J.* **2012**, *752*, 45.
20. Ho, N.; Geha, M.; Munoz, R.R.; Guhathakurta, P.; Kalirai, J.; Gilbert, K.M.; Tollerud, E.; Bullock, J.; Beaton, R.L.; Majewski, S.R. Stellar Kinematics of the Andromeda II Dwarf Spheroidal Galaxy. *Astrophys. J.* **2012**, *758*, 124.
21. Howley, K.M.; Guhathakurta, P.; van der Marel, R.; Geha, M.; Kalirai, J.; Yniguez, B.; Kirby, E.; Cuillandre, J.C.; Gilbert, K. Internal Stellar Kinematics of M32 from the SPLASH Survey: Dark Halo Constraints. *Astrophys. J.* **2013**, *765*, 65.
22. Dorman, C.E.; Guhathakurta, P.; Seth, A.C.; Weisz, D.R.; Bell, E.F.; Dalcanton, J.J.; Gilbert, K.M.; Hamren, K.M.; Lewis, A.R.; Skillman, E.D.; et al. A Clear Age-Velocity Dispersion Correlation in Andromeda's Stellar Disk. *Astrophys. J.* **2015**, *803*, 24.
23. McConnachie, A.W.; Huxor, A.; Martin, N.F.; Irwin, M.J.; Chapman, S.C.; Fahlman, G.; Ferguson, A.M.N.; Ibata, R.A.; Lewis, G.F.; Richer, H.; et al. A Trio of New Local Group Galaxies with Extreme Properties. *Astrophys. J.* **2008**, *688*, 1009–1020.
24. Dorman, C.E.; Widrow, L.M.; Guhathakurta, P.; Seth, A.C.; Foreman-Mackey, D.; Bell, E.F.; Dalcanton, J.J.; Gilbert, K.M.; Skillman, E.D.; Williams, B.F. A New Approach to Detailed Structural Decomposition from the SPLASH and PHAT Surveys: Kicked-up Disk Stars in the Andromeda Galaxy? *Astrophys. J.* **2013**, *779*, 103.
25. Gilbert, K.M.; Fardal, M.; Kalirai, J.S.; Guhathakurta, P.; Geha, M.C.; Isler, J.; Majewski, S.R.; Ostheimer, J.C.; Patterson, R.J.; Reitzel, D.B.; et al. Stellar Kinematics in the Complicated Inner Spheroid of M31: Discovery of Substructure along the Southeastern Minor Axis and Its Relationship to the Giant Southern Stream. *Astrophys. J.* **2007**, *668*, 245–267.
26. Gilbert, K.M.; Guhathakurta, P.; Kollipara, P.; Beaton, R.L.; Geha, M.C.; Kalirai, J.S.; Kirby, E.N.; Majewski, S.R.; Patterson, R.J. The Splash Survey: A Spectroscopic Portrait of Andromeda's Giant Southern Stream. *Astrophys. J.* **2009**, *705*, 1275–1297.
27. Gilbert, K.M.; Guhathakurta, P.; Beaton, R.L.; Bullock, J.; Geha, M.C.; Kalirai, J.S.; Kirby, E.N.; Majewski, S.R.; Ostheimer, J.C.; Patterson, R.J.; et al. Global Properties of M31's Stellar Halo from the SPLASH Survey. I. Surface Brightness Profile. *Astrophys. J.* **2012**, *760*, 76.
28. Williams, B.F.; Dalcanton, J.J.; Bell, E.F.; Gilbert, K.M.; Guhathakurta, P.; Lauer, T.R.; Seth, A.C.; Kalirai, J.S.; Rosenfield, P.; Girardi, L. The Panchromatic Hubble Andromeda Treasury. II. Tracing the Inner M31 Halo with Blue Horizontal Branch Stars. *Astrophys. J.* **2012**, *759*, 46.
29. Williams, B.F.; Dalcanton, J.J.; Bell, E.F.; Gilbert, K.M.; Guhathakurta, P.; Dorman, C.; Lauer, T.R.; Seth, A.C.; Kalirai, J.S.; Rosenfield, P.; et al. Tracing the Metal-poor M31 Stellar Halo with Blue Horizontal Branch Stars. *Astrophys. J.* **2015**, *802*, 49.
30. Seigar, M.S.; Barth, A.J.; Bullock, J.S. A revised Λ CDM mass model for the Andromeda Galaxy. *Mon. Not. R. Astron. Soc.* **2008**, *389*, 1911–1923.
31. Vargas, L.C.; Geha, M.C.; Tollerud, E.J. The Distribution of Alpha Elements in Andromeda Dwarf Galaxies. *Astrophys. J.* **2014**, *790*, 73.
32. Vargas, L.C.; Gilbert, K.M.; Geha, M.; Tollerud, E.J.; Kirby, E.N.; Guhathakurta, P. [α/Fe] Abundances of Four Outer M31 Halo Stars. *Astrophys. J. Lett.* **2014**, *797*, L2.

33. Kirby, E.N.; Cohen, J.G.; Guhathakurta, P.; Cheng, L.; Bullock, J.S.; Gallazzi, A. The Universal Stellar Mass-Stellar Metallicity Relation for Dwarf Galaxies. *Astrophys. J.* **2013**, *779*, 102.
34. Gilmore, G.; Wyse, R.F.G. Element Ratios and the Formation of the Stellar Halo. *Astron. J.* **1998**, *116*, 748–753.
35. Gilbert, K.M.; Font, A.S.; Johnston, K.V.; Guhathakurta, P. The Dominance of Metal-rich Streams in Stellar Halos: A Comparison Between Substructure in M31 and ΛCDM Models. *Astrophys. J.* **2009**, *701*, 776–786.
36. Gilbert, K.M.; Kalirai, J.S.; Guhathakurta, P.; Beaton, R.L.; Geha, M.C.; Kirby, E.N.; Majewski, S.R.; Patterson, R.J.; Tollerud, E.J.; Bullock, J.S.; et al. Global Properties of M31's Stellar Halo from the SPLASH Survey. II. Metallicity Profile. *Astrophys. J.* **2014**, *796*, 76.
37. Veljanoski, J.; Mackey, A.D.; Ferguson, A.M.N.; Huxor, A.P.; Côté, P.; Irwin, M.J.; Tanvir, N.R.; Peñarrubia, J.; Bernard, E.J.; Fardal, M.; et al. The outer halo globular cluster system of M31 - II. Kinematics. *Mon. Not. R. Astron. Soc.* **2014**, *442*, 2929–2950.
38. Wolf, J.; Martinez, G.D.; Bullock, J.S.; Kaplinghat, M.; Geha, M.; Muñoz, R.R.; Simon, J.D.; Avedo, F.F. Accurate masses for dispersion-supported galaxies. *Mon. Not. R. Astron. Soc.* **2010**, *406*, 1220–1237.
39. Watkins, L.L.; Evans, N.W.; An, J.H. The masses of the Milky Way and Andromeda galaxies. *Mon. Not. R. Astron. Soc.* **2010**, *406*, 264–278.
40. Guhathakurta, P.; Rich, R.M.; Reitzel, D.B.; Cooper, M.C.; Gilbert, K.M.; Majewski, S.R.; Ostheimer, J.C.; Geha, M.C.; Johnston, K.V.; Patterson, R.J. Dynamics and Stellar Content of the Giant Southern Stream in M31. I. Keck Spectroscopy of Red Giant Stars. *Astron. J.* **2006**, *131*, 2497–2513.
41. Kalirai, J.S.; Guhathakurta, P.; Gilbert, K.M.; Reitzel, D.B.; Majewski, S.R.; Rich, R.M.; Cooper, M.C. Kinematics and Metallicity of M31 Red Giants: The Giant Southern Stream and Discovery of a Second Cold Component at R = 20 kpc. *Astrophys. J.* **2006**, *641*, 268–280.
42. Fardal, M.A.; Guhathakurta, P.; Babul, A.; McConnachie, A.W. Investigating the Andromeda stream—III. A young shell system in M31. *Mon. Not. R. Astron. Soc.* **2007**, *380*, 15–32.
43. Fardal, M.A.; Guhathakurta, P.; Gilbert, K.M.; Tollerud, E.J.; Kalirai, J.S.; Tanaka, M.; Beaton, R.; Chiba, M.; Komiyama, Y.; Iye, M. A spectroscopic survey of Andromeda's Western Shelf. *Mon. Not. R. Astron. Soc.* **2012**, *423*, 3134–3147.
44. Fardal, M.A.; Weinberg, M.D.; Babul, A.; Irwin, M.J.; Guhathakurta, P.; Gilbert, K.M.; Ferguson, A.M.N.; Ibata, R.A.; Lewis, G.F.; Tanvir, N.R.; et al. Inferring the Andromeda Galaxy's mass from its giant southern stream with Bayesian simulation sampling. *Mon. Not. R. Astron. Soc.* **2013**, *434*, 2779–2802.
45. Irwin, M.J.; Ferguson, A.M.N.; Ibata, R.A.; Lewis, G.F.; Tanvir, N.R. A Minor-Axis Surface Brightness Profile for M31. *Astrophys. J. Lett.* **2005**, *628*, L105–L108, [arXiv:astro-ph/0505077].
46. Dorman, C.E.; Guhathakurta, P.; Fardal, M.A.; Lang, D.; Geha, M.C.; Howley, K.M.; Kalirai, J.S.; Bullock, J.S.; Cuillandre, J.C.; Dalcanton, J.J.; et al. The SPLASH Survey: Kinematics of Andromeda's Inner Spheroid. *Astrophys. J.* **2012**, *752*, 147.
47. Purcell, C.W.; Bullock, J.S.; Kazantzidis, S. Heated disc stars in the stellar halo. *Mon. Not. R. Astron. Soc.* **2010**, *404*, 1711–1718.
48. McCarthy, I.G.; Font, A.S.; Crain, R.A.; Deason, A.J.; Schaye, J.; Theuns, T. Global structure and kinematics of stellar haloes in cosmological hydrodynamic simulations. *Mon. Not. R. Astron. Soc.* **2012**, *420*, 2245–2262.
49. Tissera, P.B.; Scannapieco, C.; Beers, T.C.; Carollo, D. Stellar haloes of simulated Milky-Way-like galaxies: Chemical and kinematic properties. *Mon. Not. R. Astron. Soc.* **2013**, *432*, 3391–3400.

galaxies

MDPI

Article

Decoding Galactic Merger Histories

Eric F. Bell [1,*] , Antonela Monachesi [2], Richard D'Souza [1,3], Benjamin Harmsen [1],
Roelof S. de Jong [4], David Radburn-Smith [5], Jeremy Bailin [6] and Benne W. Holwerda [7]

[1] Department of Astronomy, University of Michigan, Ann Arbor, MI 48109, USA; radsouza@umich.edu (R.D.);
     benharms@umich.edu (B.H.)
[2] Departamento de Física y Astronomía, Universidad de La Serena, La Serena 1720236, Chile;
     amonachesi@userena.cl
[3] Vatican Observatory, Piazza Sabatini, 4B/5, 00041 Albano Laziale RM, Italy
[4] Leibniz-Institut für Astrophysik Potsdam (AIP), 14482 Potsdam, Germany; rdejong@iap.de
[5] Department of Astronomy, University of Washington, Seattle, WA 98195, USA; d.radburnsmith@gmail.com
[6] Department of Physics and Astronomy, University of Alabama, Tuscaloosa, AL 35487, USA; jbailin@ua.edu
[7] Department of Physics and Astronomy, University of Louisville, Louisville, KY 40292, USA;
     benne.holwerda@gmail.com
* Correspondence: ericbell@umich.edu; Tel.: +1-734-764-3408

Received: 30 June 2017; Accepted: 1 December 2017; Published: 8 December 2017

Abstract: Galaxy mergers are expected to influence galaxy properties, yet measurements of individual merger histories are lacking. Models predict that merger histories can be measured using stellar halos and that these halos can be quantified using observations of resolved stars along their minor axis. Such observations reveal that Milky Way-mass galaxies have a wide range of stellar halo properties and show a correlation between their stellar halo masses and metallicities. This correlation agrees with merger-driven models where stellar halos are formed by satellite galaxy disruption. In these models, the largest accreted satellite dominates the stellar halo properties. Consequently, the observed diversity in the stellar halos of Milky Way-mass galaxies implies a large range in the masses of their largest merger partners. In particular, the Milky Way's low mass halo implies an unusually quiet merger history. We used these measurements to seek predicted correlations between the bulge and central black hole (BH) mass and the mass of the largest merger partner. *We found no significant correlations*: while some galaxies with large bulges and BHs have large stellar halos and thus experienced a major or minor merger, half have small stellar halos and never experienced a significant merger event. These results indicate that bulge and BH growth is not solely driven by merger-related processes.

Keywords: galaxies: general; galaxies: evolution; galaxies: halos; galaxies: stellar content; galaxies: bulges; galaxies: merger history

1. Introduction

The cold dark matter paradigm predicts that the gravitational collapse and merger of dark matter halos is the prime driver of galaxy formation and growth. However, the response of stars and gas to these mergers is complex to model, and predictions of their effects on galaxy bulges, disks, and supermassive black holes are uncertain (e.g., [1,2]). Direct knowledge of the merger histories of galaxies would therefore be valuable in providing direct empirical tests of how a given merger affected a particular galaxy. Sadly, dark matter is, well... dark, and direct measurement of the merger history is impossible. Yet, because the largest dark matter subhalos are predicted to host visible satellite galaxies, we can use these satellites as visible tracers of the growth of the central galaxy's dark matter halo. The tidal disruption of these satellites is predicted to form a diffuse stellar halo (e.g., [3–6]). Consequently, study of these stellar halos gives unique—and perhaps the only available—insight into the merger and growth history of actual galaxies, and offers the possibility of understanding what mergers do to galaxies. Here we summarize the findings of [7–10] and give a brief overview of the progress that has been made towards measuring the stellar halos around nearby Milky Way mass galaxies, using them to infer the most prominent event in their merger and accretion histories, and using this knowledge to explore the role of merging in bulge and supermassive black hole growth.

This issue is the most urgent for Milky Way (MW) peers—galaxies with $M_* \sim 6 \times 10^{10} M_\odot$. In addition to holding most of the stellar mass in the present-day Universe [11], MW peers are diverse (e.g., [12]), spanning from bulgeless star-forming disks like the Milky Way or M101 to elliptical or lenticular galaxies like Centaurus A or the Sombrero galaxy. It is thought that much of this diversity should be driven by merger history (e.g., [1,2]). Galaxy merging plays an important role in the creation of elliptical galaxies (e.g., [13,14]). It is commonly argued that mergers are an important driver of the formation of at least the large "classical" galaxy bulges (e.g., [1,2,15]), although other mechanisms such as early bulge formation through chaotic collapse (e.g., [16]), violent disk instabilities (e.g., [17]), or dramatic changes in gas angular momentum [18] have been suggested. Further, given that feedback from supermassive black holes (BHs) may suppress star formation on galactic scales (e.g., [19,20]) and that merging appears to drive at least some BH growth [21,22], it is important to test the relationship between BH mass and merger history directly.

In order to assess what mergers do to galaxies (do mergers grow BHs, bulges, elliptical galaxies?), an independent and robust probe of the merger history of individual galaxies is urgently needed. Stellar halos appear to be just such a probe (see [7–10] for more details).

2. Stellar Halos Measure Merger History

Galactic outskirts are predicted to give powerful insight into their merger and accretion histories. Stars in merging galaxies are collisionless, and are torn by tides from their parent galaxy and spread out into a diffuse stellar halo (e.g., [4,6]). Because the stellar mass of satellites is a very strong function of dark matter subhalo mass, these stellar halos are predicted to accrete most of their mass from the few largest, and in most cases the *single* largest, satellite(s) (e.g., [5,10,23]).

This is illustrated in Figure 1. We use the Illustris hydrodynamical simulation [10,24] to predict the fraction of the accreted mass given by the single-most massive disrupted satellite (frac$_{Dom}$, on the x-axis) and the ratio of the second-most massive to the most massive progenitor (on the y-axis) for a sample of simulated galaxies with dark matter halo masses similar to the Milky Way ($12.05 < \log M_{DM}/M_\odot < 12.15$). A wide range of growth histories are predicted, from (relatively uncommon) stellar halos built up from multiple smaller accretions (top left) through to halos completely dominated by a single massive accretion (bottom right). The growth history broadly correlates with total accreted mass, where low-mass stellar halos are often built up from many smaller accretions, while typical and large stellar halos are dominated by a single accretion (in agreement with [23]). Most MW peer stellar halos are predicted to have more than half of their total halo mass come from the most massive progenitor, and usually this most massive progenitor is substantially more massive than the second-most massive progenitor. Because the only model ingredients required to robustly predict global stellar halo properties are realistic galaxy metallicity–mass relations and mass functions of accreted satellites [5,25], these predictions are robust and other models [23,26] give very similar results.

This framework makes an important prediction. Since galaxies show a strong relationship between their metallicity and stellar mass [27], and because stellar halos tend to be dominated by the most massive progenitor (see Figure 1 and references [10,23]), the metallicity and mass of accreted stellar halos are predicted to correlate strongly [8,10,23]. This offers a clear observational test of the accretion-driven growth of stellar halos, and we will return to it in Figures 2 and 3.

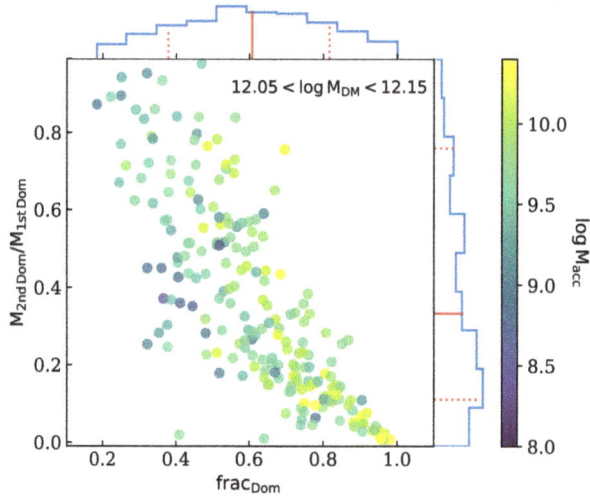

Figure 1. Stellar halos are predicted to have accreted their mass from the few largest, and in many cases the single largest, satellite(s). The *x*-axis shows frac$_{Dom}$, the fraction of the accreted mass given by the single-most massive disrupted satellite as predicted by the Illustris hydrodynamical simulation [10,24] for a sample of simulated galaxies with dark matter halo masses similar to the Milky Way (12.05 < log $M_{DM}/M_\odot$ < 12.15). The *y*-axis shows the ratio of the second-most massive to the most massive progenitor. Symbols are color-coded by the total accreted mass. Marginalized distributions of each quantity are shown, along with the 16, 50, and 84 percentiles of the distributions. The stellar halos of Milky Way (MW) peers are expected to show a wide range of properties, from the less-common halos built up from multiple smaller accretions (top left) through to the more common halos completely dominated by a single massive accretion (bottom right).

While it is not the focus of this contribution, in addition to the accreted stellar halo, there is considerable discussion about the extent to which stars formed in the main galaxy itself will exist at large radius—an in situ stellar halo. Simulators report that the prominence and origin of in situ stars varies considerably as a function of the parameters of simulations (e.g., the parameterization of stellar feedback, [28,29]). At the level of comparing observations with simulated total stellar halo masses, many simulations appear to over-produce in situ stellar halos (e.g., [8,10]).

While we await a clearer understanding of in situ stars, we have chosen to focus on the minor axis of galaxies with prominent disks. High-resolution hydrodynamical simulations predict that at minor axis distances > 15 kpc in MW peers, most stars should be accreted [30,31]. Estimates of total accreted mass and median metallicity can be inferred from minor axis measurements [10]. Using these estimates, one can measure the mass of the most massive merger partner (e.g., Figure 1 and reference [10]) and possibly when this merger occurred [10,23]. Consequently, it is important to carefully characterize the diffuse stellar halos—ideally along their minor axes—for a representative sample of galaxies.

The astronomical community is starting to assemble just such a sample. Aside from prominent individual stellar streams (e.g., [32,33]), the characterization of stellar halos has been very challenging owing to their extremely low surface brightnesses >30 *V*-band mag/arcsec^2. Recent deep diffuse light imaging has permitted detection of relatively massive halos (e.g., [34–36]). In parallel, studies of resolved red giant branch (RGB) stars in the outskirts of nearby galaxies reach fainter equivalent surface brightness limits and give estimates of both stellar halo masses and typical metallicities [7,8,37–41].

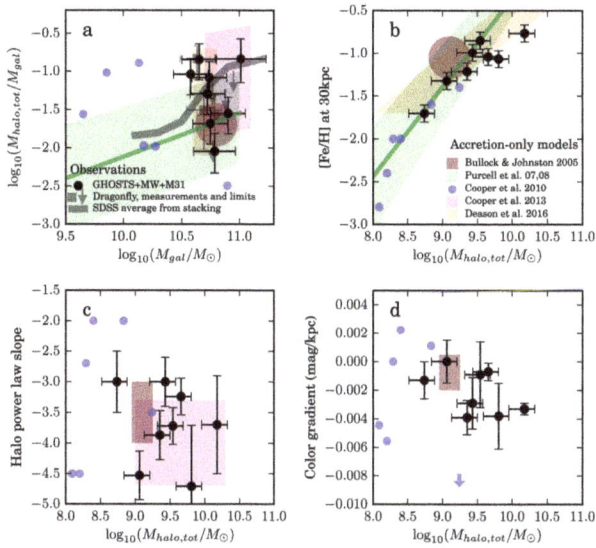

Figure 2. Both resolved star (black; [7,8,38,39,42–44]) and integrated light studies (gray; [36,45]) reveal a diversity of stellar halo masses, metallicities, density profiles, and metallicity gradients. These observational properties are in good agreement with models in which stellar halos are predominantly formed by the disruption of dwarf satellites (in colors). (**a**) Ratio of "total" stellar halo mass and total stellar mass, as a function of total stellar mass. Error bars include the uncertainty in extrapolating to "total" stellar halo mass; limits are shown with arrows. (**b**) Stellar halo metallicity at 30 kpc as a function of "total" stellar halo mass. (**c**) Inferred 3D minor axis stellar halo density power-law slope in the range 10–40 kpc as a function of "total" stellar halo mass. (**d**) Stellar halo color gradient (a proxy for metallicity gradient) as a function of "total" stellar halo mass. The observational data are shown in black and grey. Models: brick red area: [4]; light green+line: [5,25]; blue: [6]; magenta: [26]; orange: [23]. Adapted from Figure 16 from [8], reproduced with permission.

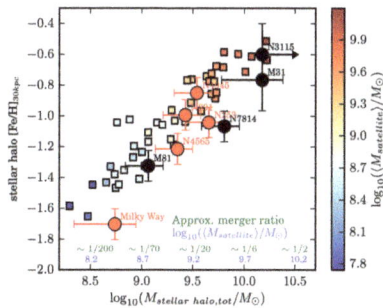

Figure 3. Circles show the observed correlation between stellar halo mass and metallicity. Galaxies with "classical" bulges are shown in black, galaxies with "pseudobulges" are shown in red. Squares show simulated stellar halo masses and metallicities from the accretion-only models of Deason et al. [23]. The squares are color-coded according to the mass weighted mean stellar mass of the contributing satellites to the halo; we indicate the approximate run of mass weighted mean accreted satellite mass in blue and corresponding approximate merger ratios for a main galaxy mass at the time of merger of $\sim 3 \times 10^{10} M_\odot$ in green. Adapted from Figure 1 of [9], reproduced with permission of the American Astronomical Society (AAS).

3. MW Peer Stellar Halos Are Diverse, Reflecting a Diversity in Merger and Accretion Histories

These observations show that MW peer stellar halos are diverse, spanning an order of magnitude or more in total mass and more than 1 dex in typical metallicity. This diversity is illustrated in Figure 2 (adapted from Figure 16 in [8]). Estimates of total stellar halo masses, metallicities at minor axis distances of 30 kpc, inferred 3D minor axis density gradients between 10 and 40 kpc, and minor axis metallicity gradients are shown in black for studies using resolved stars [7,8,38,39,42–44]. Gray symbols denote stellar halo masses or limits from integrated light studies [36,45]. Stellar halo masses have been extrapolated to give an estimate of total stellar halo masses. For this purpose, we use simulations of stellar halos including only accreted stars (e.g., [4,24,46]). For this particular analysis, we used the cosmologically-motivated particle-tagging models of [4], although we found that we recovered similar results with a particle-tagging model with more diverse accretion histories [26] and the accreted particles from two very different hydrodynamical simulations, Auriga and Illustris [9,10,24,31,47]. In these model accreted halos, most of the stellar halo mass lies at less than 10 kpc galactocentric radius. The typical correction to estimate a total stellar halo mass from "aperture" measurements (at e.g., 10–40 kpc) is a factor of several [8], with an estimated accuracy of around 40% with little systematic dependence on stellar halo properties (see also [9,10]).

Panel a of Figure 2 shows that MW peers with a factor of 4 range in total stellar mass ($3 \times 10^{10} M_\odot < M_* < 1.2 \times 10^{11} M_\odot$) show a factor of ~20 range in stellar halo fraction. We note that the correction of "aperture" stellar halo mass to a best estimate of total stellar halo mass is important, and brings the otherwise somewhat low "aperture" masses and limits presented by [36] into excellent agreement with [8,45], and in accord with predictions from accretion-only models for stellar halo formation. Panels c and d show a range in power law density profile and metallicity gradients, respectively, with little clear correlation of either with stellar halo mass (or other quantities, as discussed in [8]). Importantly, panel b shows that stellar halo metallicity (as measured on the minor axis at 30 kpc distance) and total stellar halo mass appear to correlate strongly—in agreement with the expectations of models of stellar halo formation through accretion only (as foreshadowed by [48] and explored by e.g., [10,23,49])—a result only accessible using resolved stellar populations.

This observational characterization of halos and confirmation of the predicted stellar halo metallicity–mass relation is important for a number of reasons.

- Following Section 2, it strongly suggests that the observations have indeed quantified the accreted stellar halo of galaxies.
- It shows that MW peers have a wide range of merger histories, where the Milky Way's low-mass stellar halo is very unusual. This is important because it tells us that the intuition that we have built about stellar halos from study of the Milky Way is incomplete.
- These accreted stellar halos can be used as a tool to quantify the properties of the most massive merger to have affected galaxies. The astronomical community now has quantitative access to the most massive event in individual galaxy's merger and accretion history.

The first and last of these ideas deserve some discussion.

We discuss the first point quantitatively in Figure 3 (adapted from Figure 1 of [9]), where we compare the observed stellar halo metallicity–mass relation with a set of modeled stellar halos in $10^{12} M_\odot$ dark matter halos from ([23], squares). The free parameters in these models are set *only* by the halo mass–stellar mass relation and metallicity–stellar mass relations of *galaxies*. When input model galaxies are realistic, the output accreted stellar halos closely reproduce the stellar halo metallicity–mass relation with appropriate normalization, slope, and scatter (see also [10] for an independent analysis using the Illustris hydrodynamical models with very similar results).

The relationship between stellar halo properties and the most massive accreted satellite have been articulated by [9,10,23]. In Figure 3, the models of [23] are color-coded by the mass-weighted mean stellar mass of all of the contributing satellites to the halo (termed "typical" satellite mass hereafter). The "typical" satellite mass is a strong function of the stellar halo mass (and metallicity). We provide an approximate mapping on the ordinate in blue in Figure 3. We also give an approximate merger ratio assuming a main galaxy mass at the time of merger of $\sim 3 \times 10^{10} M_\odot$ in green. Figure 3 suggests that one can broadly infer the masses of the largest satellites that were accreted by or merged into our nearby neighbors.

4. Bulges and Central Black Holes in Disk-Dominated Galaxies Correlate Poorly with Merger History

With a quantitative estimate of the most massive merger to have affected a galaxy in hand, we can now explore for the first time how MW peer merger history correlates with bulge and BH prominence. Figure 4 (Figure 2 of [9]) shows the bulge mass (left) and BH mass (right) as a function of stellar halo mass ($\sim$ mass of the most massive merger/accretion partner). No significant correlations between bulge/BH masses and stellar halo masses are detected in this dataset. Galaxies with $M_{stellar\,halo,tot} > 10^9 M_\odot$ have an order of magnitude spread in B/T ratio or bulge mass and two orders of magnitude spread in BH mass.

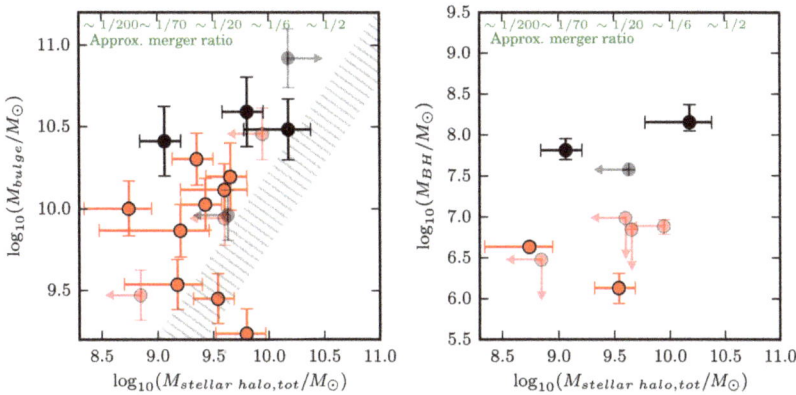

Figure 4. Bulge mass (left) and BH mass (right) as a function of stellar halo mass. Red denotes galaxies with low-mass "pseudobulges", black shows galaxies with higher-mass "classical" bulges; observational limits are shown with lighter shading. As argued in Figure 3, stellar halo mass reflects merger history, and approximate merger ratios are given in green. The shaded area in the left panel schematically illustrates what would be expected if there were a 1:1 correlation between stellar halo mass and bulge mass, as broadly expected in some simple modeling contexts (e.g., [1]). Figure 2 from [9], reproduced with permission of the AAS.

"Classical" bulges (black and grey) could have been expected to have been formed in major or minor mergers (e.g., [15]). Three galaxies with classical bulges (M31, NGC 3115, Cen A[1]) indeed have massive metal-rich stellar halos—carrying > 20% of the total galaxy stellar mass—indicative of a minor or major merger. Yet, another three galaxies with classical bulges (M81, NGC 4258, NGC 7814) have less-massive stellar halos. Most notable among these is M81, with a large classical bulge and an anemic stellar halo containing only $2 \pm 0.9\%$ of its total stellar mass. M81 shows no sign of any significant past major or minor merging activity that was expected to drive the formation of its classical bulge. A naïve one-to-one association between bulge and BH prominence and merging is ruled out, and places important constraints on models of bulge and BH formation and growth. In particular, such

---

[1] While Cen A's minor axis metallicity at 30 kpc suggests that it has a large stellar halo mass, no published estimate exists, and so it is not shown in Figure 4.

systems may be an excellent testing ground for models in which bulge and BH growth is driven by other merger-independent processes.

5. Conclusions

A central goal of observational astrophysics is to measure the merger history of individual galaxies. Models predict that stellar halos provide such a measurement and allow us to explore how merger history—and in particular the most massive merger—affects galaxy properties.

Hubble Space Telescope and ground-based observations resolve individual stars in diffuse halos along the minor axes of nearby galaxies with total stellar masses similar to the Milky Way. The mass and metallicity of these stellar halos show a considerable range and correlate with each other, agreeing with models where they are formed by the disruption of dwarf galaxies alone. These models predict that stellar halo mass and metallicity constrain the mass of the largest accreted satellite galaxy. Consequently, the range in stellar halo mass and metallicity implies a range in the mass of the largest merged/accreted satellite galaxy. Intriguingly, in having a low mass and metallicity stellar halo, the Milky Way is not normal and likely experienced an unusually quiet merger history.

Motivated by models in which bulges and their central black holes (BHs) form or grow through galaxy merging, we then explore the relationship between these features and merger history. Bulge and central BH mass correlate poorly with stellar halo mass and therefore merger history. While half of the galaxies with the largest bulges and BHs have large stellar halos and thus have experienced a major or minor merger, half have small stellar halos and have never experienced a significant merger event. These systems may be an excellent testing ground for models where bulge and BH growth are driven by merger-independent processes.

Acknowledgments: This work benefited from the support from NSF grants AST 1008342 and AST 1514835 and HST grants GO-11613, GO-12213 and GO-13696 provided by NASA through a grant from the Space Telescope Science Institute, which is operated by the Association of Universities for Research in Astronomy, Inc., under NASA contract NAS5-26555. We appreciate the useful feedback from the two referees, and helpful conversations and insights from Sarah Loebman, Monica Valluri, Kohei Hattori, Ian Roederer, Adam Smercina, Bryan Terrazas, Denija Crnojevic, Duncan Forbes, Andrei Kravtsov, Nicolas Martin, and Oleg Gnedin.

Author Contributions: D.J.R.-S., R.S.dJ. and A.M. led the data reduction and photometry; B.H., A.M., E.F.B. and R.D'S. led the analysis; E.F.B. wrote the paper with contributions from all coauthors.

Conflicts of Interest: The authors declare no conflict of interest.

References

1. Hopkins, P.F.; Bundy, K.; Croton, D.; Hernquist, L.; Keres, D.; Khochfar, S.; Stewart, K.; Wetzel, A.; Younger, J.D. Mergers and Bulge Formation in ΛCDM: Which Mergers Matter? *Astrophys. J.* **2010**, *715*, 202–229. [arXiv:astro-ph.CO/0906.5357].

2. Somerville, R.S.; Davé, R. Physical Models of Galaxy Formation in a Cosmological Framework. *Annu. Rev. Astron. Astrophys.* **2015**, *53*, 51–113. [1412.2712].

3. Bullock, J.S.; Kravtsov, A.V.; Weinberg, D.H. Hierarchical Galaxy Formation and Substructure in the Galaxy's Stellar Halo. *Astrophys. J.* **2001**, *548*, 33–46. [arXiv:astro-ph/0007295].

4. Bullock, J.S.; Johnston, K.V. Tracing Galaxy Formation with Stellar Halos. I. Methods. *Astrophys. J.* **2005**, *635*, 931–949. [arXiv:astro-ph/0506467].

5. Purcell, C.W.; Bullock, J.S.; Zentner, A.R. Shredded Galaxies as the Source of Diffuse Intrahalo Light on Varying Scales. *Astrophys. J.* **2007**, *666*, 20–33. [astro-ph/0703004].

6. Cooper, A.P.; Cole, S.; Frenk, C.S.; White, S.D.M.; Helly, J.; Benson, A.J.; De Lucia, G.; Helmi, A.; Jenkins, A.; Navarro, J.F.; et al. Galactic stellar haloes in the CDM model. *Mon. Not. R. Astron. Soc.* **2010**, *406*, 744–766. [arXiv:astro-ph.GA/0910.3211].

7. Monachesi, A.; Bell, E.F.; Radburn-Smith, D.J.; Bailin, J.; de Jong, R.S.; Holwerda, B.; Streich, D.; Silverstein, G. The GHOSTS survey - II. The diversity of halo colour and metallicity profiles of massive disc galaxies. *Mon. Not. R. Astron. Soc.* **2016**, *457*, 1419–1446. [1507.06657].

8. Harmsen, B.; Monachesi, A.; Bell, E.F.; de Jong, R.S.; Bailin, J.; Radburn-Smith, D.J.; Holwerda, B.W. Diverse stellar haloes in nearby Milky Way mass disc galaxies. *Mon. Not. R. Astron. Soc.* **2017**, *466*, 1491–1512. [1611.05448].

9. Bell, E.F.; Monachesi, A.; Harmsen, B.; de Jong, R.S.; Bailin, J.; Radburn-Smith, D.J.; D'Souza, R.; Holwerda, B.W. Galaxies Grow Their Bulges and Black Holes in Diverse Ways. *Astrophys. J.* **2017**, *837*, L8. [1702.06116].

10. D'Souza, R.; Bell, E. Accreted Metallicity-Stellar Mass Relationship. *arXiv* **2017**, arXiv:1705.08442. [1705.08442].

11. Papovich, C.; Labbé, I.; Quadri, R.; Tilvi, V.; Behroozi, P.; Bell, E.F.; Glazebrook, K.; Spitler, L.; Straatman, C.M.S.; Tran, K.V.; et al. ZFOURGE/CANDELS: On the Evolution of M* Galaxy Progenitors from z = 3 to 0.5. *Astrophys. J.* **2015**, *803*, 26. [1412.3806].

12. Blanton, M.R.; Moustakas, J. Physical Properties and Environments of Nearby Galaxies. *Annu. Rev. Astron. Astrophys.* **2009**, *47*, 159–210. [0908.3017].

13. Schweizer, F.; Seitzer, P. Correlations between UBV colors and fine structure in E and S0 galaxies—A first attempt at dating ancient merger events. *Astron. J.* **1992**, *104*, 1039–1067.

14. Rothberg, B.; Joseph, R.D. A Survey of Merger Remnants. II. The Emerging Kinematic and Photometric Correlations. *Astron. J.* **2006**, *131*, 185–207. [astro-ph/0510019].

15. Kormendy, J.; Kennicutt, R.C., Jr. Secular Evolution and the Formation of Pseudobulges in Disk Galaxies. *Annu. Rev. Astron. Astrophys.* **2004**, *42*, 603–683. [astro-ph/0407343].

16. Johansson, P.H.; Naab, T.; Ostriker, J.P. Forming Early-type Galaxies in ΛCDM Simulations. I. Assembly Histories. *Astrophys. J.* **2012**, *754*, 115. [1202.3441].

17. Ceverino, D.; Dekel, A.; Tweed, D.; Primack, J. Early formation of massive, compact, spheroidal galaxies with classical profiles by violent disc instability or mergers. *Mon. Not. R. Astron. Soc.* **2015**, *447*, 3291–3310. [1409.2622].

18. Sales, L.V.; Navarro, J.F.; Theuns, T.; Schaye, J.; White, S.D.M.; Frenk, C.S.; Crain, R.A.; Dalla Vecchia, C. The origin of discs and spheroids in simulated galaxies. *Mon. Not. R. Astron. Soc.* **2012**, *423*, 1544–1555. [arXiv:astro-ph.CO/1112.2220].

19. Terrazas, B.A.; Bell, E.F.; Henriques, B.M.B.; White, S.D.M.; Cattaneo, A.; Woo, J. Quiescence Correlates Strongly with Directly Measured Black Hole Mass in Central Galaxies. *Astrophys. J. Lett.* **2016**, *830*, L12. [1609.07141].

20. Terrazas, B.A.; Bell, E.F.; Woo, J.; Henriques, B.M.B. Supermassive Black Holes as the Regulators of Star Formation in Central Galaxies. *Astrophys. J.* **2017**, *844*, 170.

21. Koss, M.; Mushotzky, R.; Veilleux, S.; Winter, L.M.; Baumgartner, W.; Tueller, J.; Gehrels, N.; Valencic, L. Host Galaxy Properties of the Swift Bat Ultra Hard X-Ray Selected Active Galactic Nucleus. *Astrophys. J.* **2011**, *739*, 57. [arXiv:astro-ph.CO/1107.1237].

22. Ellison, S.L.; Patton, D.R.; Mendel, J.T.; Scudder, J.M. Galaxy pairs in the Sloan Digital Sky Survey - IV. Interactions trigger active galactic nuclei. *Mon. Not. R. Astron. Soc.* **2011**, *418*, 2043–2053. [1108.2711].

23. Deason, A.J.; Mao, Y.Y.; Wechsler, R.H. The Eating Habits of Milky Way-mass Halos: Destroyed Dwarf Satellites and the Metallicity Distribution of Accreted Stars. *Astrophys. J.* **2016**, *821*, 5. [1601.07905].

24. Vogelsberger, M.; Genel, S.; Springel, V.; Torrey, P.; Sijacki, D.; Xu, D.; Snyder, G.; Nelson, D.; Hernquist, L. Introducing the Illustris Project: simulating the coevolution of dark and visible matter in the Universe. *Mon. Not. R. Astron. Soc.* **2014**, *444*, 1518–1547. [1405.2921].

25. Purcell, C.W.; Bullock, J.S.; Zentner, A.R. The metallicity of diffuse intrahalo light. *Mon. Not. R. Astron. Soc.* **2008**, *391*, 550–558. [0805.2965].

26. Cooper, A.P.; D'Souza, R.; Kauffmann, G.; Wang, J.; Boylan-Kolchin, M.; Guo, Q.; Frenk, C.S.; White, S.D.M. Galactic accretion and the outer structure of galaxies in the CDM model. *Mon. Not. R. Astron. Soc.* **2013**, *434*, 3348–3367. [1303.6283].

27. Kirby, E.N.; Cohen, J.G.; Guhathakurta, P.; Cheng, L.; Bullock, J.S.; Gallazzi, A. The Universal Stellar Mass-Stellar Metallicity Relation for Dwarf Galaxies. *Astrophys. J.* **2013**, *779*, 102.

28. Zolotov, A.; Willman, B.; Brooks, A.M.; Governato, F.; Brook, C.B.; Hogg, D.W.; Quinn, T.; Stinson, G. The Dual Origin of Stellar Halos. *Astrophys. J.* **2009**, *702*, 1058–1067. [arXiv:astro-ph.GA/0904.3333].

29. Cooper, A.P.; Parry, O.H.; Lowing, B.; Cole, S.; Frenk, C. Formation of in situ stellar haloes in Milky Way-mass galaxies. *Mon. Not. R. Astron. Soc.* **2015**, *454*, 3185–3199. [1501.04630].

30. Pillepich, A.; Vogelsberger, M.; Deason, A.; Rodriguez-Gomez, V.; Genel, S.; Nelson, D.; Torrey, P.; Sales, L.V.; Marinacci, F.; Springel, V.; et al. Halo mass and assembly history exposed in the faint outskirts: The stellar and dark matter haloes of Illustris galaxies. *Mon. Not. R. Astron. Soc.* **2014**, *444*, 237–249. [1406.1174].

31. Monachesi, A.; Gómez, F.A.; Grand, R.J.J.; Kauffmann, G.; Marinacci, F.; Pakmor, R.; Springel, V.; Frenk, C.S. On the stellar halo metallicity profile of Milky Way-like galaxies in the Auriga simulations. *Mon. Not. R. Astron. Soc.* **2016**, *459*, L46–L50. [1512.03064].

32. Malin, D.; Hadley, B. HI in Shell Galaxies and Other Merger Remnants. *Publ. Astron. Soc. Aust.* **1997**, *14*, 52–58.

33. Martínez-Delgado, D.; Gabany, R.J.; Crawford, K.; Zibetti, S.; Majewski, S.R.; Rix, H.W.; Fliri, J.; Carballo-Bello, J.A.; Bardalez-Gagliuffi, D.C.; Peñarrubia, J.; et al. Stellar Tidal Streams in Spiral Galaxies of the Local Volume: A Pilot Survey with Modest Aperture Telescopes. *Astron. J.* **2010**, *140*, 962–967. [1003.4860].

34. Duc, P.A.; Cuillandre, J.C.; Karabal, E.; Cappellari, M.; Alatalo, K.; Blitz, L.; Bournaud, F.; Bureau, M.; Crocker, A.F.; Davies, R.L.; et al. The ATLAS^{3D} project - XXIX. The new look of early-type galaxies and surrounding fields disclosed by extremely deep optical images. *Mon. Not. R. Astron. Soc.* **2015**, *446*, 120–143. [1410.0981].

35. Trujillo, I.; Fliri, J. Beyond 31 mag arcsec^{-2}: The Frontier of Low Surface Brightness Imaging with the Largest Optical Telescopes. *Astrophys. J.* **2016**, *823*, 123. [1510.04696].

36. Merritt, A.; van Dokkum, P.; Abraham, R.; Zhang, J. The Dragonfly nearby Galaxies Survey. I. Substantial Variation in the Diffuse Stellar Halos around Spiral Galaxies. *Astrophys. J.* **2016**, *830*, 62. [1606.08847].

37. Radburn-Smith, D.J.; de Jong, R.S.; Seth, A.C.; Bailin, J.; Bell, E.F.; Brown, T.M.; Bullock, J.S.; Courteau, S.; Dalcanton, J.J.; Ferguson, H.C.; et al. The GHOSTS Survey. I. Hubble Space Telescope Advanced Camera for Surveys Data. *Astrophys. J.* **2011**, *195*, 18.

38. Ibata, R.A.; Lewis, G.F.; McConnachie, A.W.; Martin, N.F.; Irwin, M.J.; Ferguson, A.M.N.; Babul, A.; Bernard, E.J.; Chapman, S.C.; Collins, M.; et al. The Large-scale Structure of the Halo of the Andromeda Galaxy. I. Global Stellar Density, Morphology and Metallicity Properties. *Astrophys. J.* **2014**, *780*, 386–406. [1311.5888].

39. Gilbert, K.M.; Kalirai, J.S.; Guhathakurta, P.; Beaton, R.L.; Geha, M.C.; Kirby, E.N.; Majewski, S.R.; Patterson, R.J.; Tollerud, E.J.; Bullock, J.S.; et al. Global Properties of M31's Stellar Halo from the SPLASH Survey. II. Metallicity Profile. *Astrophys. J.* **2014**, *796*, 76. [1409.3843].

40. Rejkuba, M.; Harris, W.E.; Greggio, L.; Harris, G.L.H.; Jerjen, H.; Gonzalez, O.A. Tracing the Outer Halo in a Giant Elliptical to 25 R$_{eff}$. *Astrophys. J. Lett.* **2014**, *791*, L2. [1406.4627].

41. Peacock, M.B.; Strader, J.; Romanowsky, A.J.; Brodie, J.P. Detection of a Distinct Metal-poor Stellar Halo in the Early-type Galaxy NGC 3115. *Astrophys. J.* **2015**, *800*, 13. [1412.2752].

42. Bell, E.F.; Zucker, D.B.; Belokurov, V.; Sharma, S.; Johnston, K.V.; Bullock, J.S.; Hogg, D.W.; Jahnke, K.; de Jong, J.T.A.; Beers, T.C.; et al. The Accretion Origin of the Milky Way's Stellar Halo. *Astrophys. J.* **2008**, *680*, 295–311. [arXiv:0706.0004].

43. Xue, X.X.; Rix, H.W.; Ma, Z.; Morrison, H.; Bovy, J.; Sesar, B.; Janesh, W. The Radial Profile and Flattening of the Milky Way's Stellar Halo to 80 kpc from the SEGUE K-giant Survey. *Astrophys. J.* **2015**, *809*, 144. [1506.06144].

44. Sesar, B.; Jurić, M.; Ivezić, Ž. The Shape and Profile of the Milky Way Halo as Seen by the Canada-France-Hawaii Telescope Legacy Survey. *Astrophys. J.* **2011**, *731*, 4. [1011.4487].

45. D'Souza, R.; Kauffman, G.; Wang, J.; Vegetti, S. Parametrizing the stellar haloes of galaxies. *Mon. Not. R. Astron. Soc.* **2014**, *443*, 1433–1450. [1404.2123].

46. Rodriguez-Gomez, V.; Pillepich, A.; Sales, L.V.; Genel, S.; Vogelsberger, M.; Zhu, Q.; Wellons, S.; Nelson, D.; Torrey, P.; Springel, V.; et al. The stellar mass assembly of galaxies in the Illustris simulation: growth by mergers and the spatial distribution of accreted stars. *Mon. Not. R. Astron. Soc.* **2016**, *458*, 2371–2390. [1511.08804].

47. Grand, R.J.J.; Gómez, F.A.; Marinacci, F.; Pakmor, R.; Springel, V.; Campbell, D.J.R.; Frenk, C.S.; Jenkins, A.; White, S.D.M. The Auriga Project: the properties and formation mechanisms of disc galaxies across cosmic time. *Mon. Not. R. Astron. Soc.* **2017**, *467*, 179–207. [1610.01159].

Galaxies **2017**, *5*, 95

48. Renda, A.; Gibson, B.K.; Mouhcine, M.; Ibata, R.A.; Kawata, D.; Flynn, C.; Brook, C.B. The stellar halo metallicity-luminosity relationship for spiral galaxies. *Mon. Not. R. Astron. Soc.* **2005**, *363*, L16–L20. [astro-ph/0507281].
49. Font, A.S.; Johnston, K.V.; Bullock, J.S.; Robertson, B.E. Phase-Space Distributions of Chemical Abundances in Milky Way-Type Galaxy Halos. *Astrophys. J.* **2006**, *646*, 886–898. [astro-ph/0512611].

galaxies

MDPI

Letter

Deep MOS Spectroscopy of NGC 1316 Globular Clusters

Leandro A. Sesto [1,2,*], **Favio R. Faifer** [1,2], **Juan C. Forte** [3,4] **and Analía V. Smith Castelli** [2]

1   Facultad de Ciencias Astronómicas y Geofísicas, Universidad Nacional de La Plata , Paseo del Bosque s/n, La Plata B1900FWA, Argentina; favio@fcaglp.unlp.edu.ar
2   Instituto de Astrofísica de La Plata (IALP; CCT La Plata, CONICET-UNLP), Paseo del bosque s/n, La Plata B1900FWA, Argentina; asmith@fcaglp.unlp.edu.ar
3   The National Scientific and Technical Research Council (CONICET), CABA C1425FQB, Argentina; forte@fcaglp.unlp.edu.ar
4   Planetario de la Ciudad de Buenos Aires, CABA C1425FGC, Argentina
*   Correspondence: sesto@fcaglp.unlp.edu.ar; Tel.: +54-221-483-7324

Academic Editors: Duncan A. Forbes and Ericson D. Lopez
Received: 29 June 2017; Accepted: 9 August 2017; Published: 15 August 2017

Abstract: The giant elliptical galaxy NGC 1316 is the brightest galaxy in the Fornax cluster, and displays a number of morphological features that might be interpreted as an intermediate age merger remanent ($\sim$3 Gyr). Based on the idea that globular clusters systems (GCS) constitute genuine tracers of the formation and evolution of their host galaxies, we conducted a spectroscopic study of approximately 40 globular clusters (GCs) candidates associated with this interesting galaxy. We determined ages, metallicities, and α-element abundances for each GC present in the sample, through the measurement of different Lick indices and their subsequent comparison with simple stellar populations models (SSPs).

Keywords: elliptical galaxies; globular clusters; galaxy haloes

1. Introduction

The giant elliptical galaxy and strong radio source NGC 1316 displays a number of morphological features that might be interpreted as a merger remnant of approximately 3 Gyr ([1]). Among them, we can emphasise shells, ripples, and an unusual pattern of dust, formed by large filaments and dark structures. This galaxy is located at a distance of 20.8 Mpc ([2]), and belongs to Fornax, one of the closest and most studied galaxy clusters of the southern hemisphere.

In a previous photometric work, we detected the presence of different globular clusters (GCs) sub-populations likely associated with different merger events ([3]). In this context, we conducted a spectroscopic study of 40 globular clusters candidates belonging to NGC 1316 using the multi-object mode of the Gemini Multi-Object Spectrograph (GMOS), mounted on the Gemini South telescope (Figure 1). As a result of good quality data and a detailed reduction, we have obtained spectra with excellent signal-to-noise ratio (S/N) (some of them with S/N >50). This allowed us to determine radial velocities, ages, metallicities, and α-element abundances for each GC present in the sample (Sesto et al., in preparation).

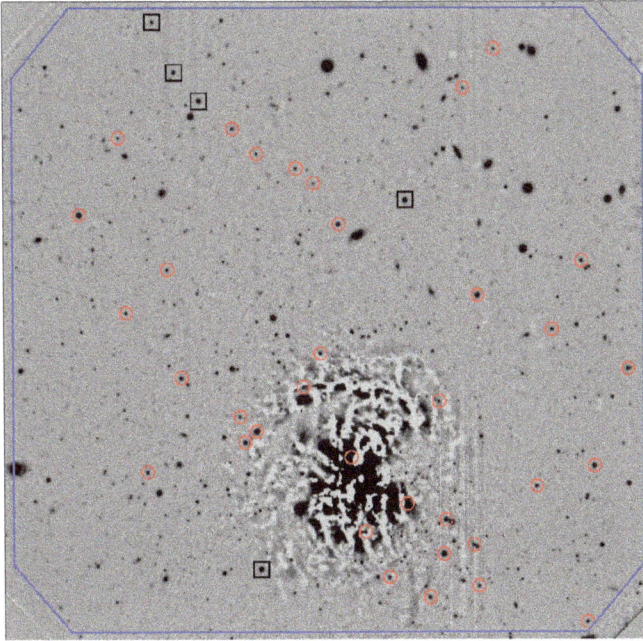

Figure 1. Image used as pre-image for spectroscopy in which the halo of NGC 1316 has been subtracted. Confirmed GCs are shown with red circles and field stars with black squares (see Section 3). The North is up and East to the left. The field of view (FOV) is 5.5′×5.5′.

2. Observations and Data Reduction

The data were obtained between August 2013 and January 2014 as part of the Gemini program GS-2013B-Q-24 (PI: Leandro Sesto). The MOS mask consisted of 40 slits of 1 arcsec width and 4–6 arcsec length. We used the B600-G5303 grating centered at 5000 and 5100 Å (to cover the CCD chip gaps), with 2 × 2 binning, and we considered exposure times of 16 × 1800 s, yielding 8 hours of on-source integration time. The obtained MOS spectra typically covers the range 3500–6500 Å, with a dispersion of 0.90 Å/pix and a spectral resolution of approximately 4.7 Å. The images were processed using the GEMINI-GMOS routines in IRAF. This process was carried out in different stages, which included corrections by bias and flat field, calibration in wavelength, and the extraction and subsequent combination of individual spectra. Finally, spectroscopic standard star observations were used to transform our instrumental spectra into flux-calibrated spectra.

3. Confirmation of GCs

The radial velocities (RVs) of the sample objects were determined through cross-correlation with different synthetic models of simple stellar populations (SSPs) using the method of [4], incorporated in the FXCOR task within IRAF. As template spectra we used stellar population synthesis models obtained from MILES libraries ([5]). A total of 19 SSP models were considered, covering a wide range of ages (2.5, 5, and 12.6 Gyr) and metallicities ([Z/H] = −2.32; −1.71; −1.31; −0.71; −0.4; 0; 0.4 dex), with a unimodal initial mass function (IMF) with a slope value of 1.3.

Thirty-five genuine globular clusters were confirmed, which present radial velocities close to 1760 km/s, adopted as the systemic velocity of NGC 1316 ([6]). Only five objects presented in the sample were field stars with heliocentric radial velocities lower than 60 km/s.

4. Lick/IDS Indices

In order to determine the age, metallicity, and α-element abundances of the GCs, the χ^2 minimization method of [7] and [8] was used. This technique simultaneously compares the different observed Lick/IDS indices with those obtained from simple stellar population models, selecting the combination that minimizes the residuals through a χ^2 fitting. In this particular case, we used the SSP models of [9,10], which have a spectral coverage between 4000 and 6500 Å, ages from 1 to 15 Gyr, and metallicities of [Z/H] = −2.25 to 0.67 dex. One of the most outstanding features of these models is the fact that they include the effects produced by the abundance relations of α-element. These models consider [α/Fe] = 0.0, 0.3, 0.5 dex.

To estimate the integrated properties of each GC, we used those spectral indexes that presented the smallest errors. These were selected from the group conformed by $H\delta_A$, $H\delta_F$, $H\gamma_A$, G4300, Fe4383, $H\beta$, Fe5015, Mgb, Fe5270, Fe5335 and Fe5406, since they provide acceptable results for the study of extragalactic GCs ([11]).

5. Ages, Metallicities, and α-element Abundances

Figure 2 shows the color-magnitude diagram of the photometric Gemini-$g'r'i'$ data (corrected for interstellar extinction) presented in [3], where colors indicate the different ages of the GCs with spectroscopic information. The figure shows the presence of an important group of young GCs with ages close to 2 Gyr.

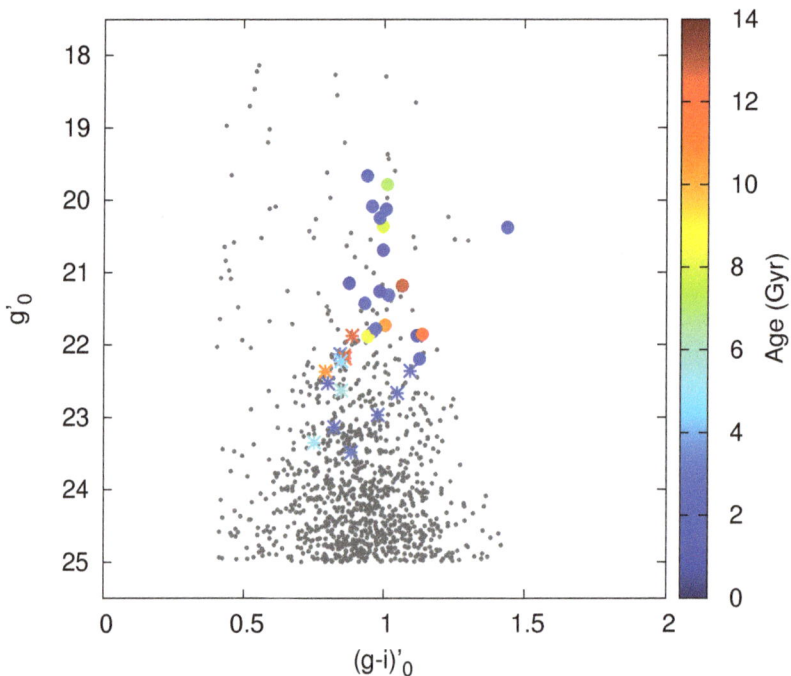

Figure 2. Color-magnitude diagram of the photometric GCs candidates ([3]). The grey dots correspond to the globular clusters (GCs) candidates brighter than g'_0 = 25 mag. The colors indicate the different ages of the of the 35 spectroscopically confirmed GCs. We distinguish between GCs with spectra with signal-to-noise ratio (S/N) <25 (stars) and S/N >25 (circles).

The ages of some GCs associated with NGC 1316 were previously established by [1]. These authors obtained spectra for three GCs with S/N good enough to determine ages and metallicities. Only one of them was present in our spectroscopic sample. A significant discrepancy was observed with respect to the age of this object, as these authors estimated in $\sim$3$\pm$0.5 Gyr, whereas in this work we measured an age of 7.1$\pm$0.4 Gyr. It is important to note that the spectrum used in this work presents a considerably higher S/N value, which allowed us to measure many more spectral indicators than those used by [1]. This result is particularly interesting, since it could be indicating that this GC was not formed in the last merger event experienced by NGC 1316.

Figure 3 shows the color-magnitude diagram of the photometric and spectroscopic data, where colors indicate the different metallicities ([Z/H] measured in dex) of the 35 spectroscopically confirmed GCs. In the particular case of objects with S/N >25, the sample is dominated by objects with relatively high metallicities; i.e., $-0.5 < [Z/H] < 0.5$ dex.

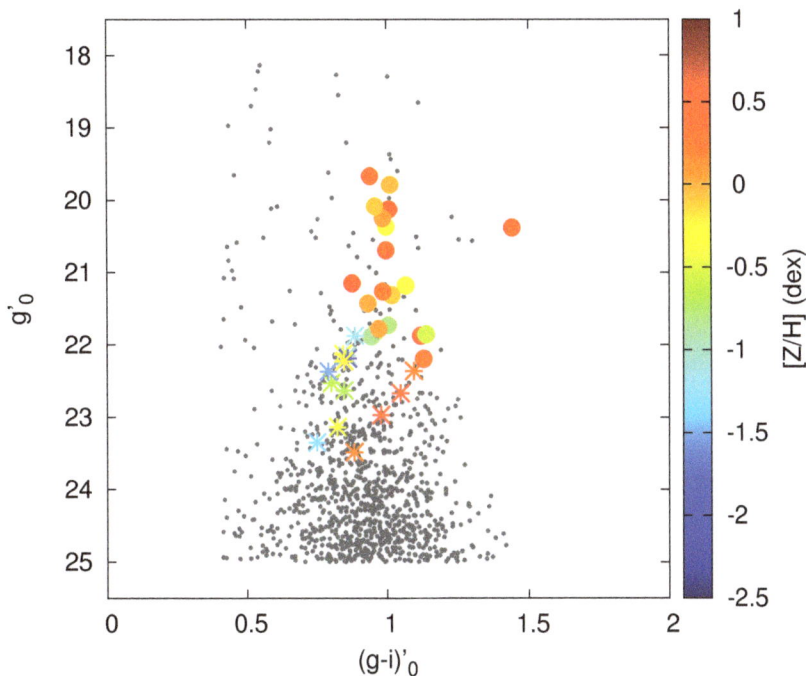

Figure 3. Color-magnitude diagram of the photometric GCs candidates ([3]). The grey dots correspond to the GCs candidates brighter than $g'_0 = 25$ mag. The colors indicate the different metallicities of the 35 spectroscopically confirmed GCs. We distinguish between GCs with spectra with S/N <25 (stars) and S/N >25 (circles).

6. Conclusions

The spectroscopic results have confirmed the presence of multiple GC populations associated with NGC 1316, among which stands out the presence of a population of young GCs with an average age of 1.7 Gyr and metallicities between $-0.5 < [Z/H] < 0.5$ dex. These results will be analyzed in a future work with the aim of describing the different episodes of star formation, and thus at obtaining a more complete picture about the evolutionary history of the galaxy.

Author Contributions: All authors contributed equally to this work.

Conflicts of Interest: The authors declare no conflict of interest.

References

1. Goudfrooij, P.; Mack, J.; Kissler-Patig, M.; Meylan, G.; Minniti, D. Kinematics, ages and metallicities of star clusters in NGC 1316: A 3-Gyr-old merger remnant. *Mon. Not. R. Astron. Soc.* **2001**, *322*, 643–657.
2. Cantiello, M.; Grado, A.; Blakeslee, J.P.; Raimondo, G.; Di Rico, G.; Limatola, L.; Brocato, E.; Della Valle, M.; Gilmozzi, R. The distance to NGC 1316 (Fornax A): Yet another curious case. *Astron. Astrophys.* **2013**, *552*, A106.
3. Sesto, L.A.; Faifer, F.R.; Forte, J.C. The complex star cluster system of NGC 1316 (Fornax A). *Mon. Not. R. Astron. Soc.* **2016**, *461*, 4260–4275.
4. Tonry, J.; Davis, M. A survey of galaxy redshifts. I-Data reduction techniques. *Astrophys. J.* **1979**, *84*, 1511–1525.
5. Vazdekis, A.; Sánchez-Blázquez, P.; Falcón-Barroso, J.; Cenarro, A.J.; Beasley, M.A.; Cardiel, N.; Gorgas, J.; Peletier, R.F. Evolutionary stellar population synthesis with MILES-I. The base models and a new line index system. *Mon. Not. R. Astron. Soc.* **2010**, *404*, 1639–1671.
6. Longhetti, M.; Rampazzo, R.; Bressan, A.; Chiosi, C. Star formation history of early-type galaxies in low density environments. II. Kinematics. *Astron. Astrophys. Suppl.* **1998**, *130*, 267–283.
7. Proctor, R.N.; Sansom, A.E. A comparison of stellar populations in galaxy spheroids across a wide range of Hubble types. *Mon. Not. R. Astron. Soc.* **2002**, *333*, 517–543.
8. Proctor, R.N.; Forbes, D.A.; Beasley, M.A. A robust method for the analysis of integrated spectra from globular clusters using Lick indices. *Mon. Not. R. Astron. Soc.* **2004**, *355*, 1327–1338.
9. Thomas, D.; Maraston, C.; Bender, R. Stellar population models of Lick indices with variable element abundance ratios. *Mon. Not. R. Astron. Soc.* **2003**, *339*, 897–911.
10. Thomas, D.; Maraston, C.; Korn, A. Higher-order Balmer line indices in α/Fe-enhanced stellar population models. *Mon. Not. R. Astron. Soc.* **2004**, *351*, L19–L23.
11. Norris, M.A.; Sharples, R.M.; Kuntschner, H. GMOS spectroscopy of the S0 galaxy NGC 3115. *Mon. Not. R. Astron. Soc.* **2006**, *367*, 815–824.

galaxies

MDPI

Article

Disk Heating, Galactoseismology, and the Formation of Stellar Halos

Kathryn V. Johnston [1,*,†], Adrian M. Price-Whelan [2,†], Maria Bergemann [3], Chervin Laporte [1], Ting S. Li [4], Allyson A. Sheffield [5], Steven R. Majewski [6], Rachael S. Beaton [7], Branimir Sesar [3] and Sanjib Sharma [8]

[1]   Department of Astronomy, Columbia University, 550 W 120th st., New York, NY 10027, USA; cfl2126@columbia.edu
[2]   Department of Astrophysical Sciences, Princeton University, 4 Ivy Lane, Princeton, NJ 08544, USA; adrn@astro.princeton.edu
[3]   Max Planck Institute for Astronomy, Heidelberg 69117, Germany; bergemann@mpia-hd.mpg.de (M.B.); bsesar@mpia.de (B.S.)
[4]   Fermi National Accelerator Laboratory, P. O. Box 500, Batavia, IL 60510, USA; sazabi@neo.tamu.edu
[5]   Department of Natural Sciences, LaGuardia Community College, City University of New York, 31-10 Thomson Ave., Long Island City, NY 11101, USA; asheffield@lagcc.cuny.edu
[6]   Department of Astronomy, University of Virginia, P.O. Box 400325, Charlottesville, VA 22904, USA; srm4n@virginia.edu
[7]   The Carnegie Observatories, 813 Santa Barbara Street, Pasadena, CA 91101, USA; rlb9n@virginia.edu
[8]   Sydney Institute for Astronomy, School of Physics, University of Sydney, NSW 2006, Australia; sanjib.sharma@gmail.com
*   Correspondence: kvj@astro.columbia.edu; Tel.: +1-212-854-3884
†   These authors contributed equally to this work.

Academic Editors: Duncan A. Forbes and Ericson D. Lopez
Received: 1 July 2017; Accepted: 14 August 2017; Published: 26 August 2017

Abstract: Deep photometric surveys of the Milky Way have revealed diffuse structures encircling our Galaxy far beyond the "classical" limits of the stellar disk. This paper reviews results from our own and other observational programs, which together suggest that, despite their extreme positions, the stars in these structures were formed in our Galactic disk. Mounting evidence from recent observations and simulations implies kinematic connections between several of these distinct structures. This suggests the existence of collective disk oscillations that can plausibly be traced all the way to asymmetries seen in the stellar velocity distribution around the Sun. There are multiple interesting implications of these findings: they promise new perspectives on the process of disk heating; they provide direct evidence for a stellar halo formation mechanism in addition to the accretion and disruption of satellite galaxies; and, they motivate searches of current and near-future surveys to trace these oscillations across the Galaxy. Such maps could be used as dynamical diagnostics in the emerging field of "Galactoseismology", which promises to model the history of interactions between the Milky Way and its entourage of satellites, as well examine the density of our dark matter halo. As sensitivity to very low surface brightness features around external galaxies increases, many more examples of such disk oscillations will likely be identified. Statistical samples of such features not only encode detailed information about interaction rates and mergers, but also about long sought-after dark matter halo densities and shapes. Models for the Milky Way's own Galactoseismic history will therefore serve as a critical foundation for studying the weak dynamical interactions of galaxies across the universe.

Keywords: galaxies: galaxy formation; galactic disks; stellar halos; Milky Way

1. Introduction

Our perspective on the Milky Way presents both unique challenges and unique opportunities within our quest to understand galaxies more generally. Because we are located inside of our own Galaxy, it is the only galaxy for which we lack a truly global perspective in a single snapshot and, instead, must survey the entire night sky to fully sample its constituents. On the other hand, it is one of the few galaxies that we can presently study by individual stars and it is the only galaxy for which we can make volume-complete samples in both position and velocity space for non-evolved stellar tracers (e.g., via Main Sequence Turnoff Stars, hereafter MSTO). Present and recent sky surveys have already considerably advanced this effort, and surveys in the near future will deliver massive datasets that will enable detailed studies of stellar structures throughout the Galactic volume.

Emerging in the 1990s, the catalogues that provided the inspiration for current and future surveys not only mapped global structures in our Galaxy, but also revealed the ubiquity of substructure within it. These revelations added an unforeseen richness to interpretations of the data sets and encouraged the development of new dynamical tools for studying ongoing interactions and formation histories. As a few examples:

- Astrometric data from the *Hipparcos* mission [1] led to the discovery of moving groups in the velocity distribution of solar-neighborhood stars [2]. Some of these likely correspond to destroyed star clusters (as expected), while others (unexpectedly) have been interpreted as signatures of resonances with the Galactic bar [3].
- Precise, large-area photometry from the Sloan Digital Sky Survey (hereafter, SDSS —[4–6]) led to the discovery of many "streams" of MSTO stars in the Galactic stellar halo. These are understood to be the remnants of long-dead satellite galaxies and dissolved globular clusters [7,8] and serve as a stunning confirmation that our Galaxy has indeed formed hierarchically (e.g., [9–11]).
- All-sky, infrared photometry from the Two Micron All Sky Survey (hereafter, 2MASS —[12]) enabled M giant stars associated with tidal debris from the Sagittarius dwarf galaxy (hereafter Sgr) to be traced around the entire sky [13], offering a new perspective on the history of its disruption [14].

From these and many other studies using large survey catalogues, it is now clear that the Milky Way is full of kinematic substructure, from the nearby regions of the Galactic disk to the distant stellar halo.

The focus of this *Article* is on three such substructures just beyond the historical "end" of the Galactic disk within the inner stellar halo. Figure 1 (reproduced with data from previous work, [15]), shows the spatial distribution of M giant stars associated with these three substructures: the so-called Monoceros Ring (Mon; also known as the Galactic Anticenter Stellar Structure, GASS), the Triangulum-Andromeda clouds (TriAnd), and A13.

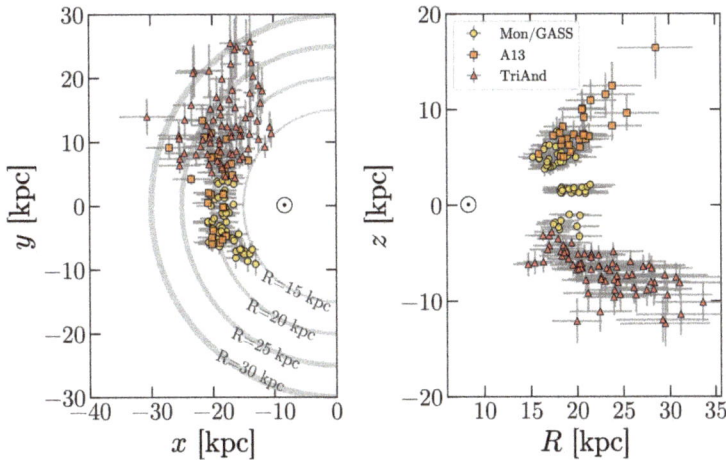

Figure 1. Summary of the spatial distribution of M giants in each of the three low-latitude structures. Note that at lower Galactic latitudes, the lack of candidate M giant members is due to selection effects and crowding near the midplane; we expect the structures to continue towards even lower latitudes, but blend with ordinary disk stars. Markers represent individual stars identified as likely members of each of the three structures discussed in this work (see figure legend). Distance estimates come from photometry alone and have expected absolute uncertainties around ≈20% for TriAnd and A13 [15,16] and ≈25% for Mon/GASS. Grey curves in left panel show Galactocentric circles with cylindrical radii, R, indicated on the figure. The position of the Sun is marked with the solar symbol ⊙.

Each of these substructures were originally identified as over-densities in stellar number counts relative to the expected global structure of the disk or inner stellar halo [1]. The same region of the sky has been shown to be richly structured on even smaller scales [19–23], but here we consider only the larger structures. Unlike most stellar streams, Mon/GASS, TriAnd, and A13 are present at a range of low to moderate Galactic latitudes and span large areas of the sky—we will hereafter refer to them collectively as the "low-latitude structures". The basic properties of the low-latitude structures are summarized below:

- *Mon/GASS* is an arc-like or partial-ring feature of stars beyond the previously-expected edge of the Galactic disk, ≈ 5 kpc beyond the Sun in cylindrical radius [24]. Stars attributed to Mon/GASS span a large area of the sky and large range in distance: Galactic longitudes ≈120° ≲ l ≲ 240°, latitudes −30° ≲ b ≲ +40°, and heliocentric distances 5 ≲ $d_\odot$ ≲ 10 kpc [17,18,25]. Radial velocity measurements of M giant stars associated with the structure follow a clear trend in mean velocity with Galactic longitude and have a velocity dispersion much smaller than the stellar halo [26].
- *TriAnd* was first discovered as a diffuse over-density of M giant stars covering the area ≈100° ≲ l ≲ 160°, ≈-35° ≲ b ≲ −15°, overlapping the Mon/GASS structure on the sky but at larger heliocentric distances of ≈15–25 kpc [27]. The M giants again exhibit a coherent radial-velocity sequence with a dispersion much smaller than the halo [27]. Deep photometry in the region revealed MSTO stars associated with the structure and proposed the existence of a second main sequence ("TriAnd 2") at larger distance, ≈30–35 kpc [22,28].
- *A13* is another tenuous association of M giants in the North Galactic Hemisphere in the area ≈125° ≲ l ≲ 210°, ≈20° ≲ b ≲ 40° at approximate distances of ≈10–20 kpc. It was initially

---

[1] Here we show just M giant stars that have been previously identified as candidate members of the structures, however, some of the structures have also been detected in MSTO stars [17,18].

discovered by applying a group finding algorithm [29] to all M giants in the 2MASS photometric catalogue [30]. Again, radial velocities of M giants in this structure have a small velocity dispersion around a roughly linear trend with Galactic longitude [15]

Several distinct scenarios have been used to explain the formation and existence of each of these structures. Mon/GASS has been attributed to the accretion of a satellite [31], a natural extension of the Galactic warp [32,33], or disturbances to the Galactic disk [21,34–38]. The extreme position of TriAnd at $(R, Z) \approx (15, -5)$ to $(25, -12)$ kpc across a large range in Galactocentric azimuth, ϕ, seemed to exclude the possibility of a disturbed Galactic disk as a possible origin and it has also been modeled as debris from a satellite on a retrograde orbit [16]. Initial abundance studies measured $[\alpha/\text{Fe}]$ and s-process values for M giants in both Mon/GASS and TriAnd and found them to be consistent with those seen in Milky Way satellite galaxies, thus unlike the disk [39,40] — see Section 2.3 for a more complete discussion.

Recent evidence points towards a more convincing, coherent picture for the nature of the three low-latitude structures: the stars in these structures likely have a common origin in the Galactic disk and have been "kicked out" to their present-day positions. This *Article* reviews recent and ongoing contributions that our own group is making to formulating this picture, which include spectroscopic surveys of the low-latitude structures to study metallicities and kinematic properties [15,16], stellar populations (Sheffield et al., in prep [41]), and detailed abundance patterns (Bergemann et al., in prep.), as well as numerical simulations (Laporte et al., in prep [16,41,42]). We summarize this observational and theoretical work in Sections 2 and 3 respectively, adding in the context of contemporary work from other groups, as well as the larger context of possible connections across the Galactic disk. Armed with this understanding of the nature of these substructures, we proceed in Section 4 to discuss prospects for mapping such structures more generally around our own and other galaxies. We end in Section 5 by outlining the motivation for making such maps, asking what they might be telling us about bigger questions: galaxy formation scenarios and the distribution of dark matter around galaxies.

2. The Nature of Structures Around the Outer Disk—Summary of Observations

From clustering in positions or distance alone, many candidate groups and over-densities of M giants have been identified in the outer disk or inner halo. Over the last five years our group has obtained spectroscopy for candidate members of these structures with the aims of (1) confirming the existence of substructure in velocities, (2) measuring chemical abundances, and (3) studying the constituent stellar populations. These goals then inform our own efforts to produce plausible dynamical formation scenarios using simulations. In particular, we *avoid* the collimated stellar streams that have been well-studied in prior work (such as Sgr, Orphan, GD1 and Pal 5—see, e.g., [14,43–45]) and instead focus on stellar structures that appear diffuse, amorphous, extended or "cloud-like" in nature, such as TriAnd, A13, and Mon/GASS. Initial interpretations of these morphologies suggested the structures could be *shells* —debris from the disruption of satellite galaxies on near-radial orbits—but as seen from an internal perspective [46]. However, our own and other recent observations instead suggest a common origin within the Galactic disk for stars associated with the low-latitude structures.

2.1. Low-resolution Spectroscopy: Metallicities and Radial Velocities

In our first study, we extended a prior sample of TriAnd M giants [27] by obtaining spectra of all candidate M giants identified by applying color-magnitude cuts to stars in the TriAnd region of the sky [16]. We identified M giants associated with the two proposed MSTO TriAnd structures (TriAnd 1 and 2, as named by [22,28]). M giant stars in both TriAnd 1 and 2 form clear over-densities in radial velocities with a small dispersion, $\sigma_v \approx 25$ km s^{-1}, compared to the background halo velocity distribution. The radial velocities of M giants in both TriAnd 1 and 2 follow the same sequence in velocity with a steady negative gradient of mean Galactic Standard-of-Rest (GSR) radial velocity (v_{GSR}) with increasing Galactic longitude, l; see Figures 2 and 3, red triangles. We initially presented a dynamical model that simultaneously and approximately reproduces TriAnd 1 and 2 as tidal debris stripped over two separate pericentric passages from a single accreted satellite on a low-eccentricity, retrograde, near-planar orbit. The debris structures in the simulation were morphologically closer to *streams* than *shells*, but still subtended large areas on the sky as observed from the Sun's position. We note, however, that because of large distance uncertainties, the M giants in TriAnd 1 and 2 are indistinguishable and overlap in distance, velocity distribution, and sky position; the existence of two distinct structures rather than a single extended structure has yet to be conclusively demonstrated (see [22] for some counter-arguments). Hereafter, we therefore refer to the TriAnd structures collectively, rather than individually.

In subsequent work, we continued this spectroscopic survey by observing M giant stars in A13 [30]. A13 overlaps the TriAnd clouds in Galactic longitude (but not latitude) at one end, and Mon/GASS at the other end, but is apparent in the Northern (rather than Southern) Galactic Hemisphere and at slightly brighter magnitudes than TriAnd. The spectra show that, like the TriAnd clouds, this structure has a coherent velocity structure with low dispersion and a steady gradient with longitude, l, confirming the genuine association of its members [15]; see Figure 2, orange squares.

Low-resolution spectroscopy has also been obtained for a sample of M giant stars that span $\approx 100°$ of the Mon/GASS structure [26]. The candidate Mon/GASS member M giants also show a clear trend in GSR velocity with Galactic longitude, and appear to form a coherent sequence with both A13 and TriAnd; see Figure 2, yellow circles.

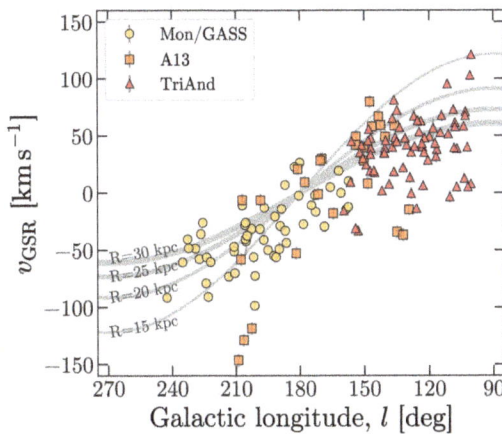

Figure 2. Summary of the velocity distribution of M giants in each of the three low-latitude structures. Markers represent individual stars identified as likely members of each of the three structures discussed in this work (see figure legend). Grey curves show the expected v_{GSR} trends for circular orbits in the Galactic disk midplane with velocity equal to 220 km s^{-1} at several Galactocentric cylindrical radii, R, as indicated on the figure. Velocity uncertainties are typically the same as or smaller than the marker sizes. Reproduced from [15].

As mentioned above, Figure 2 (reproduced with data from previous work, [15]), summarizes the line-of-sight (GSR) velocity trends of M giants in each of the three low-latitude structures. Not only do these structures all have low dispersions ($\sim$25 km/s) relative to the stellar halo ($\sim$120 km/s) — which suggests that the structures themselves are real—they also appear to *collectively* exhibit a continuous gradient with Galactic longitude, l. This suggests that the structures may also be associated with one another, as part of a larger structure in the outer Galactic disk.

2.2. Stellar Populations and Other Kinematic Tracers

Motivated by the observed, low-dispersion velocity distribution of the M giant stars in TriAnd, we sought to observe other distance tracer stars in the same region, determine their membership, and improve the distance estimates to the structure. We focused on and selected RR Lyrae stars in the TriAnd region from the Palomar Transient Factory (PTF; [47]), using a conservative distance cut to account for uncertainties in the RR Lyrae and M giant photometric distance estimates, 15 kpc $< d_\odot < 35$ kpc. We obtained spectra for $\approx 1/3$ of the total number of RR Lyrae in the M giant volume considered to be associated with TriAnd and measured radial velocities for these stars [41].

Figure 3 (reproduced from [41]) shows the results of our survey: unlike the M giants (triangles) the RR Lyrae stars (circles) show no clear, tight velocity sequence. By modeling both the RR Lyrae and M giants velocities as having been drawn from a mixture of two populations—one representing a low-dispersion sequence with varied dispersion, and one representing a halo population with large dispersion, both Gaussian—we showed that, after accounting for selection effects, the number ratio of RR Lyrae to M giants, $f_{RR:MG}$, within the overdensity is $f_{RR:MG} < 0.38$ with 95% confidence.

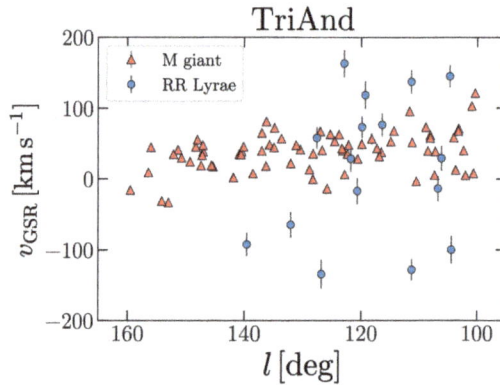

Figure 3. Comparison of the velocity distribution for M giants in the TriAnd structure (red triangles) with velocities for RR Lyrae stars in the same region of sky and distance range (blue circles). Radial velocity uncertainties of the M giant stars are typically the size of the marker or smaller. Uncertainties for the RR Lyrae stars are shown with gray error bars. Note the low-dispersion sequence in the M giant velocities, unseen in the RR Lyrae star velocities, which look like typical halo stars with a large velocity dispersion. Reproduced from [41].

Since we were unable to find any RR Lyrae clearly associated with TriAnd, our attempt to measure more accurate distances to the structure was unsuccessful. However, the upper limit on the value of $f_{RR:MG}$ was in itself interesting. RR Lyrae and M giants are tracers of populations with quite distinct metallicities: stars in the horizontal branch phase of evolution are typically only blue enough to cross the instability strip and become RR Lyrae if they have [Fe/H] $\lesssim -1.5$; and giant stars typically only evolve to colors red enough to become spectral class M if they have [Fe/H] $\gtrsim -1.5$. Hence, a stellar population has to contain a significant range of metallicities to contain both types of stars. The metallicity distributions in nearly all existing satellite galaxies orbiting the Milky Way (e.g., [48]) are typically biased towards low metallicities (i.e., [Fe/H] < -1) , such that they contain no or few M giant stars (i.e., $f_{RR:MG} = \infty$). The largest satellites (the Large Magellanic Cloud and Sgr) are exceptions as they contain substantial metal rich (i.e., [Fe/H] > -1) populations; these very metal-enriched dwarf galaxies have $f_{RR:MG} \sim 0.5$ in their still-bound stellar populations [41]. In contrast, the local Galactic disk is an overall metal-rich population and thus has very few RR Lyrae (i.e., $f_{RR:MG} \sim 0$; [49]), more consistent with our findings for the stellar population of TriAnd. Moreover, results from the APOGEE survey show the outer disk populations have [Fe/H] > −1 [50], and hence it is also likely contain a very low fraction of RR Lyrae.

Our work on the stellar populations of the TriAnd region motivated us to look at possible associations of RR Lyrae with the M giant sequences found in Mon/GASS and A13. For the surveyed regions of these structures, we find $f_{RR:MG}$ values similar to those observed for TriAnd, and therefore consistent with membership of the Galactic disk (Sheffield et al., in prep.).

2.3. High-Resolution Spectroscopy: Abundance Patterns

The origin of stellar associations can also be explored through measurements of the detailed chemical abundances of their constituent stars. It is intriguing that our finding of the low-latitude structures all having stellar populations (as indicated by $f_{RR:MG}$) that look more like the disk than known Galactic satellites (Section 2.2) appears to be at odds with prior work on abundance patterns of stars in these structures. High-resolution spectra of 21 M giants in Mon/GASS [39] showed [Ti/Fe] lower by up to 0.4 dex compared to the mean trends known for main-sequence stars of the Galactic disk (e.g., [51,52]) and a mean offset for [Y/Fe] of about 0.2 dex at [Fe/H] $\approx$ −0.5. A comparison with similar results for stars in Sgr [53] suggested that Mon/GASS may be more similar to Sgr than the disk, and therefore proposed an external origin for this structure. Subsequent work comes to similar conclusions for TriAnd stars [40].

To explore the apparent contradiction between the stellar populations and abundance work, we have recently obtained high-resolution ($R \sim 30,000$–$47,000$) and high signal-to-noise (>200 per Å) spectra of fifteen stars in the TriAnd and A13 overdensities. Fourteen stars were observed with the HIRES-S spectrograph at the Keck-1 telescope [62] and one star was observed using the UVES spectrograph at the VLT (Program ID: 097.B-0770A). Figure 4 shows the results of our preliminary analysis, with the average over all the stars in our sample presented as a black point. We find that the stars in these structures have a very narrow metallicity spread, with a value that is consistent with the prior metallicity estimates in TriAnd stars [40]. The TriAnd and A13 stars also have extremely similar chemical abundances to each other, with the abundance dispersion across the combined set of stars from both structures of ≤0.06 dex for most chemical elements. The abundances of all measured α-elements are uniformly enhanced at a level that is consistent with the abundances of the Milky Way disk stars (gray points; [52,63,64]). The [Mg/Fe] value is also consistent with the measurements of α-elements for the disk stars that have been found towards the outer disk in the APOGEE survey [50] (with the caveat that at a given [Fe/H] there is typically an offset of ∼0.2 dex between [Mg/Fe] measured in the infrared with APOGEE and other optical studies). The abundance ratios we derive are too high for the chemical abundance patterns observed in the stars of the Galactic satellites (colored points; [55,65–68]), which are known to have low, typically solar ([O/Fe]) or even sub-solar ([Mg/Fe],[Na/Fe]), ratios at [Fe/H] $\sim$ −0.5. However, note that very recent APOGEE results suggest elevated abundance levels for some elements are present in some stars in the Sgr dwarf at [Fe/H] $\lesssim$ −0.4, so further work in this area is definitely warranted (Hasselquist et al., in prep.).

We are currently exploring one explanation for the disagreement between our and prior abundance work for stars in the same structures—that the differences in interpretation are due to differences in the observed datasets and spectroscopic modeling techniques. Past work [39,40] analyzed a small wavelength region in the near-IR, limited to 150 Å from 7440 to 7590 Å. Because of this limitation, they could include only 11 Fe I and 2 Ti I lines in the determination of metallicities, and abundances. Our full analysis, through a much wider wavelength coverage and high SNR attained for the observed spectra, will allow us to include more then a hundred of Fe I and Fe II lines, as well as tens of Ti lines from both ionization stages (Ti II is important to check the influence of NLTE effects in Ti I). It has been shown that Ti I should be not be used in abundance studies because it is very sensitive to NLTE effects [69] (see also discussion in [70]). We are investigating the sensitivity of abundance diagnostics to the line selection and wavelength regimes by performing test computations on our own data, using a reduced line-list and comparing our [Fe/H] and [Ti/Fe] results between this and the

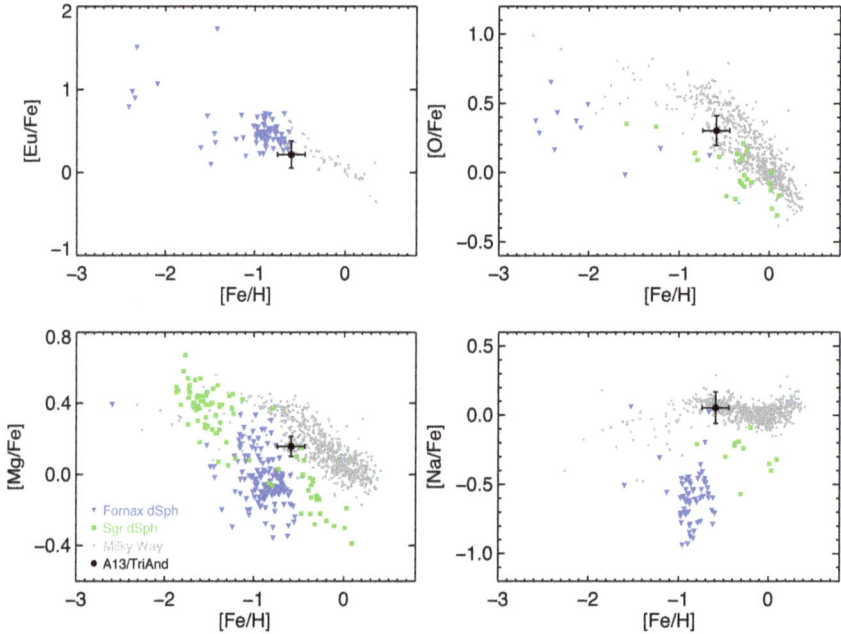

Figure 4. Abundance patterns for the thin and thick disk (grey symbols, from [52,54]), the Fornax (blue symbols, from [48,55–58]) and Sgr (green symbols, from [59–61]) dwarf spheroidal galaxies and our own results (black circle). The error bar on our measurement is the sum of the standard deviation of the stellar sample and a systematic error of the abundance measurement.

Overall, comparing to abundances of disk stars and satellite populations obtained using analogous data sets and reduction techniques, our current results indicate, in contrast to the prior work [39,40], that the birthplace of stars in TriAnd and A13 could be the outer Galactic disk (Bergemann et al., in prep.). The differences in our conclusions provide powerful motivation for a uniform survey of abundances for a much larger number of stars towards the Galactic anticenter and extending tens of kpc from the Sun (e.g., [50]), as well as at comparable metallicities in the few satellite galaxies large enough to host such metal-rich populations.

2.4. Mapping Main-Sequence Stars in the Low-Latitude Structures

While we have concentrated on follow-up studies of the known low-latitude structures as traced by M giants selected from 2MASS, knowledge of the spatial distribution of MSTO stars towards the anticenter region has been further refined using photometry from the PanSTARRS1 survey [21,25]. Recent studies using the SDSS [37] and Pan-STARRS1 (Lurie et al., in prep.) surveys employ the technique of subtracting color-magnitude diagrams (CMD's) derived from fields in their photometric data which were symmetrically placed at equal and opposite Galactic latitudes and at the same Galactic longitudes, analogous to that used for M giants in the original discovery paper for the TriAnd Clouds [27]. The vast number of MSTO stars allowed denser regions, closer to the Galactic plane and at smaller heliocentric distances to be explored. Both MSTO studies show overdense and clearly distinct arcs of stars oscillating between the Northern and Southern Hemispheres as the heliocentric distance was increased towards the anticenter of our Galaxy.

2.5. Connecting the Low-Latitude Structures to Velocity Structure Near the Solar Radius

Coincident with studies exploring structures at the very outer edge of our Galactic disk, large scale spectroscopic surveys have allowed a detailed re-examination of the *local* distribution of stellar velocities, within a few kpc of the Sun. Using data from the SDSS [71,72] and the RAdial Velocity Experiment (RAVE; [73,74]), asymmetries between the Northern and Southern Galactic Hemispheres have been seen in the density and velocity distributions of stars in the vicinity of the sun. Looking ~2 kpc out towards the Galactic anticenter, the Large Sky Area Multi-Object Fiber Spectroscopic Telescope (LAMOST; [75–77]) finds similar asymmetries in radial and vertical velocities [78]. The scale and sense of these asymmetries indicate moderate systematic motions (of order a few km/s) of stars within the disk perpendicular to the plane, suggesting both vertical movement of the midplane, and compression and rarefaction of the vertical scale (referred to as "bending" and "breathing" modes respectively—see, e.g., [79]). It is natural to associate these asymmetries in motion as a local manifestations of the oscillations traced in space over much larger scales [37,41].

3. The Nature of Structures Around the Outer Disk—Summary of Theoretical Interpretations

The observations summarized in Section 2 indicate that:

- the low-latitude structures—Mon/GASS, TriAnd, and A13 —each have low velocity dispersions supporting the genuine association of the candidate member stars;
- Mon/GASS, TriAnd, and A13 share a continuous sequence in mean GSR velocity as a function of Galactic longitude, suggestive of associations between these structures;
- the stellar populations in the structures (as indicated by $f_{RR:MG}$) are all more consistent with those in the Galactic disk rather than those observed in the stellar halo or Galactic satellites;
- the abundance patterns of stars in TriAnd and A13 are similar to those found in the disk of our Galaxy (although the discrepancy with prior work in this conclusion is still under investigation);
- the low-latitude structures (around the outer disk) may be connected to oscillating density and velocity structure on smaller scales, traced all the way back to the Solar neighborhood;

Taken together, we conclude that: (i) there is mounting evidence that Mon/GASS, TriAnd, and A13 represent populations of stars formed in the disk that now exist at extreme radii and heights above the Galactic disk; (ii) these structures are likely associated and part of a global system of vertical disk oscillations that can be traced all the way to the velocity asymmetries seen in the solar neighborhood; and (iii) the stellar populations in these structures are inconsistent with a picture in which they formed in a dwarf galaxy.

One natural interpretation of these collected observations is that the oscillations represent the response of the disk to an external perturbation—for example, the impact of a satellite galaxy that has been transmitted and amplified by its wake in the dark matter halo [38,80]. Prior work has already pointed to this as a possible explanation for the existence of Mon/GASS [34,35], with Sgr being pointed to as a plausible culprit for the perturbation [36]. It has also been demonstrated how perturbations from a satellite on an orbit perpendicular to the disk could lead to bending (at low relative impact velocity) and breathing (at higher relative velocity) modes that would be observed in the solar neighborhood as asymmetries in the local velocity distribution [79] and on larger scales as rings [81]. (Note that breathing modes can also be induced by non-axisymmetric features in the disk such as the bar and spiral arms; [82].) Simulations have also shown that Sgr could be responsible for local velocity structure [83]. Such interactions and corresponding disk features have been found to naturally occur in cosmological simulations [38].

Figure 5 illustrates these ideas with the results of simulations from our own recent work. Using simulations of a disk disturbed (separately) by satellites on orbits that mimic those expected for the Large Magellanic Cloud and Sgr [42], we extend the prior theoretical backdrop that looked at Mon/GASS to examine whether the extreme (R, Z) locations of TriAnd stars could fit within the same picture. With different masses and orbits (and consequently different interaction strength, timings, and durations) these satellites necessarily induce distinct but overlapping signatures on the global structure of the disk. In more recent work, we found a model that was capable of reproducing the scales of the observed disturbances (radial wavelength and amplitude in space, as well as magnitude of offsets in velocity locally). The model required: the interaction of Sgr with the disk of the Milky Way to be followed for several passages longer than prior work (note that the length of Sgr's streams indicate that it has been impacting the disk for several pericentric passages e.g., [14]); Sgr to have sufficient initial mass and density to impact the disk in the last Gigayear with a remaining mass of $\sim 3 \times 10^9$ M$_\odot$; and the disk to be realized with stars existing as far out as 40 kpc from the Galactic center to populate the regions corresponding to TriAnd. The interaction with the LMC modified the overall morphology of the structures induced, but was not sufficient alone to explain their properties. The full details of these results will be discussed in an upcoming paper (Laporte et al., in prep.).

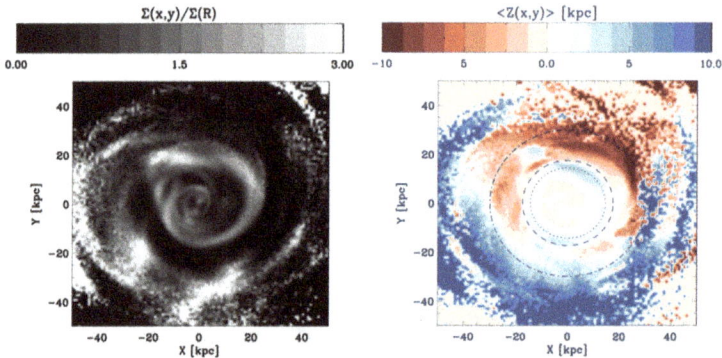

Figure 5. Visualizations at the end-point of a simulation of the impact of the Sgr dwarf galaxy on the Milky Way's disk from work currently in preparation (Laporte et al.). The simulation encompassed the last $\sim$6 Gyrs of the interaction. Sgr had encountered the disk 6 times during this time and had a final mass of $\sim 3 \times 10^9$ $M_\odot$. The left hand panel shows density fluctuations around the mean at each annulus in the disk. The right hand panel shows the average positions of particles above/below the disk at each point. Note that material is kicked to as much as 10 kpc from the midplane at distances of 20–30 kpc from the Galactic center.

4. Discussion—Observational Prospects

4.1. The Milky Way

While the connections that have already been made between the different low-latitude structures and the disk population are convincing, there are several possible directions for further observations to strengthen these claims. Kinematic and abundance information and density measurements for more tracers and over a larger volume would greatly facilitate an informative comparison to theoretical work, allowing more detailed interpretations of the history of our Galaxy.

The most obvious direction is to obtain proper motions and more accurate distance estimates to the known features, using the candidate members discussed in this work to search for other tracers. For example, proper motions for the "Anti-Center Stream" (ACS) (a thin, coherent density structure, which may or may not be part of the larger Mon/GASS system; [84]) indicate that stars in the ACS are not actually moving along the spatial extent of the stream [85]. This is inconsistent with expectations for the behavior of debris from a destroyed satellite. If similar measurements of proper motions of stars in all of the low-latitude structures showed significant motion perpendicular to the Galactic disk, this would provide even more evidence of a disk origin and connection to local oscillations (first results are just being reported in this region [86]). With precise proper motions, the velocity information would also place important constraints on dynamical models of the disk (see Section 5). In upcoming data-releases, astrometric measurements from the *Gaia* mission [87] are poised to provide these data. Expected proper motion uncertainties for the closest M giant stars ($\approx$5 kpc) in Mon/GASS correspond to tangential velocity uncertainties of $\approx$1–2 km s^{-1} for M giants with tangential velocities <50 km s^{-1}. For the farthest known M giant stars in TriAnd ($\approx$ 35 kpc), tangential velocity uncertainties will be $\approx$7–12 km s^{-1} for M giants with tangential velocities <50 km s^{-1}.

The next decade will see first light for the Large Synoptic Survey Telescope (LSST; [88]), which will survey the sky to the same depth as SDSS every three days. Over time, these data can be combined to detect MSTO stars out to 100 kpc. Hence the exquisite maps from SDSS and PanSTARRS that exploited the dense coverage of this tracer to reveal disk oscillations out to Mon/GASS can be extended to A13 and TriAnd and beyond.

Another complimentary direction for future data is to extend spectroscopic maps to fainter magnitudes and global scales. The *Gaia* catalogue will soon be enhanced by surveys from several ground-based, wide-field, multi-object spectrographs on larger telescopes capable of reaching to fainter magnitudes (e.g., WEAVE on the William Herschel Telescope, 4MOST and the Prime Focus Spectrograph on the Subaru telescope—[89–91]). However, the reach of these surveys towards lower latitudes all the way to the disk plane will be limited by extinction. An infrared spectroscopic survey, such as APOGEE [92] , could overcome this limitation. APOGEE is capable of building catalogues of radial velocities and abundance patterns for 15 chemical species across the Galactic disk, bulge, bar, halo, star clusters, and dwarf galaxies— the type of homogeneous data set needed to compare the low latitude structures to other Galactic components.

4.2. Other Galaxies

TriAnd is estimated to have a surface brightness below 32 mag/arcsec^2 [93], so studying similar features around other galaxies presents a significant challenge. Nevertheless, there are growing samples of galaxies within and beyond the Local Group being mapped to depths close to this target.

For nearby galaxies, these levels are approached through star counts studies, most spectacularly for the case of our nearest neighbor, the Andromeda Galaxy, where giant star counts have revealed a significantly extended and richly featured outer stellar disk [94,95]. Analogous studies have been carried out for galaxies up to distances of several Mpc (e.g., [96,97]), although the focus of these studies has typically been on detecting the stellar halos of these objects. The great advantage of star-count studies is the ability to reach extremely low surface brightnesses—depths below 30 mag/arcsec^2 have been estimated for the density profile of M31's and other galaxies' stellar halos [98,99].

Star count studies are limited to the galaxies in the volume within which stars can be resolved. Several dedicated surveys have made innovative advances in studying galaxies to low surface-brightness using a variety of techniques to reach limits below 30 mag/arcsec^2 in integrated light (e.g., [100–103]), although these studies face the twin challenges of calibration and Galactic cirrus to overcome.

Looking to the future, NASA's proposed Wide Field InfraRed Survey Telescope (WFIRST), with its wide field of view and high resolution, offers the possibility of extending the deep star-count sensitivities now achieved in MW and M31 to all galaxies within 10 Mpc [104]. In the next decade, images in integrated light from LSST can be combined to be sensitive to slightly shallower depth ($\sim$29 mag/arcsec^2, see [88]), but for vast numbers (many millions!) of galaxies.

5. Conclusions—What Might These Structures Tell Us about Galaxies?

The above sections summarize observational evidence for large scale vertical oscillations of the Galactic plane present in the Solar neighborhood and reaching out beyond the traditional edge of our stellar disk. We have also discussed theoretical studies that suggest that these oscillations could be caused by, and contain the signatures of, ongoing interactions of the Milky Way with its satellite system. Moreover, there are numerous observational prospects for extending this work both to map the Milky Way globally and to look for analogous features around many other galaxies.

Now that we have a physical picture of the origin of such planar undulations, as well as prospects for mapping them further within the Milky Way and detecting analogous substructures around other galaxies, we can move on to discussing how useful they are for constraining the dynamics and evolution of galaxies. While the mere existence of these low surface brightness structures around the outskirts of galaxies is interesting, they contain only a tiny fraction of the stars in galaxies and these are spread out over a large area—these properties naturally make such structures difficult to map, either because their unique signatures can be lost in the foreground star counts (e.g., in the Milky Way), or because the required surface brightness limits for detection are prohibitively low (for integrated light).

Conversely, these features may prove to be particularly powerful probes of interactions and histories, precisely because they contain so little mass: they can be modeled as test particles responding to an external perturbation.

Below are just three examples of where these structures could promise new insights into some classic questions in galactic astronomy.

- *Disk heating mechanisms* — It has been understood for a long time that disks can evolve significantly due to mergers, major or minor, and hence that their current structures bear witness to their accretion history [105–108]. This understanding has fueled a significant literature on simulations investigating the importance of the heating of galactic disks in response to encounters with other dark matter halos (that may or may not contain their own galaxies) [109–118]. In general, these studies have concentrated on the overall effects of many encounters on global properties, such as the thickness and vertical velocity dispersion in disks. The results of these simulations have traditionally been compared to the spatial and velocity scales in samples of galaxies. In contrast, the identification and mapping of vertical waves associated with ongoing interactions in the Milky Way gives us the opportunity to dissect individual disk heating events in progress (e.g., the impact of Sgr). We can use this to check our understanding of the mechanism directly and in detail rather than assessing its importance through collective effects and longterm, phase-mixed signatures.

- *Stellar halo formation processes*— The last decade has seen increasing interest in assessing how much of the content of our stellar halo could be made "in situ" along with other components of the Milky Way rather than accreted from other objects. Hydrodynamical simulations of galaxy formation suggest that tens of percent of the stars in the inner halo might be formed "in situ", either along their current orbits or "kicked-out" from orbits that were originally in the disk [119–127]. Preliminary arguments for the existence of an "in situ" population were based on transitions in the density or orbital structures of stellar halos (e.g., [128]). However, such transitions were also found to occur naturally in purely-accreted models of stellar halos [129]. More convincing observational evidence for stars in the halo that have been "kicked-out" of the disk is just beginning to emerge through studies that look for stars with halo-like kinematics, but disk-like abundances around M31 [130] and the Milky Way [70,131,132]. The work outlined above in Sections 2 and 3 adds new perspectives on the "kicked-out-disk" stellar halo formation process with the detection and modeling of disk stars that may be in transition from the disk to the halo.

- *Galactoseismic probes of interactions and dark matter* — The response of a disk to an encounter will depend on its own properties, the properties of the dark matter halo in which it is embedded, and the mass and orbit of the perturbing satellite. This leads to the suggestion that, analogous to helioseismic investigations of the structure of our Sun, maps of a disk response—such as those described in Section 2 for our Milky Way—might be similarly inverted to tell us about, for example, the structure of the dark matter halo [71]. Indeed, recent investigations into the signatures of encounters in the very outskirts of extended HI disks have successfully used simulations combined with an analytic understanding to find how the observed characteristics of the disturbed gas can be simply related to properties of the perturbing object [133–135].

Overall, the breadth and depth of studies summarized in this *Article* are indicative of the developing interest in the new discipline of Galactoseismology, fueled by the imminence of another leap in the scale of available data sets. The remaining challenge is to build the models and appropriate tools to compare to obervations that can take full advantage of these opportunities to significantly advance our understanding of galactic dynamics and evolution.

Acknowledgments: Much of the work reviewed in this paper was made possible by NSF grant AST-1312196. Kathryn V. Johnston was supported by NSF grant AST-1614743 while writing the review.

Author Contributions: This paper summarizes results from the team of listed authors. It was written by Kathryn V. Johnston and Adrian M. Price-Whelan, with section contributions from Maria Bergemann Figures were made by Adrian M. Price-Whelan (using data consolidated by Ting S. Li) and Chervin Laporte All the authors reviewed and commented on the drafts.

Conflicts of Interest: The authors declare no conflict of interest.

References

1. European Space Agency (ESA). (Ed.) *The HIPPARCOS and TYCHO Catalogues. Astrometric and Photometric Star Catalogues Derived from the ESA HIPPARCOS Space Astrometry Mission;* ESA Special Publication; ESA Publications Division: Noordwijk, The Netherland, 1997; Volume 1200.

2. Dehnen, W. The Distribution of Nearby Stars in Velocity Space Inferred from HIPPARCOS Data. *Astron. J.* **1998**, *115*, 2384–2396.

3. Dehnen, W. The Effect of the Outer Lindblad Resonance of the Galactic Bar on the Local Stellar Velocity Distribution. *Astron. J.* **2000**, *119*, 800–812.

4. York, D.G.; Adelman, J.; Anderson, J.E., Jr.; Anderson, S.F.; Annis, J.; Bahcall, N.A.; Bakken, J.A.; Barkhouser, R.; Bastian, S.; Berman, E.; et al. The Sloan Digital Sky Survey: Technical Summary. *Astron. J.* **2000**, *120*, 1579.

5. Stoughton, C.; Lupton, R.H.; Bernardi, M.; Blanton, M.R.; Burles, S.; Castander, F.J.; Connolly, A.J.; Eisenstein, D.J.; Frieman, J.A.; Hennessy, G.S.; et al. Sloan Digital Sky Survey: Early Data Release. *Astron. J.* **2002**, *123*, 485–548.

6. Abazajian, K.; Adelman-McCarthy, J.K.; Agüeros, M.A.; Allam, S.S.; Anderson, S.F.; Annis, J.; Bahcall, N.A.; Baldry, I.K.; Bastian, S.; Berlind, A.; et al. The First Data Release of the Sloan Digital Sky Survey. *Astron. J.* **2003**, *126*, 2081–2086.

7. Newberg, H.J.; Yanny, B.; Rockosi, C.; Grebel, E.K.; Rix, H.W.; Brinkmann, J.; Csabai, I.; Hennessy, G.; Hindsley, R.B.; Ibata, R.; et al. The Ghost of Sagittarius and Lumps in the Halo of the Milky Way. *Astrophys. J.* **2002**, *569*, 245–274.

8. Belokurov, V.; Zucker, D.B.; Evans, N.W.; Gilmore, G.; Vidrih, S.; Bramich, D.M.; Newberg, H.J.; Wyse, R.F.G.; Irwin, M.J.; Fellhauer, M.; et al. The Field of Streams: Sagittarius and Its Siblings. *Astrophys. J.* **2006**, *642*, L137–L140.

9. Bullock, J.S.; Kravtsov, A.V.; Weinberg, D.H. Hierarchical Galaxy Formation and Substructure in the Galaxy's Stellar Halo. *Astrophys. J.* **2001**, *548*, 33–46.

10. Bullock, J.S.; Johnston, K.V. Tracing Galaxy Formation with Stellar Halos. I. Methods. *Astrophys. J.* **2005**, *635*, 931–949.

11. Bell, E.F.; Zucker, D.B.; Belokurov, V.; Sharma, S.; Johnston, K.V.; Bullock, J.S.; Hogg, D.W.; Jahnke, K.; de Jong, J.T.A.; Beers, T.C.; et al. The Accretion Origin of the Milky Way's Stellar Halo. *Astrophys. J.* **2008**, *680*, 295–311.

12. Nikolaev, S.; Weinberg, M.D.; Skrutskie, M.F.; Cutri, R.M.; Wheelock, S.L.; Gizis, J.E.; Howard, E.M. A Global Photometric Analysis of 2MASS Calibration Data. *Astrophys. J.* **2000**, *120*, 3340–3350.

13. Majewski, S.R.; Skrutskie, M.F.; Weinberg, M.D.; Ostheimer, J.C. A Two Micron All Sky Survey View of the Sagittarius Dwarf Galaxy. I. Morphology of the Sagittarius Core and Tidal Arms. *Astrophys. J.* **2003**, *599*, 1082–1115.

14. Law, D.R.; Majewski, S.R. The Sagittarius Dwarf Galaxy: A Model for Evolution in a Triaxial Milky Way Halo. *Astrophys. J.* **2010**, *714*, 229–254.

15. Li, T.S.; Sheffield, A.A.; Johnston, K.V.; Marshall, J.L.; Majewski, S.R.; Price-Whelan, A.M.; Damke, G.J.; Beaton, R.L.; Richardson, W.; Sharma, S.; et al. Exploring Halo Substructure with Giant Stars. XV. Discovery of a Connection between the Monoceros Ring and the Triangulum-Andromeda Overdensity? *Astrophys. J.* **2017**, *844*, doi:10.3847/1538-4357/aa7a0d.

16. Sheffield, A.A.; Johnston, K.V.; Majewski, S.R.; Damke, G.; Richardson, W.; Beaton, R.; Rocha-Pinto, H.J. Exploring Halo Substructure with Giant Stars. XIV. The Nature of the Triangulum-Andromeda Stellar Features. *Astrophys. J.* **2014**, *793*, doi:10.1088/0004-637X/793/1/62.

17. Yanny, B.; Newberg, H.J.; Grebel, E.K.; Kent, S.; Odenkirchen, M.; Rockosi, C.M.; Schlegel, D.; Subbarao, M.; Brinkmann, J.; Fukugita, M.; et al. A Low-Latitude Halo Stream around the Milky Way. *Astrophys. J.* **2003**, *588*, 824–841.

18. Ibata, R.A.; Irwin, M.J.; Lewis, G.F.; Ferguson, A.M.N.; Tanvir, N. One ring to encompass them all: A giant stellar structure that surrounds the Galaxy. *Mon. Not. R. Astron. Soc.* **2003**, *340*, L21–L27.

19. Grillmair, C.J. Substructure in Tidal Streams: Tributaries in the Anticenter Stream. *Astrophys. J. Lett.* **2006**, *651*, L29–L32.

20. Grillmair, C.J.; Carlin, J.L.; Majewski, S.R. Fishing in Tidal Streams: New Radial Velocity and Proper Motion Constraints on the Orbit of the Anticenter Stream. *Astrophys. J. Lett.* **2008**, *689*, L117–L120.

21. Slater, C.T.; Bell, E.F.; Schlafly, E.F.; Morganson, E.; Martin, N.F.; Rix, H.W.; Peñarrubia, J.; Bernard, E.J.; Ferguson, A.M.N.; Martinez-Delgado, D.; et al. The Complex Structure of Stars in the Outer Galactic Disk as Revealed by Pan-STARRS1. *Astrophys. J.* **2014**, *791*, 9, doi:10.1088/0004-637X/791/1/9.

22. Martin, N.F.; Ibata, R.A.; Rich, R.M.; Collins, M.L.M.; Fardal, M.A.; Irwin, M.J.; Lewis, G.F.; McConnachie, A.W.; Babul, A.; Bate, N.F.; et al. The PAndAS Field of Streams: Stellar Structures in the Milky Way Halo toward Andromeda and Triangulum. *Astrophys. J.* **2014**, *787*, doi:10.1088/0004-637X/787/1/19.

23. Deason, A.J.; Belokurov, V.; Hamren, K.M.; Koposov, S.E.; Gilbert, K.M.; Beaton, R.L.; Dorman, C.E.; Guhathakurta, P.; Majewski, S.R.; Cunningham, E.C. TriAnd and its siblings: Satellites of satellites in the Milky Way halo. *Mon. Not. R. Astron. Soc.* **2014**, *444*, 3975–3985.

24. Robin, A.C.; Creze, M.; Mohan, V. The edge of the Galactic disk. *Astrophys. J. Lett.* **1992**, *400*, L25–L27.

25. Morganson, E.; Conn, B.; Rix, H.W.; Bell, E.F.; Burgett, W.S.; Chambers, K.; Dolphin, A.; Draper, P.W.; Flewelling, H.; Hodapp, K.; et al. Mapping the Monoceros Ring in 3D with Pan-STARRS1. *Astrophys. J.* **2016**, *825*, doi:10.3847/0004-637X/825/2/140.

26. Crane, J.D.; Majewski, S.R.; Rocha-Pinto, H.J.; Frinchaboy, P.M.; Skrutskie, M.F.; Law, D.R. Exploring Halo Substructure with Giant Stars: Spectroscopy of Stars in the Galactic Anticenter Stellar Structure. *Astrophys. J. Lett.* **2003**, *594*, L119–L122.

27. Rocha-Pinto, H.J.; Majewski, S.R.; Skrutskie, M.F.; Crane, J.D.; Patterson, R.J. Exploring Halo Substructure with Giant Stars: A Diffuse Star Cloud or Tidal Debris around the Milky Way in Triangulum-Andromeda. *Astrophys. J.* **2004**, *615*, 732–737.

28. Martin, N.F.; Ibata, R.A.; Irwin, M. Galactic Halo Stellar Structures in the Triangulum-Andromeda Region. *Astrophys. J. Lett.* **2007**, *668*, L123–L126.

29. Sharma, S.; Johnston, K.V. A Group Finding Algorithm for Multidimensional Data Sets. *Astrophys. J.* **2009**, *703*, 1061–1077.

30. Sharma, S.; Johnston, K.V.; Majewski, S.R.; Muñoz, R.R.; Carlberg, J.K.; Bullock, J. Group Finding in the Stellar Halo Using M-giants in the Two Micron All Sky Survey: An Extended View of the Pisces Overdensity? *Astrophys. J.* **2010**, *722*, 750–759.

31. Peñarrubia, J.; Martínez-Delgado, D.; Rix, H.W.; Gómez-Flechoso, M.A.; Munn, J.; Newberg, H.; Bell, E.F.; Yanny, B.; Zucker, D.; Grebel, E.K. A Comprehensive Model for the Monoceros Tidal Stream. *Astrophys. J.* **2005**, *626*, 128–144.

32. Momany, Y.; Zaggia, S.R.; Bonifacio, P.; Piotto, G.; De Angeli, F.; Bedin, L.R.; Carraro, G. Probing the Canis Major stellar over-density as due to the Galactic warp. *Astron. Astrophys.* **2004**, *421*, L29–L32.

33. Momany, Y.; Zaggia, S.; Gilmore, G.; Piotto, G.; Carraro, G.; Bedin, L.R.; de Angeli, F. Outer structure of the Galactic warp and flare: explaining the Canis Major over-density. *Astron. Astrophys.* **2006**, *451*, 515–538.

34. Kazantzidis, S.; Bullock, J.S.; Zentner, A.R.; Kravtsov, A.V.; Moustakas, L.A. Cold Dark Matter Substructure and Galactic Disks. I. Morphological Signatures of Hierarchical Satellite Accretion. *Astrophys. J.* **2008**, *688*, 254–276.

35. Younger, J.D.; Besla, G.; Cox, T.J.; Hernquist, L.; Robertson, B.; Willman, B. On the Origin of Dynamically Cold Rings around the Milky Way. *Astrophys. J. Lett.* **2008**, *676*, doi:10.1086/587099.

36. Purcell, C.W.; Bullock, J.S.; Tollerud, E.J.; Rocha, M.; Chakrabarti, S. The Sagittarius impact as an architect of spirality and outer rings in the Milky Way. *Nature* **2011**, *477*, 301–303.

37. Xu, Y.; Newberg, H.J.; Carlin, J.L.; Liu, C.; Deng, L.; Li, J.; Schönrich, R.; Yanny, B. Rings and Radial Waves in the Disk of the Milky Way. *Astrophys. J.* **2015**, *801*, doi:10.1088/0004-637X/801/2/105.

38. Gómez, F.A.; White, S.D.M.; Marinacci, F.; Slater, C.T.; Grand, R.J.J.; Springel, V.; Pakmor, R. A fully cosmological model of a Monoceros-like ring. *Mon. Not. R. Astron. Soc.* **2016**, *456*, 2779–2793.

39. Chou, M.Y.; Majewski, S.R.; Cunha, K.; Smith, V.V.; Patterson, R.J.; Martínez-Delgado, D. The Chemical Evolution of the Monoceros Ring/Galactic Anticenter Stellar Structure. *Astrophys. J. Lett.* **2010**, *720*, L5–L10.

40. Chou, M.Y.; Majewski, S.R.; Cunha, K.; Smith, V.V.; Patterson, R.J.; Martínez-Delgado, D. First Chemical Analysis of Stars in the Triangulum–Andromeda Star Cloud. *Astrophys. J. Lett.* **2011**, *731*, doi:10.1088/ 2041-8205/731/2/L30.

41. Price-Whelan, A.M.; Johnston, K.V.; Sheffield, A.A.; Laporte, C.F.P.; Sesar, B. A reinterpretation of the Triangulum-Andromeda stellar clouds: A population of halo stars kicked out of the Galactic disc. *Mon. Not. R. Astron. Soc.* **2015**, *452*, 676–685.

42. Laporte, C.F.P.; Gómez, F.A.; Besla, G.; Johnston, K.V.; Garavito-Camargo, N. Response of the Milky Way's disc to the Large Magellanic Cloud in a first infall scenario. *Mon. Not. R. Astron. Soc.* **2016**, 1–13.

43. Koposov, S.E.; Rix, H.W.; Hogg, D.W. Constraining the Milky Way Potential with a Six-Dimensional Phase-Space Map of the GD-1 Stellar Stream. *Astrophys. J.* **2010**, *712*, 260–273.

44. Küpper, A.H.W.; Balbinot, E.; Bonaca, A.; Johnston, K.V.; Hogg, D.W.; Kroupa, P.; Santiago, B.X. Globular Cluster Streams as Galactic High-Precision Scales: The Poster Child Palomar 5. *Astrophys. J.* **2015**, *803*, doi:10.1088/0004-637X/803/2/80.

45. Bovy, J.; Bahmanyar, A.; Fritz, T.K.; Kallivayalil, N. The Shape of the Inner Milky Way Halo from Observations of the Pal 5 and GD–1 Stellar Streams. *Astrophys. J.* **2016**, *833*, doi:10.3847/1538-4357/833/1/31.

46. Johnston, K.V.; Bullock, J.S.; Sharma, S.; Font, A.; Robertson, B.E.; Leitner, S.N. Tracing Galaxy Formation with Stellar Halos. II. Relating Substructure in Phase and Abundance Space to Accretion Histories. *Astrophys. J.* **2008**, *689*, 936–957.

47. Law, N.M.; Kulkarni, S.R.; Dekany, R.G.; Ofek, E.O.; Quimby, R.M.; Nugent, P.E.; Surace, J.; Grillmair, C.C.; Bloom, J.S.; Kasliwal, M.M.; et al. The Palomar Transient Factory: System Overview, Performance, and First Results. *Publ. Astron. Soc. Pac.* **2009**, *121*, doi:10.1086/648598.

48. Kirby, E.N.; Lanfranchi, G.A.; Simon, J.D.; Cohen, J.G.; Guhathakurta, P. Multi-element Abundance Measurements from Medium-resolution Spectra. III. Metallicity Distributions of Milky Way Dwarf Satellite Galaxies. *Astrophys. J.* **2011**, *727*, doi:10.1088/0004-637X/727/2/78 .

49. Amrose, S.; Mckay, T. A Calculation of the Mean Local RR Lyrae Space Density Using ROTSE. *Astrophys. J. Lett.* **2001**, *560*, L151–L154.

50. Hayden, M.R.; Bovy, J.; Holtzman, J.A.; Nidever, D.L.; Bird, J.C.; Weinberg, D.H.; Andrews, B.H.; Majewski, S.R.; Allende Prieto, C.; Anders, F.; et al. Chemical Cartography with APOGEE: Metallicity Distribution Functions and the Chemical Structure of the Milky Way Disk. *Astrophys. J.* **2015**, *808*, doi:10.1088/0004-637X/808/2/132.

51. Reddy, B.E.; Tomkin, J.; Lambert, D.L.; Allende Prieto, C. The chemical compositions of Galactic disc F and G dwarfs. *Mon. Not. R. Astron. Soc.* **2003**, *340*, 304–340.

52. Bensby, T.; Feltzing, S.; Oey, M.S. Exploring the Milky Way stellar disk. A detailed elemental abundance study of 714 F and G dwarf stars in the solar neighbourhood. *Astron. Astrophys.* **2014**, *562*, doi:10.1051/0004-6361/201322631.

53. Chou, M.Y.; Cunha, K.; Majewski, S.R.; Smith, V.V.; Patterson, R.J.; Martínez-Delgado, D.; Geisler, D. A Two Micron All Sky Survey View of the Sagittarius Dwarf Galaxy. VI. s-Process and Titanium Abundance Variations Along the Sagittarius Stream. *Astrophys. J.* **2010**, *708*, 1290–1309.

54. Battistini, C.; Bensby, T. The origin and evolution of r- and s-process elements in the Milky Way stellar disk. *Astron. Astrophys.* **2016**, *586*, doi:10.1051/0004-6361/201527385.

55. Shetrone, M.; Venn, K.A.; Tolstoy, E.; Primas, F.; Hill, V.; Kaufer, A. VLT/UVES Abundances in Four Nearby Dwarf Spheroidal Galaxies. I. Nucleosynthesis and Abundance Ratios. *Astrophys. J.* **2003**, *125*, 684–706.

56. Letarte, B.; Hill, V.; Jablonka, P.; Tolstoy, E.; François, P.; Meylan, G. VLT/UVES spectroscopy of individual stars in three globular clusters in the Fornax dwarf spheroidal galaxy. *Astron. Astrophys.* **2006**, *453*, 547–554.

57. Letarte, B.; Hill, V.; Tolstoy, E.; Jablonka, P.; Shetrone, M.; Venn, K.A.; Spite, M.; Irwin, M.J.; Battaglia, G.; Helmi, A.; et al. A high-resolution VLT/FLAMES study of individual stars in the centre of the Fornax dwarf spheroidal galaxy. *Astron. Astrophys.* **2010**, *523*, doi:10.1051/0004-6361/200913413.

58. Larsen, S.S.; Brodie, J.P.; Strader, J. Detailed abundance analysis from integrated high-dispersion spectroscopy: Globular clusters in the Fornax dwarf spheroidal. *Astron. Astrophys.* **2012**, *546*, doi:10.1051/0004-6361/201219895.

59. McWilliam, A.; Smecker-Hane, T.A. The Composition of the Sagittarius Dwarf Spheroidal Galaxy, and Implications for Nucleosynthesis and Chemical Evolution. In *Astronomical Society of the Pacific Conference Series, Proceedings of the Cosmic Abundances as Records of Stellar Evolution and Nucleosynthesis, Austin, TX, USA, 17–19 June 2004*; Barnes, T.G., III, Bash, F.N., Eds.; Springer: Berlin/Heidelberg, Germany, 2005; Volume 336; p. 221.

60. Sbordone, L.; Bonifacio, P.; Buonanno, R.; Marconi, G.; Monaco, L.; Zaggia, S. The exotic chemical composition of the Sagittarius dwarf spheroidal galaxy. *Astron. Astrophys.* **2007**, *465*, 815–824.

61. Mucciarelli, A.; Bellazzini, M.; Ibata, R.; Romano, D.; Chapman, S.C.; Monaco, L. Chemical abundances in the nucleus of the Sagittarius dwarf spheroidal galaxy. *Astron. Astrophys.* **2017**, doi:10.1051/0004- 6361/201730707 .

62. Vogt, S.S.; Allen, S.L.; Bigelow, B.C.; Bresee, L.; Brown, B.; Cantrall, T.; Conrad, A.; Couture, M.; Delaney, C.; Epps, H.W.; et al. HIRES: The high-resolution echelle spectrometer on the Keck 10-m Telescope. In *Instrumentation in Astronomy VIII*; Crawford, D.L., Craine, E.R., Eds.; SPIE: Bellingham, WA, USA, 1994; Volume 2198, Proc. SPIE, p. 362.

63. Fuhrmann, K. Nearby stars of the Galactic disk and halo. III. *Astron. Nachr.* **2004**, *325*, 3–80.

64. Bergemann, M.; Ruchti, G.R.; Serenelli, A.; Feltzing, S.; Alves-Brito, A.; Asplund, M.; Bensby, T.; Gruyters, P.; Heiter, U.; Hourihane, A.; et al. The Gaia-ESO Survey: Radial metallicity gradients and age-metallicity relation of stars in the Milky Way disk. *Astron. Astrophys.* **2014**, *565*, doi:10.1051/0004-6361/201423456.

65. Bonifacio, P.; Hill, V.; Molaro, P.; Pasquini, L.; Di Marcantonio, P.; Santin, P. First results of UVES at VLT: Abundances in the Sgr dSph. *Astron. Astrophys.* **2000**, *359*, 663–668.

66. Shetrone, M.D.; Côté, P.; Sargent, W.L.W. Abundance Patterns in the Draco, Sextans, and Ursa Minor Dwarf Spheroidal Galaxies. *Astrophys. J.* **2001**, *548*, 592–608.

67. Tolstoy, E.; Hill, V.; Tosi, M. Star-Formation Histories, Abundances, and Kinematics of Dwarf Galaxies in the Local Group. *Ann. Rev. Astron. Astrophys.* **2009**, *47*, 371–425.

68. De Boer, T.J.L.; Tolstoy, E.; Lemasle, B.; Saha, A.; Olszewski, E.W.; Mateo, M.; Irwin, M.J.; Battaglia, G. The episodic star formation history of the Carina dwarf spheroidal galaxy. *Astron. Astrophys.* **2014**, *572*, doi:10.1051/0004-6361/201424119.

69. Bergemann, M. Ionization balance of Ti in the photospheres of the Sun and four late-type stars. *Mon. Not. R. Astron. Soc.* **2011**, *413*, 2184–2198.

70. Sheffield, A.A.; Majewski, S.R.; Johnston, K.V.; Cunha, K.; Smith, V.V.; Cheung, A.M.; Hampton, C.M.; David, T.J.; Wagner-Kaiser, R.; Johnson, M.C.; et al. Identifying Contributions to the Stellar Halo from Accreted, Kicked-out, and In Situ Populations. *Astrophys. J.* **2012**, *761*, doi:10.1088/0004-637X/761/2/161.

71. Widrow, L.M.; Gardner, S.; Yanny, B.; Dodelson, S.; Chen, H.Y. Galactoseismology: Discovery of Vertical Waves in the Galactic Disk. *Astrophys. J. Lett.* **2012**, *750*, doi:10.1088/2041-8205/750/2/L4.

72. Yanny, B.; Gardner, S. The Stellar Number Density Distribution in the Local Solar Neighborhood is North-South Asymmetric. *Astrophys. J.* **2013**, *777*, doi:10.1088/0004-637X/777/2/91.

73. Steinmetz, M.; Zwitter, T.; Siebert, A.; Watson, F.G.; Freeman, K.C.; Munari, U.; Campbell, R.; Williams, M.; Seabroke, G.M.; Wyse, R.F.G.; et al. The Radial Velocity Experiment (RAVE): First Data Release. *Astrophys. J.* **2006**, *132*, 1645–1668.

74. Williams, M.E.K.; Steinmetz, M.; Binney, J.; Siebert, A.; Enke, H.; Famaey, B.; Minchev, I.; de Jong, R.S.; Boeche, C.; Freeman, K.C.; et al. The wobbly Galaxy: Kinematics north and south with RAVE red-clump giants. *Mon. Not. R. Astron. Soc.* **2013**, *436*, 101–121.

75. Cui, X.Q.; Zhao, Y.H.; Chu, Y.Q.; Li, G.P.; Li, Q.; Zhang, L.P.; Su, H.J.; Yao, Z.Q.; Wang, Y.N.; Xing, X.Z.; et al. The Large Sky Area Multi-Object Fiber Spectroscopic Telescope (LAMOST). *Res. Astron. Astrophys.* **2012**, *12*, 1197–1242.

76. Deng, L.C.; Newberg, H.J.; Liu, C.; Carlin, J.L.; Beers, T.C.; Chen, L.; Chen, Y.Q.; Christlieb, N.; Grillmair, C.J.; Guhathakurta, P.; et al. LAMOST Experiment for Galactic Understanding and Exploration (LEGUE) —The survey's science plan. *Res. Astron. Astrophys.* **2012**, *12*, 735–754.

77. Zhao, G.; Zhao, Y.H.; Chu, Y.Q.; Jing, Y.P.; Deng, L.C. LAMOST spectral survey—An overview. *Res. Astron. Astrophys.* **2012**, *12*, 723–734.

78. Carlin, J.L.; DeLaunay, J.; Newberg, H.J.; Deng, L.; Gole, D.; Grabowski, K.; Jin, G.; Liu, C.; Liu, X.; Luo, A.L.; et al. Substructure in Bulk Velocities of Milky Way Disk Stars. *Astrophys. J. Lett.* **2013**, *777*, doi:10.1088/2041-8205/777/1/L5.

79. Widrow, L.M.; Barber, J.; Chequers, M.H.; Cheng, E. Bending and breathing modes of the Galactic disc. *Mon. Not. R. Astron. Soc.* **2014**, *440*, 1971–1981.

80. Weinberg, M.D.; Blitz, L. A Magellanic Origin for the Warp of the Galaxy. *Astrophys. J. Lett.* **2006**, *641*, L33–L36.

81. D'Onghia, E.; Madau, P.; Vera-Ciro, C.; Quillen, A.; Hernquist, L. Excitation of Coupled Stellar Motions in the Galactic Disk by Orbiting Satellites. *Astrophys. J.* **2016**, *823*, doi:10.3847/0004-637X/823/1/4.

82. Monari, G.; Famaey, B.; Siebert, A.; Grand, R.J.J.; Kawata, D.; Boily, C. The effects of bar-spiral coupling on stellar kinematics in the Galaxy. *Mon. Not. R. Astron. Soc.* **2016**, *461*, 3835–3846.

83. Gómez, F.A.; Minchev, I.; O'Shea, B.W.; Beers, T.C.; Bullock, J.S.; Purcell, C.W. Vertical density waves in the Milky Way disc induced by the Sagittarius dwarf galaxy. *Mon. Not. R. Astron. Soc.* **2013**, *429*, 159–164.

84. Li, J.; Newberg, H.J.; Carlin, J.L.; Deng, L.; Newby, M.; Willett, B.A.; Xu, Y.; Luo, Z. On Rings and Streams in the Galactic Anti-Center. *Astrophys. J.* **2012**, *757*, doi:10.1088/0004-637X/757/2/151.

85. Carlin, J.L.; Casetti-Dinescu, D.I.; Grillmair, C.J.; Majewski, S.R.; Girard, T.M. Kinematics in Kapteyn's Selected Area 76: Orbital Motions Within the Highly Substructured Anticenter Stream. *Astrophys. J.* **2010**, *725*, 2290–2311.

86. De Boer, T.J.L.; Belokurov, V.; Koposov, S.E. The fall of the Northern Unicorn: Tangential motions in the Galactic Anti-centre with SDSS and Gaia. *Mon. Not. R. Astron. Soc.* **2017**, 1–16.

87. Gaia Collaboration.; Prusti, T.; de Bruijne, J.H.J.; Brown, A.G.A.; Vallenari, A.; Babusiaux, C.; Bailer-Jones, C.A.L.; Bastian, U.; Biermann, M.; Evans, D.W.; et al. The Gaia mission. *Astron. Astrophys.* **2016**, *595*, doi:10.1051/0004-6361/201629272.

88. Ivezic, Z.; Axelrod, T.; Brandt, W.N.; Burke, D.L.; Claver, C.F.; Connolly, A.; Cook, K.H.; Gee, P.; Gilmore, D.K.; Jacoby, S.H.; et al. Large Synoptic Survey Telescope: From Science Drivers To Reference Design. *Serbian Astron. J.* **2008**, *176*, 1–13.

89. Dalton, G.; Trager, S.; Abrams, D.C.; Bonifacio, P.; Aguerri, J.A.L.; Middleton, K.; Benn, C.; Dee, K.; Sayède, F.; Lewis, I.; et al. Final design and progress of WEAVE: The next generation wide-field spectroscopy facility for the William Herschel Telescope. In Proceedings of the SPIE Ground-based and Airborne Instrumentation for Astronomy VI, Edinburgh, UK, 26–30 June 2016; Volume 9908, p. 99081G.

90. Walcher, C.J.; de Jong, R.S.; Dwelly, T.; Bellido, O.; Boller, T.; Chiappini, C.; Feltzing, S.; Irwin, M.; McMahon, R.; Merloni, A.; et al. 4MOST: Science operations for a large spectroscopic survey program with multiple science cases executed in parallel. In Proceedings of the SPIE Observatory Operations: Strategies, Processes, and Systems VI, Edinburgh, UK, 26 June–1 July 2016; Volume 9910, p. 99101N.

91. Tamura, N.; PFS Collaboration. Prime Focus Spectrograph (PFS): A Very Wide-Field, Massively Multi-Object, Optical and Near-Infrared Fiber-Fed Spectrograph on the Subaru Telescope. In *Multi-Object Spectroscopy in the Next Decade: Big Questions, Large Surveys, and Wide Fields*; Skillen, I., Barcells, M., Trager, S., Eds.; Astronomical Society of the Pacific Conference Series; Astronomical Society of the Pacific: San Francisco, CA, USA, 2016; Volume 507, p. 387.

92. Majewski, S.R.; APOGEE Team; APOGEE-2 Team. The Apache Point Observatory Galactic Evolution Experiment (APOGEE) and its successor, APOGEE-2. *Astron. Nachr.* **2016**, *337*, doi:10.1002/asna.201612387.

93. Majewski, S.R.; Ostheimer, J.C.; Rocha-Pinto, H.J.; Patterson, R.J.; Guhathakurta, P.; Reitzel, D. Detection of the Main-Sequence Turnoff of a Newly Discovered Milky Way Halo Structure in the Triangulum-Andromeda Region. *Astrophys. J.* **2004**, *615*, 738–743.

94. Ferguson, A.M.N.; Irwin, M.J.; Ibata, R.A.; Lewis, G.F.; Tanvir, N.R. Evidence for Stellar Substructure in the Halo and Outer Disk of M31. *Astrophys. J.* **2002**, *124*, 1452–1463.

95. Ibata, R.; Chapman, S.; Ferguson, A.M.N.; Lewis, G.; Irwin, M.; Tanvir, N. On the Accretion Origin of a Vast Extended Stellar Disk around the Andromeda Galaxy. *Astrophys. J.* **2005**, *634*, 287–313.

96. Monachesi, A.; Bell, E.F.; Radburn-Smith, D.J.; Vlajić, M.; de Jong, R.S.; Bailin, J.; Dalcanton, J.J.; Holwerda, B.W.; Streich, D. Testing Galaxy Formation Models with the GHOSTS Survey: The Color Profile of M81's Stellar Halo. *Astrophys. J.* **2013**, *766*, doi:10.1088/0004-637X/766/2/106 .

97. Crnojević, D.; Sand, D.J.; Spekkens, K.; Caldwell, N.; Guhathakurta, P.; McLeod, B.; Seth, A.; Simon, J.D.; Strader, J.; Toloba, E. The Extended Halo of Centaurus A: Uncovering Satellites, Streams, and Substructures. *Astrophys. J.* **2016**, *823*, doi:10.3847/0004-637X/823/1/19.

98. Ibata, R.; Martin, N.F.; Irwin, M.; Chapman, S.; Ferguson, A.M.N.; Lewis, G.F.; McConnachie, A.W. The Haunted Halos of Andromeda and Triangulum: A Panorama of Galaxy Formation in Action. *Astrophys. J.* **2007**, *671*, 1591–1623.

99. Harmsen, B.; Monachesi, A.; Bell, E.F.; de Jong, R.S.; Bailin, J.; Radburn-Smith, D.J.; Holwerda, B.W. Diverse stellar haloes in nearby Milky Way mass disc galaxies. *Mon. Not. R. Astron. Soc.* **2017**, *466*, 1491–1512.

100. Martínez-Delgado, D.; Gabany, R.J.; Crawford, K.; Zibetti, S.; Majewski, S.R.; Rix, H.W.; Fliri, J.; Carballo-Bello, J.A.; Bardalez-Gagliuffi, D.C.; Peñarrubia, J.; et al. Stellar Tidal Streams in Spiral Galaxies of the Local Volume: A Pilot Survey with Modest Aperture Telescopes. *Astrophys. J.* **2010**, *140*, 962–967.

101. Van Dokkum, P.G.; Abraham, R.; Merritt, A. First Results from the Dragonfly Telephoto Array: The Apparent Lack of a Stellar Halo in the Massive Spiral Galaxy M101. *Astrophys. J. Lett.* **2014**, *782*, doi:10.1088/2041-8205/782/2/L24.

102. Duc, P.A.; Cuillandre, J.C.; Karabal, E.; Cappellari, M.; Alatalo, K.; Blitz, L.; Bournaud, F.; Bureau, M.; Crocker, A.F.; Davies, R.L.; et al. The ATLAS^{3D} project - XXIX. The new look of early-type galaxies and surrounding fields disclosed by extremely deep optical images. *Mon. Not. R. Astron. Soc.* **2015**, *446*, 120–143.

103. Spavone, M.; Capaccioli, M.; Napolitano, N.R.; Iodice, E.; Grado, A.; Limatola, L.; Cooper, A.; Cantiello, M.; Forbes, D.A.; Paolillo, M.; et al. VEGAS: A VST Early-type GAlaxy Survey. II. Photometric study of giant ellipticals and their stellar halos. *Astron. Astrophys.* **2017**, doi:10.1051/0004-6361/201629111.

104. Spergel, D.; Gehrels, N.; Breckinridge, J.; Donahue, M.; Dressler, A.; Gaudi, B.S.; Greene, T.; Guyon, O.; Hirata, C.; Kalirai, J.; et al. WFIRST-2.4: What Every Astronomer Should Know. *ArXiv* **2013**, arXiv:1305.5425.

105. Toth, G.; Ostriker, J.P. Galactic disks, infall, and the global value of Omega. *Astrophys. J.* **1992**, *389*, 5–26.

106. Quinn, P.J.; Hernquist, L.; Fullagar, D.P. Heating of galactic disks by mergers. *Astrophys. J.* **1993**, *403*, 74–93.

107. Walker, I.R.; Mihos, J.C.; Hernquist, L. Quantifying the Fragility of Galactic Disks in Minor Mergers. *Astrophys. J.* **1996**, *460*, doi:10.1086/176956.

108. Velazquez, H.; White, S.D.M. Sinking satellites and the heating of galaxy discs. *Mon. Not. R. Astron. Soc.* **1999**, *304*, 254–270.

109. Font, A.S.; Navarro, J.F.; Stadel, J.; Quinn, T. Halo Substructure and Disk Heating in a Λ Cold Dark Matter Universe. *Astrophys. J. Lett.* **2001**, *563*, L1–L4.

110. Ardi, E.; Tsuchiya, T.; Burkert, A. Constraints of the Clumpiness of Dark Matter Halos through Heating of the Disk Galaxies. *Astrophys. J.* **2003**, *596*, 204–215.

111. Benson, A.J.; Lacey, C.G.; Frenk, C.S.; Baugh, C.M.; Cole, S. Heating of galactic discs by infalling satellites. *Mon. Not. R. Astron. Soc.* **2004**, *351*, 1215–1236.

112. Stewart, K.R.; Bullock, J.S.; Wechsler, R.H.; Maller, A.H.; Zentner, A.R. Merger Histories of Galaxy Halos and Implications for Disk Survival. *Astrophys. J.* **2008**, *683*, 597–610.

113. Hopkins, P.F.; Hernquist, L.; Cox, T.J.; Younger, J.D.; Besla, G. The Radical Consequences of Realistic Satellite Orbits for the Heating and Implied Merger Histories of Galactic Disks. *Astrophys. J.* **2008**, *688*, 757–769.

114. Villalobos, Á.; Helmi, A. Simulations of minor mergers–I. General properties of thick discs. *Mon. Not. R. Astron. Soc.* **2008**, *391*, 1806–1827.

115. Purcell, C.W.; Kazantzidis, S.; Bullock, J.S. The Destruction of Thin Stellar Disks Via Cosmologically Common Satellite Accretion Events. *Astrophys. J. Lett.* **2009**, *694*, L98–L102.

116. Kazantzidis, S.; Zentner, A.R.; Kravtsov, A.V.; Bullock, J.S.; Debattista, V.P. Cold Dark Matter Substructure and Galactic Disks. II. Dynamical Effects of Hierarchical Satellite Accretion. *Astrophys. J.* **2009**, *700*, 1896–1920.

117. Sachdeva, S.; Saha, K. Survival of Pure Disk Galaxies over the Last 8 Billion Years. *Astrophys. J. Lett.* **2016**, *820*, doi:10.3847/2041-8205/820/1/L4.

118. Moetazedian, R.; Just, A. Impact of cosmological satellites on the vertical heating of the Milky Way disc. *Mon. Not. R. Astron. Soc.* **2016**, *459*, 2905–2924.

119. Abadi, M.G.; Navarro, J.F.; Steinmetz, M. Stars beyond galaxies: The origin of extended luminous haloes around galaxies. *Mon. Not. R. Astron. Soc.* **2006**, *365*, 747–758.

120. Zolotov, A.; Willman, B.; Brooks, A.M.; Governato, F.; Brook, C.B.; Hogg, D.W.; Quinn, T.; Stinson, G. The Dual Origin of Stellar Halos. *Astrophys. J.* **2009**, *702*, 1058–1067.

121. Zolotov, A.; Willman, B.; Brooks, A.M.; Governato, F.; Hogg, D.W.; Shen, S.; Wadsley, J. The Dual Origin of Stellar Halos. II. Chemical Abundances as Tracers of Formation History. *Astrophys. J.* **2010**, *721*, 738–743.

122. Font, A.S.; McCarthy, I.G.; Crain, R.A.; Theuns, T.; Schaye, J.; Wiersma, R.P.C.; Dalla Vecchia, C. Cosmological simulations of the formation of the stellar haloes around disc galaxies. *Mon. Not. R. Astron. Soc.* **2011**, *416*, 2802–2820.

123. McCarthy, I.G.; Font, A.S.; Crain, R.A.; Deason, A.J.; Schaye, J.; Theuns, T. Global structure and kinematics of stellar haloes in cosmological hydrodynamic simulations. *Mon. Not. R. Astron. Soc.* **2012**, *420*, 2245–2262.

124. Tissera, P.B.; Scannapieco, C.; Beers, T.C.; Carollo, D. Stellar haloes of simulated Milky-Way-like galaxies: Chemical and kinematic properties. *Mon. Not. R. Astron. Soc.* **2013**, *432*, 3391–3400.

125. Tissera, P.B.; Beers, T.C.; Carollo, D.; Scannapieco, C. Stellar haloes in Milky Way mass galaxies: From the inner to the outer haloes. *Mon. Not. R. Astron. Soc.* **2014**, *439*, 3128–3138.

126. Pillepich, A.; Madau, P.; Mayer, L. Building Late-type Spiral Galaxies by In-situ and Ex-situ Star Formation. *Astrophys. J.* **2015**, *799*, doi:10.1088/0004-637X/799/2/184.

127. Cooper, A.P.; Parry, O.H.; Lowing, B.; Cole, S.; Frenk, C. Formation of in situ stellar haloes in Milky Way-mass galaxies. *Mon. Not. R. Astron. Soc.* **2015**, *454*, 3185–3199.

128. Carollo, D.; Beers, T.C.; Lee, Y.S.; Chiba, M.; Norris, J.E.; Wilhelm, R.; Sivarani, T.; Marsteller, B.; Munn, J.A.; Bailer-Jones, C.A.L.; et al. Two stellar components in the halo of the Milky Way. *Nature* **2007**, *450*, 1020–1025.

129. Deason, A.J.; Belokurov, V.; Evans, N.W.; Johnston, K.V. Broken and Unbroken: The Milky Way and M31 Stellar Halos. *Astrophys. J.* **2013**, *763*, 113.

130. Dorman, C.E.; Widrow, L.M.; Guhathakurta, P.; Seth, A.C.; Foreman-Mackey, D.; Bell, E.F.; Dalcanton, J.J.; Gilbert, K.M.; Skillman, E.D.; Williams, B.F. A New Approach to Detailed Structural Decomposition from the SPLASH and PHAT Surveys: Kicked-up Disk Stars in the Andromeda Galaxy? *Astrophys. J.* **2013**, *779*, doi:10.1088/0004-637X/779/2/103.

131. Hawkins, K.; Kordopatis, G.; Gilmore, G.; Masseron, T.; Wyse, R.F.G.; Ruchti, G.; Bienaymé, O.; Bland-Hawthorn, J.; Boeche, C.; Freeman, K.; et al. Characterizing the high-velocity stars of RAVE: The discovery of a metal-rich halo star born in the Galactic disc. *Mon. Not. R. Astron. Soc.* **2015**, *447*, 2046–2058.

132. Bonaca, A.; Conroy, C.; Wetzel, A.; Hopkins, P.F.; Keres, D. Gaia reveals a metal-rich in-situ component of the local stellar halo. *Astrophys. J.* **2017**, doi:10.3847/1538-4357/aa7d0c.

133. Chakrabarti, S.; Blitz, L. Tidal imprints of a dark subhalo on the outskirts of the Milky Way. *Mon. Not. R. Astron. Soc.* **2009**, *399*, L118–L122.

134. Chakrabarti, S.; Bigiel, F.; Chang, P.; Blitz, L. Finding Dwarf Galaxies from Their Tidal Imprints. *Astrophys. J.* **2011**, *743*, doi:10.1088/0004-637X/743/1/35.

135. Chang, P.; Chakrabarti, S. Dark subhaloes and disturbances in extended H I discs. *Mon. Not. R. Astron. Soc.* **2011**, *416*, 618–628.

galaxies

MDPI

Conference Report

Dissecting Halo Components in IFU Data

Michael Merrifield [1,*], Evelyn Johnston [2] and Alfonso Aragón-Salamanca [1]

1   School of Physics & Astronomy, University of Nottingham, Nottingham NG7 2RD, UK;
    alfonso.aragon@nottingham.ac.uk
2   European Southern Observatory, Alonso de Córdova 3107 Vitacura, Casilla 19001, Santiago de Chile, Chile;
    ejohnsto@eso.org
*   Correspondence: michael.merrifield@nottingham.ac.uk; Tel.: +44-115-951-5186

Academic Editors: Duncan A. Forbes and Ericson D. Lopez
Received: 20 April 2017; Accepted: 23 May 2017; Published: 25 May 2017

Abstract: While most astronomers are now familiar with tools to decompose images into multiple components such as disks, bulges, and halos, the equivalent techniques for spectral data cubes are still in their infancy. This is unfortunate, as integral field unit (IFU) spectral surveys are now producing a mass of data in this format, which we are ill-prepared to analyze effectively. We have therefore been developing new tools to separate out components using this full spectral data. The results of such analyses will prove invaluable in determining not only whether such decompositions have an astrophysical significance, but, where they do, also in determining the relationship between the various elements of a galaxy. Application to a pilot study of IFU data from the cD galaxy NGC 3311 confirms that the technique can separate the stellar halo from the underlying galaxy in such systems, and indicates that, in this case, the halo is older and more metal poor than the galaxy, consistent with it forming from the cannibalism of smaller satellite galaxies. The success of the method bodes well for its application to studying the larger samples of cD galaxies that IFU surveys are currently producing.

Keywords: galaxy formation and evolution; integral field unit spectroscopy; cD galaxies

---

1. Introduction

Our understanding of galactic structure and its evolution has been driven forward on a tidal wave of data, which has allowed us to systematically quantify the properties of these beautiful systems on the basis of very large surveys. The Sloan Digital Sky Survey (SDSS) produced imaging data on millions of galaxies, which, though an amazing resource, presented the additional challenge of how to quantify so much information in a meaningful form. Fortunately, tools such as GALFIT had been developed that allow the properties of the components that make up galaxies to be objectively quantified by simultaneously fitting a number of simple functions such as two-dimensional Sersic profiles to their images to quantify each component [1]. Such automated techniques can readily be applied even to surveys on the scale of SDSS, producing a mass of summarizing data on the disks, bulges, and other components that make up galaxies in the nearby Universe [2].

However, such imaging data has its limitations, as there are degeneracies and ambiguities when seeking to understand the stellar populations that make up each component if only broad-band colours are available. To address this problem, several large spectroscopic galaxy surveys such as SAMI [3] and MaNGA [4] are currently under way, using integral field units (IFUs) that produce a spectrum at each point on the sky across the face of each galaxy, generating a cube of data with a spectrum in the z-direction for every position $\{x, y\}$. From such spectral data, one can derive properties such as age and metallicity, and how they vary with position in each galaxy. Unfortunately, tools similar to those in common usage for imaging data do not exist for such data cubes, so we have been developing new ones specifically to decompose spectral data cubes of galaxies into their constituent components. Here, we describe the first application of such a technique to a cD galaxy in order to separate the spectral properties of its halo from the main elliptical component.

2. Materials and Methods

The essence of the new method we have developed is very simple. At every wavelength, an IFU data cube contains a narrow-band image of the galaxy at that wavelength, so the conventional machinery of image fitting and decomposition can be applied, yielding the amount of light in each component at that wavelength. Repeating the analysis for each such wavelength slice then gives the amount of light in each component as a function of wavelength, more conventionally described as the spectrum of each component. Since a complete data cube has been reduced to a few one-dimensional spectra, the resulting signal-to-noise ratios of the derived component spectra are very high, allowing them to be readily analyzed in terms of their stellar populations.

Of course, in practice things are more complicated. For one thing, each image slice must correspond to the same *rest* wavelength across the whole galaxy, so the shifts and broadening due to the galaxy's kinematics have to be allowed for. In addition, the fitting process occasionally produces unphysical results, particularly on the relatively low signal-to-noise ratio 'images' of single wavelength slices, so the fitting process can be quite unstable. Fortunately, there is the additional physical constraint that the structural parameters of a galaxy's components will not vary dramatically between adjacent wavelengths, so the fitting process can be regularized by imposing this additional constraint.

We have recently successfully implemented this approach in analyzing IFU data, and applied it to a study of S0 galaxies, deriving spectra for their bulge and disk components [5]. The results showed that the bulges were systematically younger than the disks, implying that the 'last gasp' of star formation occurred when gas was channeled into the centres of the systems as they were being stripped, meaning that, at least in a luminosity-weighted sense, the starlight of the bulges is younger than that of the surrounding disk [5]. Although initially applied to separating bulges and disks, the technique is completely general, and can be used to derive spectra for any photometrically-distinct components in a galaxy, so here we turn to separating elliptical cores from their surrounding stellar halos in cD systems, to explore their interrelationship.

An excellent target for this pilot study is offered by NGC 3311, the classic cD galaxy at the centre of the Hydra I Cluster. The centre of this cluster was the target for a set of four pointings with the large-area Multi-Unit Spectroscopic Explorer (MUSE) [6] at the Very Large Telescope (VLT) obtained under ESO Program 094.B-0711(A) (PI Arnaboldi); this archival dataset covered a significant fraction of the light from NGC 3311. Such an extensive observation of a nearby cD galaxy is complete overkill for the analysis that we wish to undertake here, but it provides the ideal test case as the high-quality data means that the results of our simple summarizing analysis can be compared with a more sophisticated study of the detailed data, to provide some confidence in the method when such unusually high-quality observations are not available.

3. Results

A sample image slice of the data set is presented in Figure 1, which shows the footprint of the four MUSE pointings, the image data in this single wavelength slice, the elliptical and halo components fitted as a de Vaucouleurs law and an exponential respectively, and the residuals once they are subtracted. There are significant negative residuals at small radii, associated with the documented unusually low surface brightness of the centre of this galaxy [7], but tests showed that the results were robust whether or not this region was included in the fit. There are also still minor artifacts in the data associated with the flatfielding of individual pointings, but again not at a level that should compromise the results unduly.

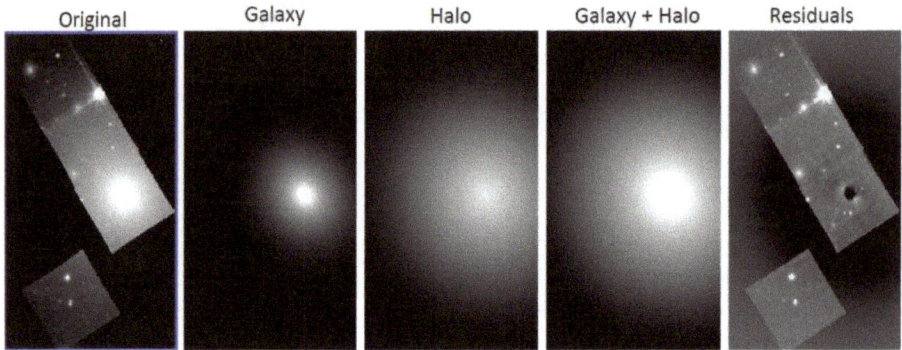

Figure 1. A sample image slice through the MUSE observations of NGC 3311, showing the footprint of the observations and image at a single wavelength, the model fit to these data, and the residuals left after the fitted model has been subtracted.

Having repeated this fitting process for all wavelength slices, as set out in Section 2, we analyzed the resulting preliminary spectra of the disentangled galaxy and halo components by determining various Lick indices from them. The results, together with an indicative grid of values characteristic of various single stellar population models, is presented in Figure 2. As is clear from this figure, both components are very old, with the halo apparently containing slightly older stars than the main body of the galaxy. More clear-cut, the halo is of significantly lower metallicity with [Fe/H] ~−0.4, as compared to the galaxy's value of ~+0.1. The latter value is characteristic of a massive elliptical galaxy, but the halo value is what one would expect of a galaxy that is a factor of ~10 less luminous [8] (Figure 8). This fits well with a scenario in which all the stars that make up this galaxy formed early, but the halo component was originally in smaller satellite galaxies that formed their stars slightly earlier than the central system, which were subsequently accreted by the central galaxy in a series of minor mergers.

Further evidence for the early formation of all the stars that ultimately make up NGC 3311 comes from the alpha enhancement measured in both components (see Figure 2b). Such enhancement occurs when star formation is complete on a timescale shorter than the point at which Type 1a supernovae start to occur, so the only heavy nuclei that have been recycled into stars are from the Type 2 supernovae that produce alpha elements in abundance. Thus, we know that the stars making up both galaxy and halo must have been formed in a single short burst rather than being spread over the lifetime of the system.

In this test case, the data are of high enough quality that one can see crudely how these indices vary with projected radius in the raw spectra. The results are shown as the small points in Figure 2; reassuringly, these results are consistent with previous study of the variation in the projected stellar population with radius [9]. As can be seen, the Lick indices generally show the kind of gradient with a radius that one would expect if the observed light were made up of the derived galaxy and halo components, whose relative weight changes with the radius such that the total spectrum transitions smoothly from galaxy-dominated at small radii to halo-dominated at large radii. The only clear exception to this consistent picture is at very small radii where the Hβ absorption-line index is consistently low, reflecting the presence of gas at the centre of this system, whose emission lines partially fill in the Hβ absorption line. Note, however, that the fitting process effectively rejects this discrepant data, making the analysis presented here robust against such contaminants, and would do so even in cases where we did not have data of the quality required to identify them explicitly, as we benefit from here.

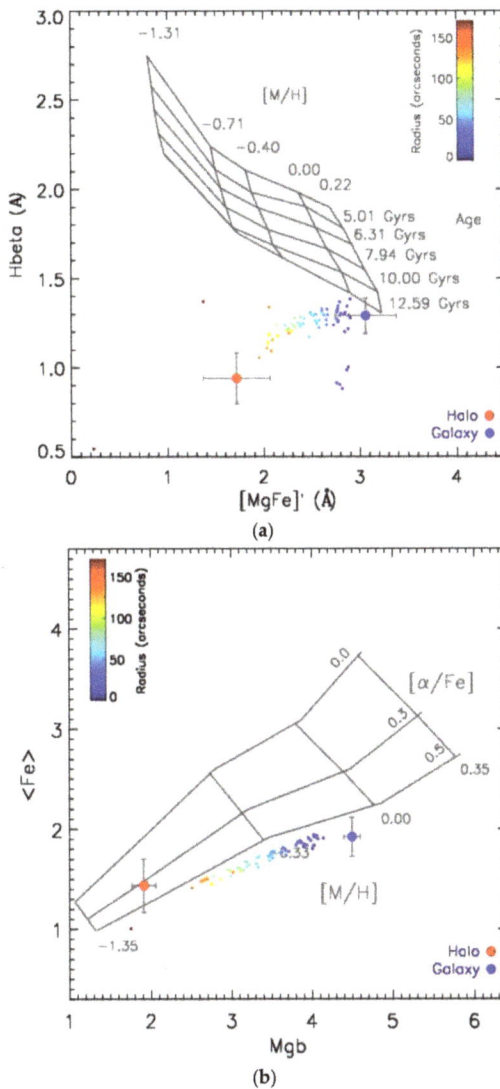

Figure 2. Stellar population models superimposed on Lick indices measured from decomposed galaxy and halo: (**a**) age and metallicity; (**b**) metallicity and alpha-element enhancement. The large points show the values from decomposed components; the small points show values derived as a function of projected radius directly from the data.

4. Discussion

In this paper, we have shown how spectral data cubes can be analyzed in a manner analogous to image data, decomposing a galaxy into separate spectra of its photometric components. A preliminary pilot study of IFU spectral observations of the cD galaxy NGC 3311 indicates how this system can be decomposed into a main galaxy component and a surrounding stellar halo; both components are made up of old rapidly-formed stellar populations, whose properties are consistent with the main component forming in situ, while the halo was acquired from mergers with multiple smaller satellite systems.

More generally, the success of this analysis in reproducing the results of more detailed analysis of high-quality data bodes well for its application to smaller, lower-quality data cubes. In addition, the automated robust nature of the fitting process means that it is well suited to an objective analysis of larger samples of cD galaxies. We therefore look forward to its application to more extensive survey data in the coming years, to see how the formation histories of the extended stellar halos around cD galaxies might depend on other properties of these systems.

Acknowledgments: The authors wish to thank the organizers of the excellent conference, "On the Origin (and Evolution) of Baryonic Halos", which motivated this work, and at which it was presented. The results presented here are based on observations collected at the European Organisation for Astronomical Research in the Southern Hemisphere under ESO programme 094.B-0711(A).

Author Contributions: E.J. carried out the analysis presented here; A.A.-S. contributed his expertise on the interpretation of line indices; M.M. suggested the project and wrote the paper.

Conflicts of Interest: The authors declare no conflict of interest.

References

1. Peng, C.Y.; Ho, L.C.; Impey, C.D.; Rix, H.-W. Detailed structural decomposition of galaxy images. *Astron. J.* **2002**, *124*, 266–293. [CrossRef]

2. Meert, A.; Vikram, V.; Bernardi, M. A catalogue of 2D photometric decompositions in the SDSS-DR7 spectroscopic main galaxy sample: Preferred models and systematics. *Mon. Not. R. Astron. Soc.* **2015**, *446*, 3943–3974. [CrossRef]

3. Bryant, J.J.; Owers, M.S.; Robotham, A.S.G.; Croom, S.M.; Driver, S.P.; Drinkwater, M.J.; Lorente, N.P.F.; Cortese, L.; Scott, N.; Colless, M.; et al. The SAMI Galaxy Survey: Instrument specification and target selection. *Mon. Not. R. Astron. Soc.* **2015**, *447*, 2857–2879. [CrossRef]

4. Yan, R.; Bundy, K.; Law, D.R.; Bershady, M.A.; Andrews, B.; Cherinka, B.; Diamond-Stanic, A.M.; Drory, N.; MacDonald, N.; Sánchez-Gallego, J.R.; et al. SDSS-IV MaNGA IFS Galaxy Survey—Survey Design, Execution, and Initial Data Quality. *Astron. J.* **2016**, *152*, 197. [CrossRef]

5. Johnston, E.J.; Häußler, B.; Aragón-Salamanca, A.; Merrifield, M.R.; Bamford, S.; Bershady, M.A.; Bundy, K.; Drory, N.; Fu, H.; Law, D.; et al. SDSS-IV MaNGA: Bulge-disc decomposition of IFU data cubes (BUDDI). *Mon. Not. R. Astron. Soc.* **2017**, *465*, 2317–2341. [CrossRef]

6. Bacon, R.; Accardo, M.; Adjali, L.; Anwand, H.; Bauer, S.; Biswas, I.; Blaizot, J.; Boudon, D.; Brau-Nogue, S.; Brinchmann, J.; et al. The MUSE second-generation VLT instrument. *Proc. SPIE* **2010**, *7735*. [CrossRef]

7. Vasterberg, A.R.; Lindblad, P.O.; Jorsater, S. An optical study of the cD galaxy NGC3311 and the giant elliptical galaxy NGC3309 in the cluster Hydra I. *Astron. Astrophys.* **1991**, *247*, 335–347.

8. Chilingarian, I.V.; Mieske, S.; Hilker, M. Dynamical versus stellar masses of ultracompact dwarf galaxies in the Fornax cluster. *Mon. Not. R. Astron. Soc.* **2011**, *412*, 1627–1638. [CrossRef]

9. Barbosa, C.E.; Arnaboldi, M.; Coccato, L.; Hilker, M.; Mendes de Oliveira, C.; Richtler, T. The Hydra I cluster core. I. Stellar populations in the cD galaxy NGC 3311. *Astron. Astrophys.* **2016**, *589*, A139. [CrossRef]

galaxies

MDPI

Letter

Distribution and Evolution of Metals in the Magneticum Simulations

Klaus Dolag [1,2,*], Emilio Mevius [1] and Rhea-Silvia Remus [1]

[1] Universitäts-Sternwarte München, Faculty of Physics, Ludwig-Maximilians-Universität, Scheinerstr. 1, D-81679 München, Germany; mevius84@gmail.com (E.M.); rhea@usm.uni-muenchen.de (R.-S.R.)

[2] Max Planck Institut for Astrophysics, D-85748 Garching, Germany

* Correspondence: dolag@usm.lmu.de

Academic Editors: Duncan A. Forbes and Ericson D. Lopez

Received: 30 June 2017; Accepted: 31 July 2017; Published: 4 August 2017

Abstract: Metals are ideal tracers of the baryonic cycle within halos. Their composition is a fossil record connecting the evolution of the various stellar components of galaxies to the interaction with the environment by in- and out-flows. The *Magneticum* simulations allow us to study halos across a large range of masses and environments, from massive galaxy clusters containing hundreds of galaxies, down to isolated field galaxies. They include a detailed treatment of the chemo-energetic feedback from the stellar component and its evolution, as well as feedback from the evolution of supermassive black holes. Following the detailed evolution of various metal species and their relative composition due to continuing enrichment of the IGM and ICM by SNIa, SNII and AGB winds of the evolving stellar population is revealed the complex interplay of local star-formation processes, mixing, global baryonic flows, secular galactic evolution and environmental processes. We present results from the Magneticum simulations on the chemical properties of simulated galaxies and galaxy clusters, carefully comparing them to observations. We show that the simulations already reach a very high level of realism within their complex descriptions of the chemo-energetic feedback, successfully reproducing a large number of observed properties and scaling relations. Our simulated galaxies clearly indicate that there are no strong secondary parameters (such as star-formation rates at a fixed redshift) driving the scatter in these scaling relations. The remaining differences clearly point to detailed physical processes, which have to be included in future simulations.

Keywords: galaxy clusters; intracluster medium; galaxies; stellar population; numerical simulation

1. Introduction

In cosmological hydrodynamical simulations, star formation is typically treated as based on a sub-grid model. Star particles are formed when the gas locally exceeds a certain density threshold, when it is Jeans unstable, and the local gas flow converges at a gas particle. Each star particle represents a simple stellar population that is characterized by an initial mass function (IMF).

In addition, each star particle emits a feedback that mimics the winds that are launched by stars in reality by averaging over the whole stellar population included in each simulated stellar particle. In most models, the considered winds are composed of three main stellar sources: supernovae Type II (SNII), supernovae Type Ia (SNIa), and asymptotic giant branch (AGB) stars. These three sources are considered to be the most important drivers of observed stellar outflows and thus the main contributors to metal enrichment of the interstellar medium.

Stars with masses larger than 8 $M_\odot$ are believed to end their lives through a "core collapse" (SNII), typically releasing an energy of 10^{51} erg per supernova together with their chemical imprint into the surrounding gas. These supernovae usually occur shortly after the formation of the simulated star particle, as the most massive stars only live for a short time, and thus this part of the stellar population of the star particle emits first. SNIa, on the other hand, are believed to arise from thermonuclear explosions of white dwarfs within binary stellar systems. These supernovae are delayed relative to the SNII and occur later in a star particle's life, as the white dwarf in the according binary system has to first be formed, and second, needs to accrete matter from its companion until it reaches the mass threshold for the onset of thermonuclear burning. Therefore, a detailed modeling of the evolution of the stellar population has to be included in the simulations to follow the chemo-energetic imprint of the SNIa on the surrounding gas. AGB stars contribute emitting strong winds, exhibiting strong mass losses during their life and importantly contribute to the nucleosynthesis of heavy elements.

To include the effects of these sources in the sub-grid models of simulated stellar particles raises the need to integrate a set of complicated equations describing the evolution of a simple stellar population. Such a set of integral equations then allows us to compute at each time the rate at which the current AGB stars pollute their environment by stellar winds and the rate at which the SNIa and SNII are exploding. Therefore they allow us to properly treat the chemo-energetic imprint on the surrounding inter-galactic medium (IGM) and the inter-cluster medium (ICM). In the following, we will give a short, schematic description of such calculations on the basis of the description presented by Dolag et al., 2017 [1] (for a more detailed review, see Matteucci, 2003 [2] and Borgani et al., 2008 [3], and references therein). Then we will introduce the cosmological hydrodynamical simulation set used in this work, and present a comparison of the metal enrichment in galaxies and clusters of galaxies caused by this model feedback, along with observations. Finally, we will summarize and discuss the results of this comparison in the light of model improvements needed for the future.

2. Chemical Enrichment

To describe the continuous enrichment of the IGM and ICM through winds from SNIa, SNII and AGB stars in the evolving stellar population of each star particle, several ingredients are needed. In the following, we will shortly present the ingredients used in many state-of-the-art cosmological simulations of galaxy formation.

2.1. Initial Mass Function

One of the most important quantities in models of chemical evolution is the IMF. It directly determines the relative ratio between SNII and SNIa, and therefore the relative abundance of α-elements and Fe-peak elements. The shape of the IMF also determines the ratio between low-mass long-living and massive short-living stars. This ratio directly affects the amount of energy released by the SNe, as well as the present luminosity of galaxies, which is dominated by low-mass stars and the (metal) mass-locking in the stellar phase.

The IMF $\phi(m)$ describes the number of stars of a given mass per unit logarithmic mass interval. Historically, a commonly used form is the Salpeter IMF (Salpeter, 1955 [4]), which follows a single power-law with an index of $x = 1.35$. However, different expressions of the IMF have been proposed more recently in order to model a flattening in the low-mass regime of the stellar mass function that is currently favoured by a number of observations. Among the newer, often-used models is the Chabrier IMF (Chabrier, 2003 [5]), which has a continuously changing slope and is more top-heavy than the Salpeter IMF:

$$\phi(m) \propto \begin{cases} m^{-1.3} & m > 1\,\mathrm{M_\odot} \\ e^{\frac{-(\log(m)-\log(m_c))^2}{2\sigma^2}} & m \leq 1\,\mathrm{M_\odot} \end{cases} \tag{1}$$

However, the questions of whether there is a global IMF or if the IMF is changing with galaxy mass, morphology or cosmological time, and which IMF has to be chosen, are still unsolved problems and matters of heavy debate.

2.2. Lifetime Functions

To model the evolution of a simple stellar population, a detailed knowledge of the lifetimes of stars with different masses is required. Different choices for the mass-dependence of the lifetime function have been proposed in the literature (e.g., Padovani and Matteucci, 1993 [6]; Maeder and Meynet, 1989 [7]; Chiappini et al., 1997 [8]), where the latest reads:

$$\tau(m) = \begin{cases} 10^{-0.6545 \log m + 1} & m \leq 1.3 \text{ M}_\odot \\ 10^{-3.7 \log m + 1.351} & 1.3 < m \leq 3 \text{ M}_\odot \\ 10^{-2.51 \log m + 0.77} & 3 < m \leq 7 \text{ M}_\odot \\ 10^{-1.78 \log m + 0.17} & 7 < m \leq 15 \text{ M}_\odot \\ 10^{-0.86 \log m - 0.94} & 15 < m \leq 53 \text{ M}_\odot \\ 1.2 \times m^{-1.85} + 0.003 & \text{otherwise.} \end{cases} \tag{2}$$

2.3. Stellar Yields

The ejected masses of the different metal species i produced by a star of mass m are called stellar yields $p_{Z_i}(m,Z)$. These yields depend on the metallicity Z with which the star originally formed, and on the type of outflow ejected from the star. Therefore, detailed predictions for the main three sources of enrichment are needed, namely for the mass loss of AGB stars, of SNII, and of SNIa. To date, such predictions still suffer from significant uncertainties, mainly due to the still poorly understood mass loss through stellar winds in stellar evolution models, which depends on multiple additional physical processes.

For the mass loss through AGB stars, the most recent predictions can be found in Karakas, 2007 [9]. Predictions for the mass loss from massive stars driving SNII are presented by Nomoto, Kobayashi & Tominaga, 2013 [10]. The most complete table for SNIa to date is presented by Thielmann, 2003 [11].

2.4. Modeling the Enrichment Process

As summarized by Dolag, 2017 [1], the assumption of a generic star-formation history represented by an arbitrary function of time $\psi(t)$ allows us to compute the rates for the different contributions in the form of a set of integral equations as shown in the following (for more details, see also Matteucci, 2003 [2], Borgani et al., 2008 [3], and references therein). This formalism can be individually applied to the large number of particles representing the continuous star-formation process within cosmological simulations. Every star particle here represents a stellar population born in a single burst. The combination of all the stellar components within the simulated galaxies then results in a model that describes the legacy of the detailed star-formation history of any simulated galaxy.

2.4.1. Type Ia Supernovae

SNIa occur in binary systems with masses in a mass range of 0.8–8 M$_\odot$. Letting m_B be the total mass of the binary system and m_2 be the mass of the secondary companion, we can now use $f(\mu)$ as the distribution function of binary systems with $\mu = m_2/m_B$, and define A as the fraction of stars in binary systems that are progenitors of SNIa. Therefore, A has to be given or be obtained by a model. Constructing such detailed models for SNIa progenitors is particularly difficult; see, for example, Greggio & Renzini, 1983 [12] or Greggio, 2005 [13]. On the basis of such kinds of models, typical values for A are inferred to be in the range from 0.05 to 0.1, on the basis of comparisons of chemical enrichment models with observed iron metallicities within galaxy clusters (e.g., Matteucci & Gibson, 1995 [14]) or within the solar neighbourhood (e.g., Matteucci & Greggio, 1986 [15]). Within the current simulations we are using a value of $A = 0.1$. With these ingredients and the mass-dependent lifetime functions $\tau(m)$, we can model the rate of SNIa as

$$R_{\mathrm{SN\,Ia}}(t) = A \int_{M_{\mathrm{B,inf}}}^{M_{\mathrm{B,sup}}} \phi(m_{\mathrm{B}}) \int_{\mu_{\mathrm{m}}}^{\mu_{\mathrm{M}}} f(\mu)\,\psi(t - \tau_{m_2})\,d\mu\,dm_{\mathrm{B}} \tag{3}$$

where M_{Bm} and M_{BM} are the smallest and largest values, respectively, that are allowed for the progenitor binary mass m_{B}. Then, the integral over m_{B} runs in the range between $M_{\mathrm{B,inf}}$ and $M_{\mathrm{B,sup}}$, which represent the minimum and the maximum value of the total mass of the binary system that is allowed to explode at the time t. These values in general are functions of M_{Bm}, M_{BM}, and $m_2(t)$, which in turn depend on the star-formation history $\Psi(t)$. In simulations, where the individual stellar particles are commonly modeled as an impulsive star-formation event, $\psi(t)$ can therefore be approximated with a Dirac δ-function. The sum of all stellar particles and their individual formation time then represent the complex star-formation history of the galaxies within the simulation.

2.4.2. Supernovae Type II, Low-, and Intermediate-Mass Stars

Computing the rates of SNII, low-mass stars (LMS), and intermediate-mass stars (IMS) is conceptually simpler than calculating the rates of SNIa, as they are purely driven by the lifetime function $\tau(m)$ convolved with the star-formation history $\psi(t)$ and multiplied by the IMF $\phi(m = \tau^{-1}(t))$. Because $\psi(t)$ is a delta-function for the simple stellar populations used in simulations, the SNII, LMS and IMS rates read

$$R_{\mathrm{SNII|LMS|IMS}}(t) = \phi(m(t)) \times \left(-\frac{d\,m(t)}{d\,t} \right) \tag{4}$$

where $m(t)$ is the mass of the star that dies at time t. For AGB rates, the above expression must be multiplied by a factor of $(1 - A)$ if the mass $m(t)$ falls in the range of masses, which is relevant for the secondary stars of SNIa binary systems.

2.4.3. The Equations of Chemical Enrichment

In order to compute the total metal release from the simple stellar population, we have to fold the above rates with the yields $p_{Z_i}^{\mathrm{SNIa|SNII|AGB}}(m, Z)$ from SNIa, SNII and AGB stars for a given element i for stars born with initial metallicity Z_i, and compute the evolution of the density $\rho_i(t)$ for each element i at each time t. As shown by Borgani et al., 2008 [3], this reads

$$\dot\rho_i(t) = \quad - \quad \psi(t)Z_i(t) \tag{5}$$

$$+ \quad \int_{M_{BM}}^{M_U} \psi(t - \tau(m))p_{Z_i}^{\mathrm{SNII}}(m, Z)\varphi(m)\,dm \tag{6}$$

$$+ \quad A \int_{M_{Bm}}^{M_{BM}} \phi(m) \left[\int_{\mu_m}^{\mu_M} f(\mu)\psi(t - \tau_{m_2})p_{Z_i}^{\mathrm{SNIa}}(m, Z)\,d\mu \right] dm \tag{7}$$

$$+ \quad (1 - A) \int_{M_{Bm}}^{M_{BM}} \psi(t - \tau(m))p_{Z_i}^{\mathrm{AGB}}(m, Z)\varphi(m)\,dm \tag{8}$$

$$+ \quad \int_{M_L}^{M_{Bm}} \psi(t - \tau(m))p_{Z_i}^{\mathrm{AGB}}(m, Z)\varphi(m)\,dm \tag{9}$$

where M_L and M_U are the minimum and maximum mass of a star in the simple stellar population, respectively. Commonly adopted choices for these limiting masses are $M_L \simeq 0.1\ M_\odot$ and $M_U \simeq 100\ M_\odot$.

In the above equation, the first line describes the locking of metals in newborn stars through the currently ongoing star formation $\psi(t)$, which, for the assumed sub-grid model case, vanishes as $\psi(t)$ is a delta-function. For a comprehensive review of the analytic formalism we refer the reader to Greggio, 2005 [13].

3. The Magneticum Simulations

The *Magneticum* simulation set covers a huge dynamical range, from very large cosmological volumes as shown in Figure 1, which can be used for statistical studies of clusters and voids (e.g., Bocquet et al., 2016 [16], Pollina et al., 2017 [17]), to very high-resolution simulations of smaller cosmological volumes, which allow for a morphological classification and a detailed analysis of galaxies and their properties (e.g., Teklu et al., 2015 [18], Remus et al., 2017 [19]). Table 1 lists the detailed properties, such as size and stellar mass-resolution, of the different simulations. These simulations treat the metal-dependent radiative cooling, heating from a uniform time-dependent ultraviolet background, star formation and the chemo-energetic evolution of the stellar population as traced by SNIa, SNII and AGB stars with the associated feedback processes and stellar evolution details described before. They also include the formation and evolution of supermassive black holes and the associated quasar and radio-mode feedback processes. For a detailed description of the simulation sample, see Dolag et al., (in prep); Hirschmann et al., 2014 [20]; and Teklu et al., 2015 [18].

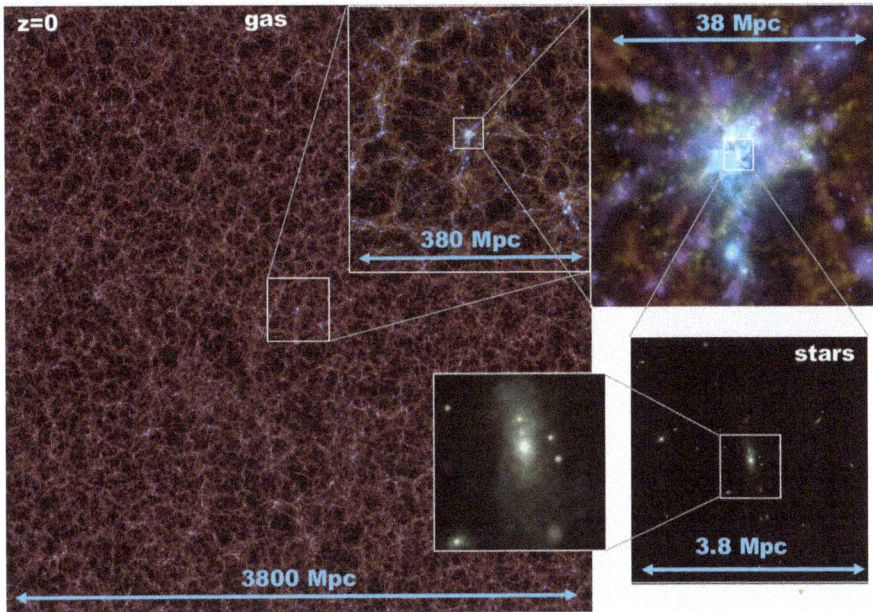

Figure 1. Visualization of the large-scale distribution of the gas and stellar components within *Box0* (see Bocquet et al., 2016 [16]) of the *Magneticum* simulation set at redshift $z = 0$. The inlays show a consecutive zoom onto the most massive galaxy cluster, where the individual galaxies become visible.

Table 1. Size and number of particles for the different simulations. The last two rows list the average mass and softening of the star particles for the different resolution levels.

Simulation	Box0	Box1	Box2b	Box2	Box3	Box4	m_{star} $[M_\odot/h]$	ϵ_{star} [kpc/h]
Size [Mpc]	3820	1300	910	500	180	68		
mr	2×4536^3	2×1512^3	–	2×594^3	2×216^3	2×81^3	4.7×10^8	5
hr	–	–	2×2880^3	2×1564^3	2×576^3	2×216^3	2.6×10^7	2
uhr	–	–	–	–	2×1536^3 (z = 2)	2×576^3	1.3×10^6	0.7

4. Metallicities from Magneticum in Comparison to Observations

As previously shown from re-simulations of massive galaxy clusters, the observed radial profiles of iron within the ICM can be well reproduced in simulations: Biffi et al., 2017, [21] demonstrated that, especially at high redshifts, the implemented AGN feedback is the main driver for enhancing the metal enrichment in the ICM at large cluster-centric distances to the observed level.

The Magneticum simulations now allow such investigations across a much larger range of halo masses. For this study, we use a large simulation volume (Box2 hr) with enough resolution to resolve mean properties of galaxies (and AGNs), resulting in the same resolution as used by Biffi et al., 2017 [21] and Hirschmann et al., 2014 [20]. For galaxies, the star-formation efficiency (SFE) is a measure of the SFR per unit of gas mass. To evaluate whether the SF activity in the simulations matches the observational data is a key step to proceed towards reproducing the observed mass–metallicity relation (MZR). In Figure 2, the evolution of the observed SFE up to a redshift of $z \approx 2$ is shown in comparison to the evolution of the SFE of star-forming galaxies in Magneticum. As can clearly be seen, the simulations are in excellent agreement with the observed trends up to $z \approx 2$.

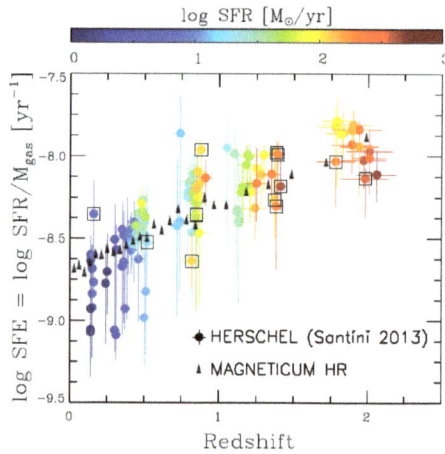

Figure 2. Redshift evolution of the star-formation efficiency (SFE; coloured points with error bars) from Herschel observations presented by Santini et al., 2014 [22] overlayed with the predicted median values at different redshifts from star-forming galaxies in the Magneticum simulations (black triangles).

4.1. Galaxy Clusters: ICM Metallicities

X-ray observations of the intracluster medium allow for a detailed study of the distribution of different metal species within the ICM via their line emissions within the X-ray band. Here, the composition of the individual metal species allows us to interpret their abundances as an imprint of the relative contributions to the enrichment from SNIa and SNII, hence keeping a record on when and where these metals are injected into the ICM (e.g., De Plaa et al., 2007 [23]). The lower left panel of

Figure 3 shows a comparison of such observational data for various individual galaxy clusters using simulations, clearly demonstrating the ability of the Magneticum simulations to reproduce the correct absolute iron abundance within the ICM self-consistently.

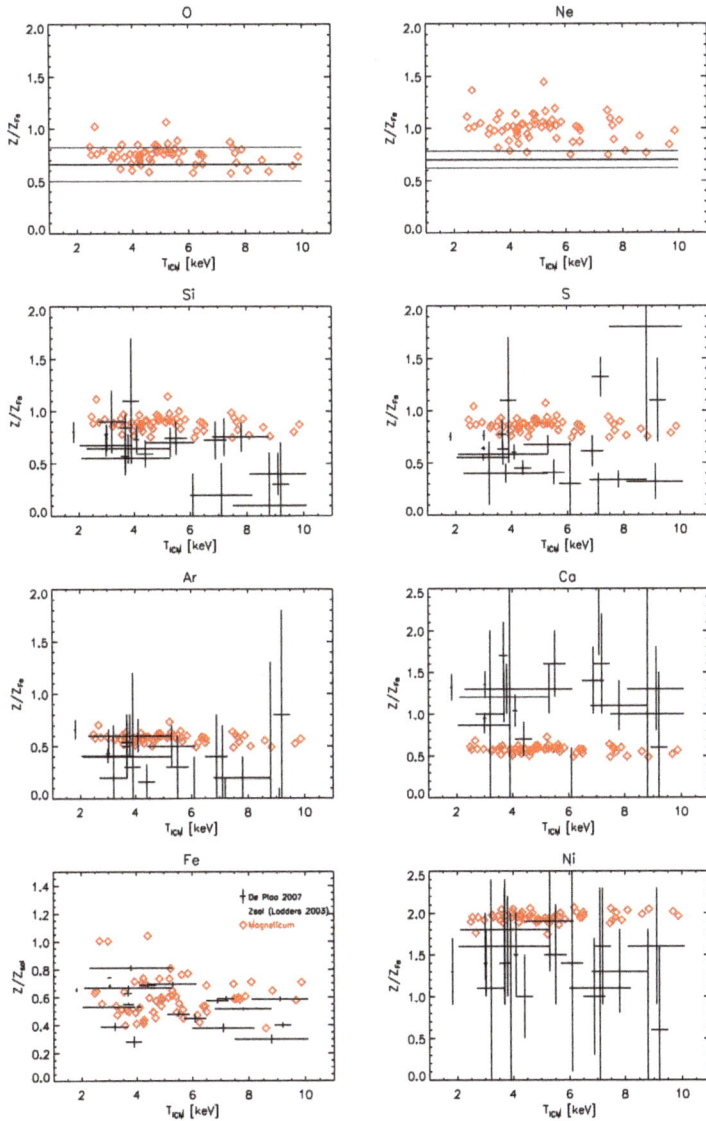

Figure 3. Comparison between the observed inter-cluster medium (ICM) metallicity within R_{2500} (black symbols with error bars) and the metallicities of galaxy clusters from the Magneticum simulations (red diamonds) as a function of the ICM temperature. Different panels show the ratios for different metal types, as indicated in the panels. From top left to bottom right, the relative contribution of supernovae Type Ia (SNIa) is expected to increase, while the relative contribution of supernovae Type II (SNII) is expected to decrease. Observational data are taken from De Plaa et al., 2007 [23].

Here, the scattering of the absolute iron abundance within the simulated clusters shows a similar spread (factor of two) as the observed values, and a very mild trend of increasing iron abundance for low-mass (e.g., lower ICM temperature) clusters. In addition, the simulations also broadly reproduce the chemical composition footprint of various elements' species, as shown in the various panels. This strongly indicates that the simulations predict the correct ratio between the contributions to the metal enrichment from SNIa and SNII and its interplay with the AGN feedback. The simulations also predict typically less spread in the composition of the metals for the individual clusters than they show in their absolute iron metallicity.

4.2. Galaxies: Gas Metallicities

To estimate the gas-phase metallicity of galaxies, it is important to consider that observationally, the measurements are obtained only from star-forming regions. Therefore, it can be misleading to only calculate the mean metallicity of all gas particles inside a simulated galaxy. Thus, after selecting star-forming galaxies (see Teklu et al., 2017 [24]) we can either calculate their mean metallicity by averaging over all particles that are currently star-forming, or alternatively, calculate the mean metallicity of the newborn stars. We tested both methods and found that the latter gives slightly better results because of the fact that it is difficult to catch the metallicity of the gas phase in the moment of star formation within simulations, given the large timespan between the simulation outputs, while the young stellar population freezes the record of the metallicity of the gas from which it was formed. This leads to a MZR that is in good agreement with observations, as shown in the left panel of Figure 4, where we used the predicted oxygen abundances from our calculations to be consistent with observations (Sanders et al., 2015 [25] at $z = 2.3$ and Bresolin et al., 2016 [26] at $z = 0$). Interestingly, even the overall evolution of the MZR is well captured by the simulations and matches the observations, as demonstrated in the right panel of Figure 4. However, when calculating gas-phase metallicity gradients within the galaxies, the simulations predict a steeper profile than the observations. Given that the prediction of the mean metallicities is in good agreement with observations, this indicates that the simulations either still lack the resolution to properly describe the mixing of the enriched gas within galaxies, or that on these scales, the diffusion of metals might have to be modeled more explicitly. Nevertheless, we clearly showed that including detailed modeling of the stellar population, combined with current AGN feedback models, significantly improves the predicted ICM and IGM metallicities, and is needed to successfully reproduce various aspects of the observations.

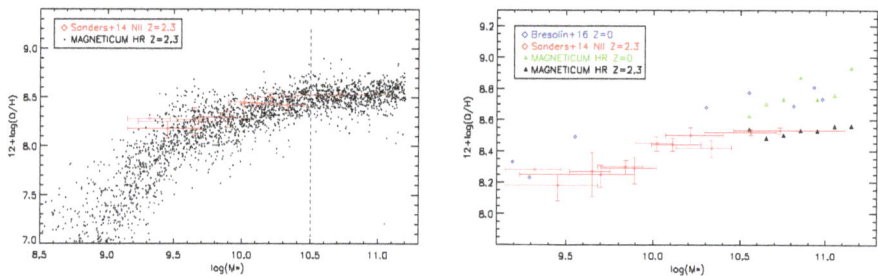

Figure 4. Gas-phase metallicity from Magneticum galaxies in comparison to observational data by Sanders et al., 2015 [25] at redshift $z = 2.3$ (red datapoints with error bars) and Bresolin et al., 2016 [26] at redshift $z = 0$ (blue diamonds). **Left panel:** Each black point represents a galaxy from the simulation. **Right panel:** Median values for each mass bin at $z = 2.3$ (black) and $z = 0$ (green) are shown as triangles.

4.3. Galaxies: Stellar Metallicities

In general, we assume that the mean metallicity of a stellar particle in the simulation represents the mean stellar metallicity of the stellar population represented by the stellar particle, neglecting any self-enrichment within stars. To be consistent with the observations, we based all calculations on the iron abundance, as predicted by the simulations. For this part of the study, we used a smaller cosmological volume with a higher resolution, as this allows for a classification of the galaxies due to their morphological type (see Teklu et al., 2015 [18] for more details on the classification and this particular simulation). This higher resolution also allows for a more detailed resolution of the metallicity gradients with the radius for individual galaxies.

Although the obtained mean metallicities of our galaxies are close to observational results, the stellar MZR obtained from the simulations is somewhat shallower than that obtained from CALIFA observations by Gonzalez Delgado et al., 2014 [27], as shown in the left panel of Figure 5. This again indicates that the treatment of the mixing between accreted, more pristine material from outside the galaxy and the enriched gas within the galaxies is not fully captured yet. At this point, it is unclear whether this will be resolved by further enhancing the resolution of simulations, or whether explicit diffusion of metals has to be taken into account.

For a proper comparison of the radial metallicity gradients between simulations and observational data, it is necessary to calculate the effective radius R_{eff} for each galaxy. This is done by selecting all stellar particles within 10% of the virial radius and then inferring the according half-mass radius, which we associate to the effective radius R_{eff}. This method has already been used to demonstrate that the Magneticum simulations successfully reproduce the mass–radius relation of galaxies for different morphological types (e.g., Remus et al., 2015 [28]) and at different redshifts (e.g., Remus et al., 2017 [19]).

Interestingly, the radial stellar metallicity gradients obtained from the simulations match the CALIFA observations very well, as shown in the right panel of Figure 5. We note also that the increasing spread towards a larger radius is an intrinsic, point-by-point spread within the individual galaxies, and that it agrees well with the behavior measured for individual galaxies by Pastorello et al., 2014 [29]. However, while the simulations successfully reproduce the observed iron abundances and radial gradients, they still produce a flat ratio of oxygen (or magnesium or silicium) over iron, in contrast to the observations, and here, the sub-grid model clearly needs to be advanced.

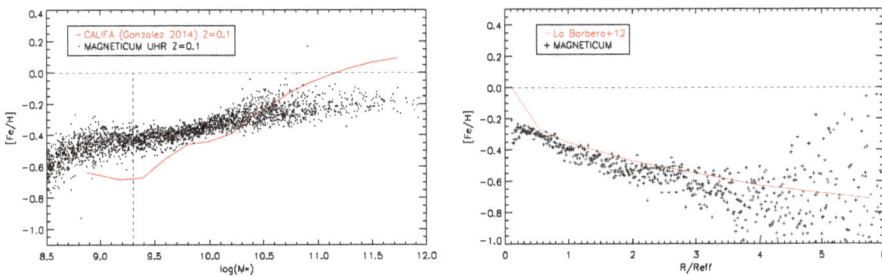

Figure 5. Left panel: Stellar metallicity versus stellar mass obtained from the Magneticum galaxies (black dots), in comparison with observational data for galaxies independent of their morphology from Gonzalez Delgado et al., 2014 [27] at redshift $z = 0$. **Right panel:** Mean stellar radial metallicity gradient from 100 Magneticum galaxies in 100 radial bins normalized to the half-mass radius (black dots), compared to the observed profile from La Barbera et al., 2012 [30] (red line).

5. Discussion and Conclusions

Metals are measured in all phases of the baryonic universe, starting from the ICM in galaxy clusters down to gas and stars in individual galaxies. As the radial distribution of different types of metals encode a record of the star-formation process over the whole period of a galaxy's evolution, comparing the chemical footprint, as seen in simulations, with observations reflects an excellent test on the reliability of the numerical sub-grid models included in modern state-of-the-art simulations of galaxy formation in a cosmological context in reproducing the complex processed involved in galaxy formation.

We demonstrated that the Magneticum simulations are able to reproduce a large variety of observational findings over a large range of halo masses, indicating that the star-formation and feedback processes included start to be highly realistic. At the galaxy-cluster scale, the simulations show an excellent agreement with both the absolute value and the cluster-by-cluster variation of the the iron abundance, when compared to X-ray observations, and they also broadly reproduce the chemical composition of the ICM. On galaxy scales, the tight mass–metallicity relations found for our simulated galaxies indicate that there are no strong secondary parameters (such as star-formation rates at a fixed redshift) driving the scatter in these relations. The remaining differences between the observed properties and the simulation results indicate, however, that the incorporation of physical processes such as diffusion and mixing have to be improved within the next generation of simulations to successfully reproduce the detailed distribution of metals on individually resolved galaxy scales.

Acknowledgments: The Magneticum Pathfinder simulations were performed at the Leibniz-Rechenzentrum with CPU time assigned to the Projects "pr86re" and "pr83li". This work was supported by the DFG Cluster of Excellence "Origin and Structure of the Universe". We are especially grateful for the support by M. Petkova through the Computational Center for Particle and Astrophysics (C2PAP).

Author Contributions: K.D. performed the simulation and wrote the paper; E.M. analysed the data; R.-S.R. contributed to the physical interpretation of the results, the supervising of E.M. during his master, and the writing of the paper.

Conflicts of Interest: The authors declare no conflict of interest.

References

1. Dolag, K. Hydrodynamic Methods for Cosmological Simulations. In *The Encyclopedia of Cosmology*; Fazio, G.G., Ed; World Scientific: Singapore, 2017; Volume 2.
2. Matteucci, F. *The Chemical Evolution of the Galaxy*; Springer: Houten, Netherlands, 2003.
3. Borgani, S.; Fabjan, D.; Tornatore, L.; Schindler, S.; Dolag, K.; Diaferio, A. The Chemical Enrichment of the ICM from Hydrodynamical Simulations. *Space Sci. Rev.* **2008**, *134*, 379–403.
4. Salpeter, E.E. The Luminosity Function and Stellar Evolution. *Astrophys. J.* **1955**, *121*, 161–167.
5. Chabrier, G. Galactic Stellar and Substellar Initial Mass Function. *Publ. Astron. Soc. Pac.* **2003**, *115*, 763–795,
6. Padovani, P.; Matteucci, F. Stellar Mass Loss in Elliptical Galaxies and the Fueling of Active Galactic Nuclei. *Astrophys. J.* **1993**, *416*, 26.
7. Maeder, A.; Meynet, G. Grids of evolutionary models from 0.85 to 120 solar masses—Observational tests and the mass limits. *Astron. Astrophys.* **1989**, *210*, 155–173.
8. Chiappini, C.; Matteucci, F.; Gratton, R. The Chemical Evolution of the Galaxy: The Two-Infall Model. *Astrophys. J.* **1997**, *477*, 765–780.
9. Karakas, A.; Lattanzio, J.C. Stellar Models and Yields of Asymptotic Giant Branch Stars. *Publ. Astron. Soc. Pac.* **2007**, *24*, 103–117.
10. Nomoto, K.; Kobayashi, C.; Tominaga, N. Nucleosynthesis in Stars and the Chemical Enrichment of Galaxies. *Ann. Rev. Astron. Astrophys.* **2013**, *51*, 457–509.
11. Thielemann, F.K.; Argast, D.; Brachwitz, F.; Hix, W.R.; Höflich, P.; Liebendörfer, M.; Martinez-Pinedo, G.; Mezzacappa, A.; Nomoto, K.; Panov, I. Supernova Nucleosynthesis and Galactic Evolution. In *From Twilight to Highlight: The Physics of Supernovae*; Hillebrandt, W., Leibundgut, B., Eds.; Springer: Berlin, Germany, 2003; p. 331.

12. Greggio, L.; Renzini, A. The binary model for type I supernovae—Theoretical rates. *Astron. Astrophys.* **1983**, *118*, 217–222.

13. Greggio, L. The rates of type Ia supernovae. I. Analytical formulations. *Astron. Astrophys.* **2005**, *441*, 1055–1078.

14. Matteucci, F.; Gibson, B.K. Chemical abundances in clusters of galaxies. *Astron. Astrophys.* **1995**, *304*, 11.

15. Matteucci, F.; Greggio, L. Relative roles of type I and II supernovae in the chemical enrichment of the interstellar gas. *Astron. Astrophys.* **1986**, *154*, 279–287.

16. Bocquet, S.; Saro, A.; Dolag, K.; Mohr, J.J. Halo mass function: baryon impact, fitting formulae, and implications for cluster cosmology. *Mon. Not. R. Astron. Soc.* **2016**, *456*, 2361–2373.

17. Pollina, G.; Hamaus, N.; Dolag, K.; Weller, J.; Baldi, M.; Moscardini, L. On the linearity of tracer bias around voids. *Mon. Not. R. Astron. Soc.* **2017**, *469*, 787–799.

18. Teklu, A.F.; Remus, R.S.; Dolag, K.; Beck, A.M.; Burkert, A.; Schmidt, A.S.; Schulze, F.; Steinborn, L.K. Connecting Angular Momentum and Galactic Dynamics: The Complex Interplay between Spin, Mass, and Morphology. *Astrophys. J.* **2015**, *812*, 29.

19. Remus, R.S.; Dolag, K.; Naab, T.; Burkert, A.; Hirschmann, M.; Hoffmann, T.L.; Johansson, P.H. The co-evolution of total density profiles and central dark matter fractions in simulated early-type galaxies. *Mon. Not. R. Astron. Soc.* **2017**, *464*, 3742–3756.

20. Hirschmann, M.; Dolag, K.; Saro, A.; Bachmann, L.; Borgani, S.; Burkert, A. Cosmological simulations of black hole growth: AGN luminosities and downsizing. *Mon. Not. R. Astron. Soc.* **2014**, *442*, 2304–2324.

21. Biffi, V.; Planelles, S.; Borgani, S.; Fabjan, D.; Rasia, E.; Murante, G.; Tornatore, L.; Dolag, K.; Granato, G.L.; Gaspari, M.; et al. The history of chemical enrichment in the intracluster medium from cosmological simulations. *Mon. Not. R. Astron. Soc.* **2017**, *468*, 531–548.

22. Santini, P.; Maiolino, R.; Magnelli, B.; Lutz, D.; Lamastra, Z.; Li Causi, G.; Eales, S.; Andreani, P.; Berta, S.; Buat, V.; et al. The evolution of the dust and gas content in galaxies. *Astron. Astrophys.* **2014**, *562*, A30.

23. De Plaa, J.; Werner, N.; Bleeker, J.A.M.; Vink, J.; Kaastra, J.S.; Méndez, M. Constraining supernova models using the hot gas in clusters of galaxies. *Astron. Astrophys.* **2007**, *465*, 345–355.

24. Teklu, A.F.; Remus, R.S.; Dolag, K.; Burkert, A. The Morphology-Density-Relation: Impact on the Satellite Fraction. *ArXiv* **2017**, arXiv:1702.06546.

25. Sanders, R.L.; Shapley, A.E.; Kriek, M.; Reddy, N.A.; Freeman, W.R.; Coil, A.L.; Siana, B.; Mobasher, B.; Shivaei, I.; Price, S.H.; et al. The MOSDEF Survey: Mass, Metallicity, and Star-formation Rate at z $\sim$ 2.3. *Astrophys. J.* **2015**, *799*, 138.

26. Bresolin, F.; Kudritzki, R.; Urbaneja, M.A.; Gieren, W.; Ho, I.; Pietrzyński, G. Young stars and ionized nebulae in M83: Comparing chemical abundances at high metallicity. *Astrophys. J.* **2016**, *830*, 64–85.

27. González Delgado, R.M.; Cid Fernandes, R.; Garcia-Benito, R.; Perez, E.; de Amorim, A.L.; Cortijo-Ferrero, C.; Lacerda, E.A.D.; Lopez Fernandez, R.; Sanchez, S.F.; Vale Asari, N.; et al. Insights on the Stellar Mass-Metallicity Relation from the CALIFA Survey. *Astrophys. J.* **2014**, *791*, L16.

28. Remus, R.S.; Dolag, K.; Bachmann, L.K.; Beck, A.M.; Burkert, A.; Hirschmann, M.; Teklu, A. Disk Galaxies in the Magneticum Pathfinder Simulations. In *Galaxies in 3D across the Universe, Proceedings of the International Astronomical Union (IAU) Symposium, Vienna, Austria, 7–11 July 2014*; Ziegler, B.L., Combes, F., Dannerbauer, H., Verdugo, M., Eds.; Cambridge University Press: Cambridge, UK, 2015; Volume 309, pp. 145–148.

29. Pastorello, N.; Forbes, D.A.; Foster, C.; Brodie, J.P.; Usher, C.; Romanowsky, A.J.; Strader, J.; Arnold, J.A. The SLUGGS survey: exploring the metallicity gradients of nearby early-type galaxies to large radii. *Mon. Not. R. Astron. Soc.* **2014**, *442*, 1003–1039.

30. La Barbera, F.; Ferreras, I.; de Carvalho, R.R.; Bruzual, G.; Charlot, S.; Pasquali, A.; Merlin, E. SPIDER—VII. Revealing the stellar population content of massive early-type galaxies out to 8R_e. *Mon. Not. R. Astron. Soc.* **2012**, *426*, 2300–2317.

galaxies

MDPI

Article

Dust Deficiency in the Interacting Galaxy NGC 3077

Jairo Armijos-Abendaño [1,*], Ericson López [1], Mario Llerena [1], Franklin Aldás [1] and Crispin Logan [2]

1   Observatorio Astronómico de Quito, Escuela Politécnica Nacional, Av. Gran Colombia S/N, Quito 170403, Ecuador; ericson.lopez@epn.edu.ec (E.L.); mario.llerena01@epn.edu.ec (M.L.); franklin.aldas@epn.edu.ec (F.A.)
2   H.H. Wills Physics Laboratory, University of Bristol, Tyndall Ave, Bristol BS8 1TL, UK; crispin.logan@bristol.ac.uk
*   Correspondence: jairo.armijos@epn.edu.ec

Academic Editor: Emilio Elizalde
Received: 24 July 2017; Accepted: 13 September 2017; Published: date

Abstract: Using 70 µm observations taken with the PACS instrument of the Herschel space telescope, the dust content of the nearby and interacting spiral galaxy NGC 3077 has been compared with the dust content of the isolated galaxies such as NGC 2841, NGC 3184 and NGC 3351. The dust content has allowed us to derive dust-to-gas ratios for the four spiral galaxies of our sample. We find that NGC 2841, NGC 3184 and NGC 3351 have dust masses of 6.5–9.1 $\times$ 10^7 M$_\odot$, which are a factor of ~10 higher than the value found for NGC 3077. This result shows that NGC 3077 is a dust deficient galaxy, as was expected, because this galaxy is affected by tidal interactions with its neighboring galaxies M81 and M82. NGC 3077 reveals a dust-to-gas ratio of 17.5%, much higher than the average ratio of 1.8% of the isolated galaxies, evidencing that NGC 3077 is also deficient in H$_2$ + HI gas. Therefore, it seems that, in this galaxy, gas has been stripped more efficiently than dust.

Keywords: spiral galaxies; dust mass; dust-to-gas ratio

---

1. Introduction

Several studies have shown that galaxies located in high density environments lose atomic neutral hydrogen (HI) due to tidal interactions [1], simultaneous ram pressure and tidal interactions [2], among other mechanisms such as viscous stripping [3] and thermal evaporation [4]. Galaxies in high density environments have less HI content than isolated galaxies [5,6]. There are few works devoted to the study of environmental effects of spiral galaxies on dust content; as examples, we have the studies of [7,8]. Not only HI gas but also dust is stripped in spiral galaxies located in a cluster environment [7,8], where HI gas is stripped more efficiently than dust [7]. The authors of [9] have studied the emission of dust from a large sample of nearby galaxies, including the galaxies addressed in this paper. Their work focused on the dust-to-gas ratios derived from maps of dust mass surface density, obtained from pixel-by-pixel modeling of infrared data.

To investigate possible effects of the environment on the dust of a nearby spiral galaxy, we study the dust content and the dust-to-gas ratio of the galaxy NGC 3077, that is part of a galaxy triplet [10] and therefore affected by tidal interactions. For comparison purposes, we include in our sample the spiral galaxies NGC 2841, NGC 3184 and NGC 3351, which we have considered as isolated galaxies. The positions, morphology and distances of the galaxies of our sample are given in Table 1. In a recent work, dedicated to the study of the environmental effects on dust of nearby galaxies [7], our sources have not been considered.

Table 1. Galaxy sample, morphology and positions.

Galaxy Name	RA [1] (hh:mm:ss.s)	DEC [1] (dd:mm:ss.s)	Morphology [1]	Distance [1] (Mpc)
NGC 2841	09:22:02.7	+50:58:35.3	SAa C	14.6
NGC 3077	10:03:19.1	+68:44:02.2	S0 C	3.8
NGC 3184	10:18:17.0	+41:25:27.8	SAc C	11.3
NGC 3351	10:43:57.7	+11:42:13.0	SBb C	10.5

[1] Information taken from the SIMBAD Astronomical Database.

2. Infrared Data

As it was mentioned above, we aim to study the dust content of four nearby spiral galaxies. For this, we use 70 μm archival maps that can be downloaded from the SIMBAD Astronomical Database[1]. These data were observed with the Photoconductor Array Camera and Spectrometer (PACS) instrument of the Herschel space telescope[2] and obtained thanks to the KINGFISH (Key Insights on Nearby Galaxies: a Far-Infrared Survey with Herschel) survey. These data were first published by [11]. The PACS maps of our four galaxies are shown in Figure 1.

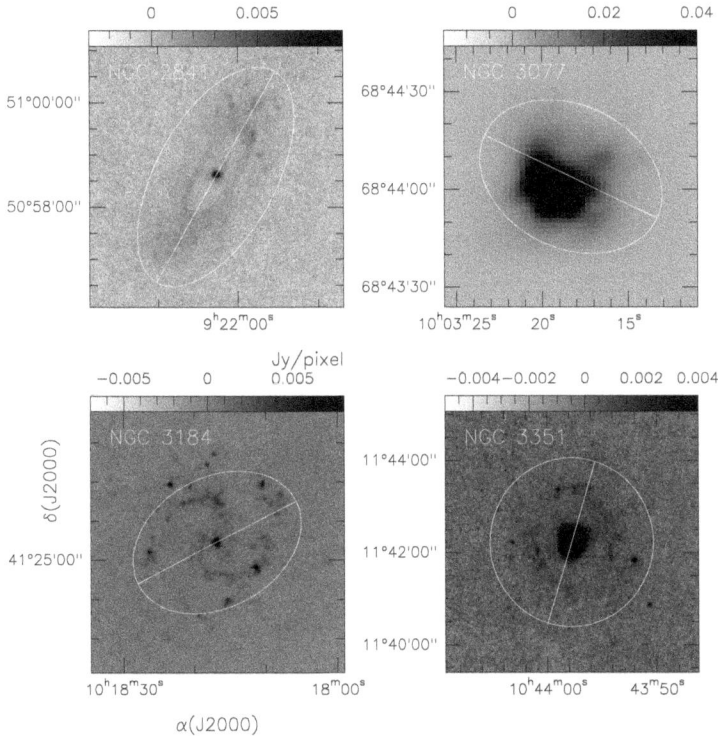

Figure 1. Images at 70 μm from our galaxy sample. The white ellipse is used to derive the infrared flux density (see Section 3.1). The white line indicates the major-axis of each galaxy.

---

[1] http://simbad.u-strasbg.fr/simbad/
[2] Herschel is an ESA space observatory with science instruments provided by European-led Principal Investigator consortia with an important NASA participation

3. Results and Discussion

3.1. Infrared Flux Density

To estimate the infrared flux density (S_λ) for the galaxies in our sample, we define an ellipse which encompasses almost all the infrared emission from each galaxy disk (see Figure 1). The S_λ will be used to estimate the dust mass in Section 3.2. The sizes of the ellipse used to enclose the galaxy disks are the same as those used to derive CO(2-1) luminosities by [12], who used CO luminosities to derive H_2 + HI masses. The similarity in the size of the ellipses in both studies is due to the fact that a good correlation exists in the spatial distribution of the CO(2-1) line emission and the 70 μm emission. This correlation can be tested by comparing the NGC 2841, NGC 3077, NGC 3184 and NGC 3351 maps given in Figure 1 of [12] with the maps of galaxies given in Figure 1 of this work. The derived values of S_λ are given in Table 2, where we also listed the semi-major axis, semi-minor axis and the position angle (PA) of the ellipses.

3.2. Dust Mass

Once the S_λ flux density is derived, we can estimate the dust mass (M_d) using the expression given by [13]:

$$M_d = \frac{S_\lambda d^2}{k_\lambda B_\lambda(T_d)} \tag{1}$$

where d is the distance to the source, k_λ is the dust mass absorption coefficient and B_λ is the Planck function at a given dust temperature (T_d). We use Equation 1 to estimate the M_d values, listed in Table 2 for the galaxies in our sample. For our mass estimates, we assumed a T_d of 25 K and k_λ of 48 cm^2 g^{-1}. This T_d is a compromise value derived from values found for a sample of nearby galaxies [14].

We found dust masses in the range of $\sim$6.5–9.1$\times 10^7$ M$_\odot$ for NGC 2841, NGC 3184 and NGC 3351. NGC 3077 has a dust mass that is a factor of $\sim$10 lower than the dust masses of the other galaxies included in our study. NGC 3077 is the nearest galaxy in our sample and its average size (observed at 70 μm) is a factor of 4.2–4.8 smaller than those (observed at 70 μm) of NGC 2841, NGC 3184 and NGC 3351. Based on this fact, the dust mass of NGC 3077 is expected to be a factor of 4.2–4.8 lower than the other galaxies, which contrasts with what has been previously estimated. This result implies that NGC 3077 is dust deficient, which may be caused by the tidal interactions that this galaxy suffers. Dust deficiency is also observed in cluster galaxies [5,6].

Table 2. Ellipse parameters and derived physical parameters for our sample of galaxies.

Galaxy Name	Semi-Major Axis Arcsec	Semi-Minor Axis Arcsec	PA[1] Degrees	$S_{70\,\mu m}$ Jy Arcsec[2]	M_d $\times 10^7$ M$_\odot$	$M^2_{H_2+HI}$ $\times 10^9$ M$_\odot$	Dust-to-Gas Ratio %
NGC 2841	140	70	150	24.4	6.6	9.0	0.7
NGC 3077	30	22	65	36.1	0.7	0.04	17.5
NGC 3184	140	100	117	34.5	6.5	3.7	1.8
NGC 3351	110	110	163	65.3	9.1	3.2	2.8

[1] Parameter taken from the SIMBAD Astronomical Database. [2] This mass is taken from the work of [12].

3.3. Dust-To-Gas Ratio

As mentioned in Section 3.1, the size of the ellipse used to derive the S_λ flux density is the same as that used to derive the CO luminosity by [12], who used this parameter to find the H_2 + HI mass (M_{H_2+HI}) for the galaxies in our study. These masses are listed in Table 2. Thanks to the similarity in the size of the ellipses, we are able to derive dust-to-gas ratios for the studied galaxies; values are given in Table 2. NGC 2841, NGC 3184 and NGC 3351 show an average dust-to-gas ratio of 1.8% that is consistent with the average value of $\sim$1% found in a sample of nearby star-forming galaxies [9]. On the other hand, NGC 3077 has a dust-to-gas ratio of 17.5%, which is much higher than the average value found for the other studied galaxies. The 17.5% ratio suggests that NGC 3077 is also deficient in H_2 + HI gas, in addition to being more deficient in H_2 + HI gas than dust. The authors of [7] found that HI gas is stripped more efficiently than dust in their study of a sample of spiral galaxies in a cluster environment, which is consistent with the scenario observed in NGC 3077.

4. Conclusions

We studied the dust content and dust-to-gas ratios of four nearby spiral galaxies. The main conclusions of our work are the following:

- For the isolated NGC 2841, NGC 3184 and NGC 3351 galaxies, we find dust masses in the range of 6.5–9.1 × 10^7 $M_\odot$. The dust masses of these galaxies are a factor of ∼10 higher than the dust mass found for NGC 3077 affected by the tidal interactions of galaxies M81 and M82, indicating that NGC 3077 is a dust-deficient galaxy.
- NGC 3077 shows a dust-to-gas ratio of 17.5% that is much higher than the average value of 1.8% found for NGC 2841, NGC 3184 and NGC 3351. The ratio of 17.5% suggests that NGC 3077 is also deficient in H_2 + HI gas, which has been stripped more efficiently than dust in this galaxy.

Author Contributions: J. Armijos-Abendaño, E. López, M. Llerena and F. Aldás performed the data analysis. J. Armijos-Abendaño and C. Logan wrote the manuscript. All authors contributed to the discussion and interpretation of the results.

Conflicts of Interest: The authors declare no conflict of interest.

References

1. Rasmussen, J.; Ponman, T.; Verdes-Montenegro, L.; Yun, M.S.; Borthakur, S. Galaxy evolution in Hickson compact groups: The role of ram-pressure stripping and strangulation. *Mon. Not. R. Astron. Soc.* **2008**, *388*, 1245–1264.
2. Mayer, L.; Mastropietro, C.; Wadsley, J.; Stadel, J; Moore, B. Simultaneous ram pressure and tidal stripping; how dwarf spheroidals lost their gas. *Mon. Not. R. Astron. Soc.* **2006**, *369*, 1021–1038.
3. Nulsen, P.E.J. Transport processes and the stripping of cluster galaxies. *Mon. Not. R. Astron. Soc.* **1982**, *198*, 1007–1016.
4. Cowie, L.L.; Songaila, A. Thermal evaporation of gas within galaxies by a hot intergalactic medium. *Nature* **1977**, *266*, 501–503.
5. Giovanelli, R.; Haynes, M.P. Gas deficiency in cluster galaxies: a comparison of nine clusters. *Astrophys. J.* **1985**, *292*, 404–425.
6. Solanes, J.M.; Manrique, A.; García-Gómez, C.; González-Casado, G.; Giovanelli, R.; Haynes, M.P. The HI content of spirals. II. Gas deficiency in cluster galaxies. *Astrophys. J.* **2001**, *548*, 97–113.
7. Pappalardo, C.; Bianchi, S.; Corbelli, E.; Giovanardi, C.; Hunt, L.; Bendo, G.J.; Boselli, A.; Cortese, L.; Magrini, L.; Zibetti, S.; et al. The Herschel Virgo Cluster Survey. XI. Environmental effects on molecular gas and dust in spiral disks. *Astron. Astrophys.* **2012**, *545*, A75–A91.
8. Cortese, L.; Bekki, K.; Boselli, A.; Catinella, B.; Ciesla, L.; Hughes, T.M.; Baes, M.; Bendo, G.J.; Boquien, M.; de Looze, I.; et al. The selective effect of environment on the atomic and molecular gas-to-dust ratio of nearby galaxies in the Herschel reference survey. *Mon. Not. R. Astron. Soc.* **2016**, *459*, 3574–3584.
9. Sandstrom, K.M.; Leroy, A.K.; Walter, F.; Bolatto, A.D.; Crowall, K.V.; Draine, B.T.; Wilson, C.D.; Wolfire, M.; Calzetti, D.; Kennicutt, R.C.; et al. The CO-to-H_2 Conversion Factor and Dust-to-Gas Ratio on Kiloparsec Scales in Nearby Galaxies. *Astrophys. J.* **2013**, *777*, 5, doi:10.1088/0004-637X/777/1/5.
10. Walter, F.; Weiss, A.; Martin, C.; Scoville, N. The Interacting Dwarf Galaxy NGC 3077: The Interplay of Atomic and Molecular Gas with Violent Star Formation. *Astron. J.* **2002**, *123*, 225–237.
11. Kennicutt, R.C.; Calzetti, D.; Aniano, G.; Appleton, P.; Armus, L.; Beirão, P.; Bolatto, A.D.; Brandl, B.; Crocker, A.; Croxall, K. KINGFISH-Key Insights on Nearby Galaxies: A Far-Infrared Survey with Herschel: Survey Description and Image Atlas. *Publ. Astron. Soc. Pac.* **2011**, *123*, 1347–1369.
12. López, E.; Armijos-Abendaño, J.; Llerena, M.; Aldás, F. Upper limits to magnetic fields in the outskirts of galaxies. *Galaxies* **2017**, in press.

13. Hildebrand, R.H. The Determination of Cloud Masses and Dust Characteristics from Submillimetre Thermal Emission. *Q. J. R. Astron. Soc.* **1983**, *24*, 267–282.
14. Trewhella, M.; Davies, J.I.; Alton, P.B.; Bianchi, S.; Madore, B.M. ISO Long Wavelength Spectrograph Observations of Cold Dust in Galaxies. *Astrophys. J.* **2000**, *543*, 153–160.

galaxies

MDPI

Article

Galaxies with Shells in the Illustris Simulation: Metallicity Signatures

Ana-Roxana Pop [1,*], Annalisa Pillepich [1,2], Nicola C. Amorisco [1,3] and Lars Hernquist [1]

[1]   Harvard-Smithsonian Center for Astrophysics, 60 Garden St., Cambridge, MA 02138, USA;
    pillepich@mpia-hd.mpg.de (A.P.); nicola.amorisco@cfa.harvard.edu (N.C.A.);
    lhernquist@cfa.harvard.edu (L.H.)
[2]   Max-Planck-Institut für Astronomie, Königstuhl 17, 69117 Heidelberg, Germany
[3]   Max Planck Institute for Astrophysics, Karl-Schwarzschild-Strasse 1, D-85740 Garching, Germany
*   Correspondence: ana-roxana.pop@cfa.harvard.edu

Academic Editors: Duncan A. Forbes and Ericson D. Lopez
Received: 1 July 2017; Accepted: 1 August 2017; Published: 4 August 2017

Abstract: Stellar shells are low surface brightness arcs of overdense stellar regions, extending to large galactocentric distances. In a companion study, we identified 39 shell galaxies in a sample of 220 massive ellipticals ($M_{200crit} > 6 \times 10^{12} \, M_\odot$) from the Illustris cosmological simulation. We used stellar history catalogs to trace the history of each individual star particle inside the shell substructures, and we found that shells in high-mass galaxies form through mergers with massive satellites (stellar mass ratios $\mu_{stars} \gtrsim 1:10$). Using the same sample of shell galaxies, the current study extends the stellar history catalogs in order to investigate the metallicity of stellar shells around massive galaxies. Our results indicate that outer shells are often times more metal-rich than the surrounding stellar material in a galaxy's halo. For a galaxy with two different satellites forming $z = 0$ shells, we find a significant difference in the metallicity of the shells produced by each progenitor. We also find that shell galaxies have higher mass-weighted logarithmic metallicities ($[Z/H]$) at 2–4 R_{eff} compared to galaxies without shells. Our results indicate that observations comparing the metallicities of stars in tidal features, such as shells, to the average metallicities in the stellar halo can provide information about the assembly histories of galaxies.

Keywords: cosmology: theory; galaxies: evolution; galaxies: interactions; galaxies: kinematics and dynamics; galaxies: structure; methods: numerical; stellar shells; stellar metallicities

1. Introduction

Stellar halos are diffuse, low surface brightness regions that contain unique morphological features, such as tidal streams, stellar shells, rings, or plumes (e.g., [1,2]). Shells are a particular type of tidal debris that appear as interleaved caustics with large opening angles, which have been identified in numerous deep surveys probing the faint outskirts of galaxies [3–10]. Interest in the detailed study of stellar substructures such as shells has been stimulated by previous work indicating that tidal features are powerful tracers of galaxies' assembly histories (see e.g., [11–16]). Due to the long dynamical timescales in the outskirts of galaxies, information about the accretion histories of galaxies is well preserved at the present epoch. Previous studies indicate that stellar shells are very long-lived structures, with average lifetimes of ~2–3 Gyr according to a recent paper [17].

Several theories have been proposed for the formation of shells, ranging from models invoking star formation in shocked galactic winds [18–21] to tidal interaction theories in which the shells are density waves induced by a passing galaxy [22,23]. Nevertheless, extensive numerical and observational studies support the merger scenario, in which shells are composed of stars stripped from accreted satellites (e.g., [24–29]). The increasing richness of stellar halo observations will give us a unique look into the past merger histories of galaxies, yet the interpretation of the data requires detailed model predictions for the formation of stellar substructures. Most of the previous studies on shell galaxies have focused on idealized mergers involving small satellites on radial orbits, (e.g., [25–28,30–33]), yet major mergers have also been shown to produce shells, (e.g., [34–37]). In a companion paper, Pop et al. ([17], hereafter P17) found that shells in high-mass galaxies are often the result of mergers with massive satellites (relative to the mass of the host galaxy). Since the efficiency of dynamical friction is higher for major mergers compared to minor mergers, massive satellites can form shells for a wide range of impact parameters; small satellites, however, need almost perfectly radial orbits in order to produce stellar shells. As a result, in P17, we show that most satellites producing shells in massive ellipticals have stellar mass ratios $\mu_{stars} \gtrsim 1 : 10$.

Previous studies suggest that metallicity gradients can be used to characterize the relative importance of major and minor mergers in the accretion history of a galaxy (e.g., [38,39]). During the initial formation phase of massive galaxies, in-situ star formation will leave a steep metallicity gradient, with stars on the outskirts of galaxies forming out of lower density gas and metals being further removed from the outer regions of the stellar halo by stellar winds [39,40]. Overall, subsequent accretion events ($z \lesssim 1$) have the net effect of flattening the metallicity profiles (see e.g., [41–43]). In a previous study of Illustris galaxies, Cook et al. [44] found that galaxies with higher fractions of accreted material tend to have less steep gradients than galaxies with a lower fraction of ex-situ stars. Based on these results, it has been suggested that metallicity gradients could provide additional information about the mass ratios of mergers that produce shells. Several groups have found that shell galaxies tend to have lower central Mg_2 values than non-shell galaxies (e.g., [45,46]), with galaxies with shells lying below the Mg_2–σ relation from [47].

Studies of shell galaxies using deep imaging face several challenges, such as the extremely low surface brightness of shells ($\sim$28 mag arcsec^{-2} or fainter in the V band, e.g., [8,48]), the varied and sometimes irregular shell morphologies, and the fact that outer shells extend far out in the outskirts of the stellar halo, at galactocentric distances $\sim$100 kpc. For sufficiently nearby elliptical galaxies ($\lesssim$20 Mpc), direct stellar photometry has successfully provided strong constraints on the age and metallicity of stars extending to large galactocentric distances in shell galaxies (e.g., [49–52]). In this paper, we investigate the metallicity of stellar shells in the Illustris simulation, studying how different metallicity signatures can set constraints on the number and mass of shell-forming progenitors. As metallicity measurements are reaching larger radii through slit spectroscopy (e.g., [53–57]), integral-field spectroscopy (e.g., [58–62]), and deep photometric studies (e.g., [51,52,63,64]), as well as probing increasingly larger samples of galaxies [55,65], the study of stellar metallicities of tidal features will provide previously inaccessible information about the galaxies' accretion histories.

2. Simulations and Methods

2.1. Illustris

This project uses the Illustris simulation [66–68], a cosmological hydrodynamical simulation run using AREPO [69]—a moving-mesh code based on a quasi-Lagrangian finite volume method. We are working with the highest resolution run (Illustris-1), which corresponds to a simulation box with a periodic volume of $(106.5 \text{ Mpc})^3$ and mass resolution of $m_{DM} = 6.26 \times 10^6 \, M_\odot$. The Illustris simulations include a broad range of astrophysical processes (e.g., feedback from supermassive black holes, supernovae, and AGNs) in order to self-consistently reproduce galaxy formation and evolution [70,71].

2.2. The Galaxy Sample

Similar to P17, we start with an initial sample consisting of the 220 most massive central galaxies in Illustris at redshift $z = 0$. This corresponds to a virial mass cut $M_{200crit} > 6 \times 10^{12} \, M_\odot$, where $M_{200crit}$ is defined as the total mass inside a radius enclosing a sphere with mean density 200 times greater than the critical density of the Universe. In Illustris, central galaxies are the most massive galaxies in a given halo/group, and satellites inside each halo are identified using the SUBFIND algorithm [72–74].

2.3. Stellar History Catalogs and Shell Galaxies Identification

The present study uses the Illustris shell galaxies previously identified in P17 using a two-step approach:

- *Step 1*: Galaxies with stellar shells are visually identified using stellar surface density maps. Each galaxy in the sample is studied using all three projections (x–y, y–z, z–x) and two different contrast levels, and each image stamp received a score ranging from 0–2 from three different members of our team. A score of 2 indicates a galaxy with two or more well-defined shells, a score of 1 corresponds to galaxies with one or two shell-like structures, and a score of 0 indicates no shell detection. We order candidate shell galaxies based on the total scores given to all their corresponding image stamps. As discussed in P17, most shell galaxies exhibit shell-like structures in at least 2/3 projections, and we use the second identification step to verify the presence of shells in each of the candidate galaxies.
- *Step 2*: We develop stellar history catalogs in order to identify the shell-forming progenitors and separate stars in shells from all other stars in the galaxy. In these catalogs, we trace the birth, trajectory and progenitors of all star particles inside $z = 0$ halos, saving information about each individual star particle at three key moments during its life: formation, accretion (when the parent satellite enters the virial radius of the host) and stripping time (when the star becomes gravitationally bound to the new host). In P17, we show that shells in Illustris are composed of ex-situ stars, which is in agreement with merger models predicting that shell-like structures correspond to overdensities of stripped stars accumulating at the apocenters of their orbits (see analytical and numerical studies by, e.g., [24–29,34,75,76]). By tracing the configuration and phase space (v_r vs. r) distribution of star particles with a common parent satellite, we obtain a systematic sample of all the satellites that are responsible for $z = 0$ stellar shells.

Out of 220 massive ellipticals in Illustris (with $M_{200crit} > 6 \times 10^{12}\,M_\odot$), Pop et al. [17] find that 39 galaxies exhibit stellar shells visible at $z = 0$, corresponding to a fraction of 18% $\pm$ 3% of all galaxies in the sample.[1] Moreover, the fraction of galaxies with shells increases with increasing mass cut, with as many as 34% $\pm$ 11% of galaxies with $M_{200crit} > 3 \times 10^{13}\,M_\odot$ having shells. Despite significant uncertainties in the number of observed shell galaxies, the incidence of galaxies with shells in Illustris is in overall agreement with current observational limits (e.g., [4–8]).

In this paper, we take advantage of the versatility of stellar history catalogs, which can be extended to study a wide range of properties for stars inside tidal features like shells, streams or plumes (for more details about the construction of the catalogs, see P17). In particular, the current study focuses on the metallicity signatures of several shell galaxies from the Illustris simulation. We use stellar surface density maps such as those presented in Figure 1, in which we weigh each star particle in the $z = 0$ halo by its total metallicity (Z), normalized to that of the Sun ($Z_\odot$). Each bin in the 2D histograms in Figure 1 is colored based on the median total metallicity of all star particles therein. In Illustris, the metallicity of each individual star particle is based on the metallicity of the gas cell at the star's time of birth (i.e., when the gas cell is converted into a star particle).

---

[1] We provide Poisson errors for the fractions of galaxies with shells in Illustris.

However, the actual uncertainties in the f_{shells} could be higher due to environment (rich groups and clusters vs. field galaxies), mass distribution of galaxies in the sample, redshift evolution $f_{shells}(z)$, surface brightness limits, and projection effects. For more details about the impact of these effects, see the discussion in Pop et al. [17].

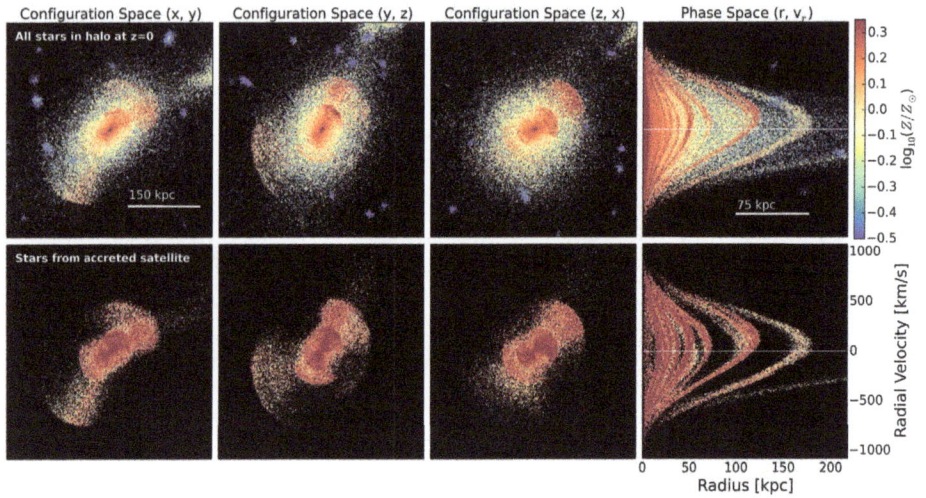

Figure 1. Stellar maps in configuration space (first three columns) and phase space, i.e., radial velocity vs. radial distance (last column). The color of each bin of area $A_{bin} = (1.5\,\text{kpc})^2$ in the 2D histograms above corresponds to the median total metallicity (Z) of all star particles therein. **Top row**: entire halo of a redshift $z = 0$ shell galaxy with total stellar mass $M_{stars} = 6.3 \times 10^{11}\,M_\odot$, virial mass $M_{200crit} = 1.4 \times 10^{13}\,M_\odot$, and effective radius $R_{eff} = 16.3\,\text{kpc}$ (where R_{eff} is defined as the radius that contains half of the total light in a galaxy). **Bottom row**: redshift $z = 0$ stars that have been accreted from the same common progenitor, which is responsible for the shell structures in the top row. We find that stellar shells in this galaxy have higher total metallicities than stars in the outer regions of the stellar halo. Moreover, we find a gradient in the average metallicity of individual shells, with outer shells having lower average Z that inner shells.

In configuration space, we identify shells as interleaved stellar overdensities, often observed on both sides of the galaxy center and extending to very large galactocentric distances, in the low surface brightness regions of the stellar halo. Stellar shells also have a specific signature in phase space, i.e., in the space of galactocentric distance vs. radial velocity (r–v_r). Stars spend longer times near the apocenters of their orbits, where their radial velocities are close to zero, thus creating phase wraps peaking at $v_r \simeq 0$, as exemplified in the right column of Figure 1. Using stellar history catalogs, we identify satellites responsible for $z = 0$ shells, and we investigate the metallicity distribution of stars inside these shells (see bottom rows of Figures 1 and 2).

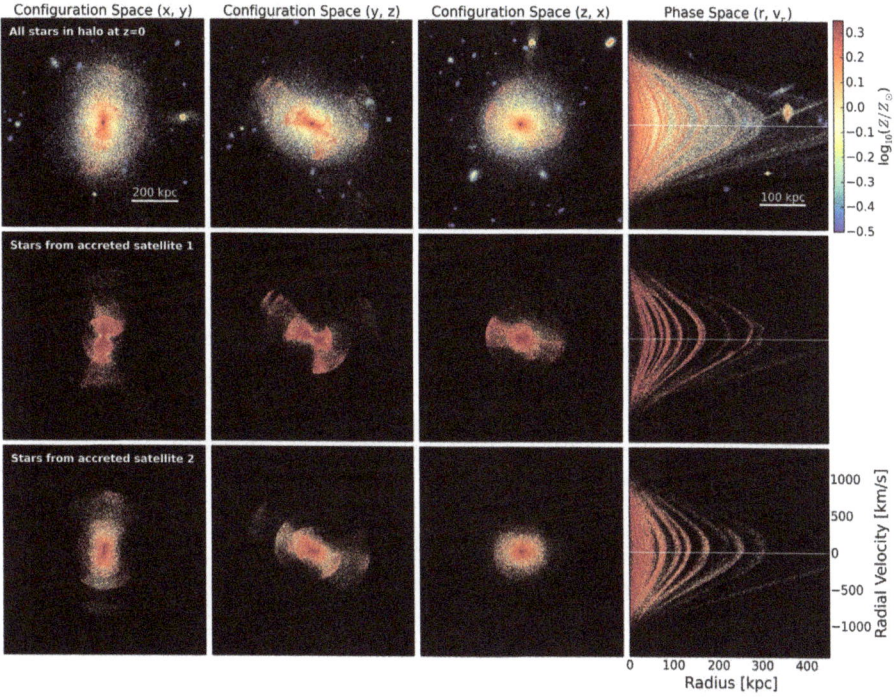

Figure 2. Stellar maps in configuration space (first three columns) and phase space, i.e., radial velocity vs. radial distance (last column). The color of each bin of area $A_{bin} = (1.5\,\text{kpc})^2$ in the 2D histograms above corresponds to the median total metallicity (Z) of all star particles therein. **Top row**: entire halo of a $z = 0$ shell galaxy with total stellar mass $M_{stars} = 1.3 \times 10^{12}\,M_\odot$, virial mass $M_{200crit} = 4.6 \times 10^{13}\,M_\odot$, and $R_{eff} = 33.5\,\text{kpc}$. This galaxy had two different shell-forming progenitors responsible for the shells visible at $z = 0$. **Middle and bottom rows**: redshift $z = 0$ stars accreted from each of the two shell-forming satellites with stellar mass ratios $\mu_{stars} = M^{stars}_{satellite}/M^{stars}_{host} = 0.15$ and 0.24, measured at accretion time (i.e., when the satellite comes inside the virial radius of the host). Stellar shells in this galaxy have higher average metallicity than the other stars in the outskirts of the stellar halo. Moreover, the shells produced by the first satellite are on average more metal-rich than those produced by the second satellite. Similar to Figure 1, we find that the outer shells have lower total metallicities that inner shells.

3. Results and Implications

We begin by investigating the distribution of stellar metallicities for two shell galaxies, one for which the $z = 0$ shells are composed of stars accreted from a single satellite, while the shells in the second galaxy are the result of mergers with two different satellites. In Figure 1, we study the configuration and phase space distribution of all star particles inside the halo of a galaxy with total stellar mass $M_{stars}(z = 0) = 6.3 \times 10^{11}\,M_\odot$ and virial mass $M_{200crit}(z = 0) = 1.4 \times 10^{13}\,M_\odot$. The shell-forming satellite depicted in the bottom row corresponds to a close to 1 : 1 merger (in total mass) and a stellar mass ratio of $\mu_{stars} = M^{stars}_{satellite}/M^{stars}_{host} = 0.22$. The satellite was accreted $\sim$4 Gyr ago on a low angular momentum orbit, with a radial velocity ratio[2] at accretion time of $v_r/|v| = -0.97$.

---

[2] The radial velocity ratio $(v_r/|v|)$ is defined as the radial component of the relative velocity of the satellite (v_r), normalized to the modulus of the relative velocity (v).

This shell-forming satellite has properties that agree with the overall trends for shell formation found in P17: most of the shell galaxies in the current sample form through mergers with stellar mass ratios $\mu_{stars} \gtrsim 1:10$, and the shell-forming satellites are accreted on low angular momentum orbits, at intermediate accretion times ($\sim$4–8 Gyr ago). We determine the color of each bin in the stellar 2D histograms in Figure 1 by computing the median of the total metallicity (Z) for all star particles in that bin. We find metal-rich shells visible in all three projections and that extend to large galactocentric distances (>50 kpc). While in the top panels of Figure 1 we determine the color of each bin by considering both shell stars and other halo stars that are inside the same spatial bin, in the bottom panels, we show only those stars accreted from the shell-forming satellite (identified as described in Section 2.3). Thus, since in the bottom row we are isolating the stellar shells, it becomes more evident that, for the galaxy in Figure 1, stars in shells are more metal-rich than the overall stellar population of the host galaxy.

For the second study case, we consider a shell galaxy that exhibits $z = 0$ shells composed of stars accreted from two different satellites. The host galaxy in Figure 2 has a total stellar mass $M_{stars} = 1.3 \times 10^{12}\,M_\odot$ and virial mass $M_{200crit} = 4.6 \times 10^{13}\,M_\odot$. As before, we color each bin based on the median total metallicity of all star particles therein, and we find significantly more stellar shells at large radii, compared to the example in Figure 1. However, similarly to the previous study case, stars in shells have higher total metallicities than the other halo stars at similarly large radii. The two shell-forming satellites in Figure 2 were accreted between $\sim$6–7 Gyr ago, and they were stripped $\sim$3 and $\sim$2 Gyr ago, for satellites 1 and 2, respectively. Both of these merger events correspond to relatively major mergers with $\mu_{stars(t_{acc})} = 0.15$ and 0.24, and the satellites had radial infall trajectories, with $v_r/|v| = -0.99$ and -0.84 at t_{acc}. The middle and bottom rows of Figure 2 depict only those stars accreted from each of these two satellites. On average, both satellites contribute stars with higher total metallicities than the average metallicity of stars in the host galaxy. Moreover, we find that the shells created by the first satellite are more metal-rich than those created by the second satellite. This opens an interesting avenue for future observational studies—measurements of the metallicities of stars in shells could be used to identify shell galaxies with more than one shell-forming progenitor.

In both Figures 1 and 2, we find a gradient in the metallicity of stars in shells, with outer shells having lower average Z than inner shells. Studies of the satellites' infall trajectories show that shells in the current sample are often composed of stars stripped while the parent satellite passes close to the center of the host several times (see P17). Stars stripped during the satellite's first pericenter passage form the first generation of shells, with subsequent stripping events forming several generations of shells that are located at decreasing galactocentric distances compared to previous generations, (e.g., [31,32,77,78]). As a result, we expect the outer-most shells to be composed of stars that were initially located on the outskirts of the parent satellite, and, thus, they had lower average metallicities than stars closer to the satellite's core. This explains the trends observed in Figures 1 and 2, with the first generation of shells having lower metallicities than shells located closer to the galaxy center.

In Figure 3, we show total metallicity Z-weighted stellar surface density maps for nine galaxies with shells in our sample. The virial masses of the host galaxies at $z = 0$ span one order of magnitude: $M_{200crit} \in (6.5 \times 10^{12}\,M_\odot,\ 6.3 \times 10^{13}\,M_\odot)$, and total stellar masses are in the range $M_{stars} \in (3.1 \times 10^{11}\,M_\odot,\ 2.2 \times 10^{12}\,M_\odot)$. The examples presented in Figure 3 represent a selection of shell galaxies in Illustris that have outer shells more metal-rich than the average metallicity of surrounding stars at similar galactocentric distances in the host galaxy. Nonetheless, we do not exclude that some galaxies could have shells composed of stars with similar or even lower metallicities than halo stars located at similar radii.

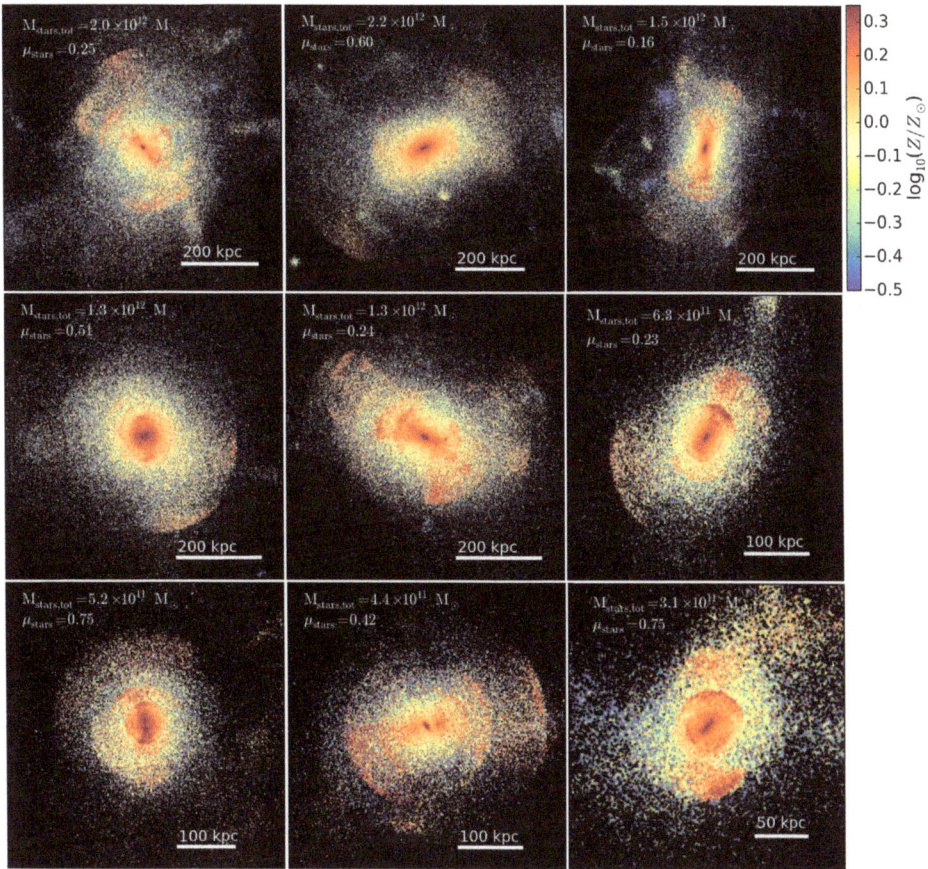

Figure 3. Examples of nine different massive elliptical galaxies with shells from the Illustris simulation. The color of each bin of area $A_{bin} = (1.5\,\text{kpc})^2$ in the 2D histograms above corresponds to the median total metallicity (Z) of all star particles therein. The total stellar masses of these galaxies range from $M_{stars} = 3.1 \times 10^{11}\,M_\odot$ to $M_{stars} = 2.2 \times 10^{12}\,M_\odot$, while the virial masses are in the range $M_{200crit} \in (6.5 \times 10^{12}\,M_\odot, 6.3 \times 10^{13}\,M_\odot)$. For several of the shell galaxies in our sample, stellar shells are metal-rich compared to the total metallicity (Z) of stars in the outer stellar halo.

In order to estimate the difference in total metallicity Z between outer shells and stars at similar galactocentric distances, and the resolution levels necessary to detect metal-rich shells such as those presented in Figure 3, we next study a series of Z-weighted stellar maps at increasingly lower resolutions. In the top row of Figure 4, we show the (y–z) projection of the same galaxy previously presented in Figure 2, and we vary the area of the 2D histogram bins from $A_{bin} = (1\,\text{kpc})^2$ to $A_{bin} = (5\,\text{kpc})^2$, $(10\,\text{kpc})^2$, and finally, $(15\,\text{kpc})^2$. This galaxy has an effective radius $R_{eff} = 33.5\,\text{kpc}$, so the right-most stellar map in Figure 4 corresponds to a low-resolution image, with the length of each bin being $L_{bin} \approx R_{eff}/2$. Some of the fine structure associated with tidal debris is completely erased for $A_{bin} > (5\,\text{kpc})^2$, and the number of detectable shells drops significantly as we lower the resolution. Nonetheless, there is sufficiently high contrast between the metallicities of some of the outer shells and the metallicities of stars located at similar radii in the halo, such that several outer shells can still be observed even for $A_{bin} \simeq (15\,\text{kpc})^2$. In the bottom row of Figure 4, we present a larger version of the image stamp with $A_{bin} = (10\,\text{kpc})^2$, on which we mark six different targets, each having an area

$A_{target} = (50\,\text{kpc})^2$. In the table accompanying the figure, we provide galactocentric distances for the centers of each of the six targets in units of kpc and R_{eff}, as well as the median $\log_{10}(Z/Z_\odot)$ computed for all star particles located inside each target area[3]. The median $\log_{10}(Z/Z_\odot)$ is highest for target area 1, which is centered on the core of the galaxy. Areas 2–4 are targeting three different shells located at galactocentric distances: $R_2 = 4.2\,R_{eff}$, $R_3 = 8.1\,R_{eff}$, and $R_4 = 7.1\,R_{eff}$, respectively. The median $\log_{10}(Z/Z_\odot)$ of all stars in these target areas varies between $\log_{10}(Z/Z_\odot) = 0.18$ to -0.021. Despite the median total metallicities of stars in outer shells being lower than the median Z of stars near the galaxy center (where $\log_{10}(Z/Z_\odot) = 0.33$), there is still a significant contrast in metallicity between targets at similar radial distances depending on whether they are centered on a shell or not. For example, for target 2, we measure a median $\log_{10}(Z/Z_\odot) = 0.18$, while target 5 is positioned at the same galactocentric distance ($R = 4.2\,R_{eff}$), in a region of the stellar halo where we detect no tidal features, and it has a median $\log_{10}(Z/Z_\odot) = -0.22$. This corresponds to a metallicity contrast $\Delta\log_{10}(Z/Z_\odot) = 0.4$ between targets 2 and 5. We can similarly compare targets 3 and 6, with the former being centered on a shell in the upper left quadrant and having median $\log_{10}(Z/Z_\odot) = 0.036$, while target 6 is positioned at a similar radial distance ($R = 8.1\,R_{eff}$) but in a low stellar density region of the halo, where the median $\log_{10}(Z/Z_\odot) = -0.64$, leading to a contrast $\Delta\log_{10}(Z/Z_\odot) \approx 0.7$ between targets 3 and 6. In addition, we find that the median total metallicities of outer shells ($\log_{10}(Z/Z_\odot) = 0.036$ and -0.021 for targets 3 and 4, respectively) are lower than the median metallicity for the shell located at a smaller radius, near target number 2 ($\log_{10}(Z/Z_\odot) = 0.18$). These results are in agreement with the metallicity gradients in shells observed in Figures 1 and 2: outer shells are less metal-rich than inner shells. Due to the extremely low surface brightness of shells, reaching the precision levels above is challenging for current observations, though not technically impossible.

Next, we quantify the difference in the median stellar metallicities of shell vs. non-shell galaxies by measuring the mass-weighted logarithmic metallicities [Z/H] of stars around all 220 galaxies in our sample (see Figure 5). Following the definitions adopted in [44], we measure [Z/H] within three radial ranges: the inner galaxy (0.1–1 R_{eff}), outer galaxy (1–2 R_{eff}), and stellar halo (2–4 R_{eff}). The average effective radius of the entire sample is 20.9 kpc, with shell galaxies having a higher average ($\bar{R}_{eff} = 24.5\,\text{kpc}$) than galaxies without shells ($\bar{R}_{eff} = 20.1\,\text{kpc}$). We do not find a significant difference in the average metallicities of stars within $\sim$2 R_{eff} of the galaxies' centers. However, in the outer regions of the stellar halo (2–4 R_{eff}), shell galaxies have higher average metallicities than galaxies without shells: the median values for the two distributions are $[Z/H]_{shells} = -0.28$ and $[Z/H]_{non\text{-}shells} = -0.36$. In turn, if stars on the outskirts of shell galaxies are more metal-rich, this implies that the metallicitiy gradients of galaxies with shells could be shallower than those of non-shell galaxies. Despite the higher average [Z/H] for shell galaxies, we find that there is significant scatter in the metallicity distributions for galaxies in our sample. Most galaxies do not show azimuthal symmetry, and, therefore, by computing [Z/H] averaged over radial profiles, we are ignoring the relative influence of different stellar substructures present at similar radii in the halo. We thus expect measurements of stellar metallicities targeted towards regions with significant stellar overdensities or clumpy substructures (see, for example, Figure 4) to be better suited at identifying new shells and characterizing the mergers producing them than measurements of overall metallicity gradients or [Z/H] averaged over large radial bins.

---

[3] We assign stars to each target area by selecting all star particles gravitationally bound to the galaxy and located inside a rectangular box with (y–z) square cross-section of area $A_{target} = (50\,\text{kpc})^2$ and infinite length.

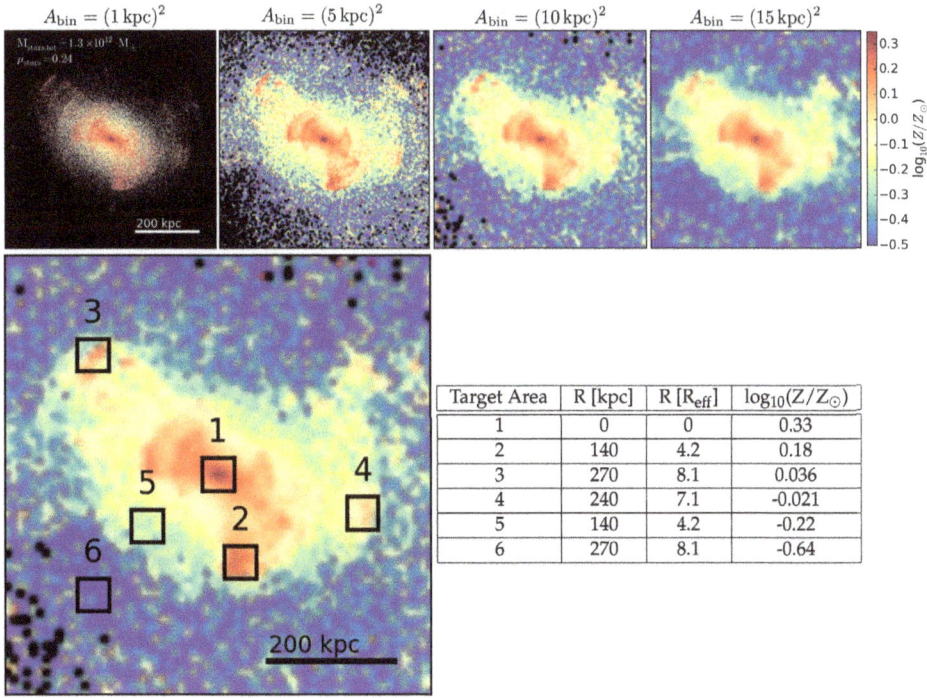

Figure 4. Top row: Z-weighted stellar maps corresponding to the $(y$–$z)$ projection of all star particles inside the $z = 0$ halo of a galaxy with shells with $M_{stars} = 1.3 \times 10^{12} \, M_\odot$ and $R_{eff} = 33.5$ kpc (same galaxy as in Figure 2). The resolution of the image stamps decreases from left to right, with the areas of 2D histogram bins given by: $A_{bin} \in [(1 \, \text{kpc})^2, (5 \, \text{kpc})^2, (10 \, \text{kpc})^2, (15 \, \text{kpc})^2]$. **Bottom row:** Zoom-in of the third image stamp in the top row, corresponding to a histogram with $A_{bin} = (10 \, \text{kpc})^2$. We mark six different target regions, each having a cross-section $A_{target} = (50 \, \text{kpc})^2$. **Table:** We provide the distance R from the center of each of the target areas to the center of the galaxy in units of kpc and R_{eff}, respectively, as well as the median $\log_{10}(Z/Z_\odot)$ measured over all star particles inside each target area. Targets centered around outer shells (e.g., 2 and 3) have significantly higher median $\log_{10}(Z/Z_\odot)$ than targets at the same galactocentric distances but centered on regions without tidal features (e.g., targets 5 and 6).

In P17, we show that most merger events responsible for forming $z = 0$ shells in massive elliptical Illustris galaxies can be described by an order-zero recipe, involving only three parameters: stellar mass ratio (μ_{stars}), radial velocity ratio ($v_r/|v|$), and accretion time (t_{acc}). Satellites that successfully form $z = 0$ shells correspond to mergers with stellar mass ratios $\mu_{stars} \gtrsim 1 : 10$ and are accreted on very low angular momentum orbits, about $\sim$4–8 Gyr ago. While the mean total mass ratio ($\mu_{total} = M_{satellite}^{total}/M_{host}^{total}$) of shell-forming merger events in P17 is lower than the mean stellar mass ratio (μ_{stars}), both μ_{total} and μ_{stars} are biased towards close-to-major mergers. Due to dynamical friction being more efficient at radializing the orbits of satellites involved in close to 1 : 1 mergers (see, e.g., [79,80]), massive satellites are allowed to probe a wider range of impact parameters and still be successful at forming shells (see discussion in P17). On the other hand, in order to produce shells in massive ellipticals, small satellites need very fine-tuned, almost perfectly radial orbits. Thus, P17 find that shells in massive galaxies are produced more frequently by major or close-to-major mergers than by minor mergers.

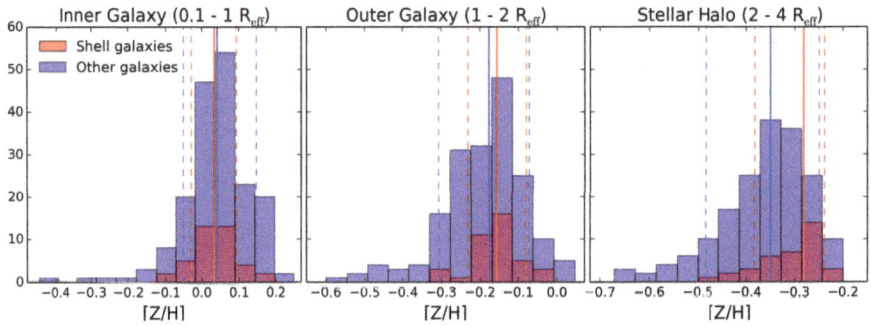

Figure 5. Histograms of the average logarithmic metallicities of stars in a sample of 220 massive elliptical galaxies. The shell galaxies identified in P17 are shown in red, with the other galaxies in the sample shown in blue. We measure [Z/H] within three radial ranges, based on the effective radius of each galaxy, R_{eff}, defined as the radius that contains half of the total light. The median [Z/H] for all shell/non-shell galaxies is marked by a continuous line, while dashed lines mark the 10% and 90% percentiles. The average total stellar metallicities are similar for shell and non-shell galaxies at galactocentric distances $< 2 R_{eff}$. In the extended stellar halo, we find that shell galaxies in our sample have higher average metallicities (shallower metallicity gradients) than galaxies without shells.

Previous studies have shown that major mergers can deposit material closer to the center of the host, unlike minor mergers which mostly contribute to the ex-situ stellar fraction at large galactocentric radii [80,81]. Major mergers with shell-forming satellites containing metal-rich stars will thus deposit high-metallicity material over a wide range of radii—from the inner regions of the halo out to distances $\gtrsim 100$ kpc. As a result, there are two competing effects that need to be considered when estimating the metallicity of outer shells relative to the metallicity of halo stars at similar radii. On one hand, the outer-most shells (or the first generation of shells) are composed of stars stripped from the outskirts of the satellite's stellar halo during its first pericenter approach, while subsequent shells at lower radii correspond to more metal-rich stars that were located closer to the center of the shell-forming progenitor. This has the effect of generating a metallicity gradient in the shells, with outer shells being more metal-poor than inner shells (as observed in Figures 1 and 2). On the other hand, shells extend to extremely large distances in the low surface brightness and metal-poor outskirts of the host's stellar halo. Despite the fact that the very first $\sim$1–2 shells at large radii are generally not very metal-rich, the next shells will start to contain relatively high-Z stars stripped from closer to the center of the satellite. After the merger event, these stars are located in shells extending to much larger effective radii in the host galaxy compared to their initial galactocentric distances with respect to the center of the shell-forming progenitor. Therefore, in the case of major mergers, many of the outer shells can be on average more metal-rich than halo stars located at similar radii in the host galaxy, and we observe this effect in several of the shell galaxies in our sample (see Figure 3).

When averaging stellar metallicities over large radial bins, we find that shell galaxies have higher median [Z/H] than non-shell galaxies in the outer regions of the stellar halo (2–4 R_{eff}). Hirschmann et al. [38] used zoom hydrodynamical simulations of 10 galaxies with masses above 6×10^{12} M$_\odot$, probing a similar mass range as our current study. They found that systems that have undergone major mergers have significantly shallower metallicity gradients at galactocentric distances $> 2 R_{eff}$ compared to systems dominated by minor mergers. For Illustris galaxies, Cook et al. [44] showed that galaxies with high ex-situ fractions tend to have less steep metallicity gradients (in the radial range 2–4 R_{eff}) than those with smaller accreted fractions, yet the metallicity gradients are not strongly correlated with the mean merger mass-ratio. As a result, it can be difficult to draw conclusions based solely on the distribution of overall metallicity gradients in shell galaxies, and precision metallicity measurements targeted around individual shells can provide more information about the shell-forming progenitor.

Based on the results presented above, we expect even outer shells (formed from material stripped from the outskirts of satellites) to be relatively metal-rich if the shells are produced by a massive parent satellite. Therefore, observations of metal-rich shells could indicate that a high fraction of the shells observed in massive early-type galaxies have a major merger origin. In this way, precision metallicity measurements targeting stars in and around the shells can provide additional information about the formation processes of shell galaxies.

4. Summary

In the current study, we investigate the average total metallicity of stellar shells around massive elliptical galaxies from the Illustris simulation. We present several shell galaxies for which we identify shells in both configuration space (x–y, y–z, z–x) and phase space (r–v_r). Stars in the outer shells of these galaxies have higher total metallicities than other stars at similar radii in the host galaxies. This result suggests that the shells studied in this paper are the by-product of major mergers. Moreover, in the particular case of a galaxy with two different shell-forming progenitors, we find a significant difference in the metallicity of the shells produced by each of the two satellites. This indicates that high-precision metallicity measurements could potentially identify shell galaxies with multiple progenitors based on their different metallicity signatures.

Previous analytical and numerical studies (e.g., [78,82]) have indicated that several generations of shells are necessary to explain the wide range of shell radii measured in observations of early-type galaxies [83]. In agreement with these results, we find that shells in Illustris are often formed by stripping stars while the parent satellite performs several pericenter passages. Stars in the first generation of shells (the outer-most shells) correspond to the least bound stars on the outskirts of the progenitor galaxy and, thus, they are the first stars stripped from the satellite. We find a gradient in the average metallicity of shells, with outer shells in our two study cases having lower metallicities than inner shells. Since stars on the outskirts of galaxies tend to have lower metallicities, first generation shells would be expected to be more metal-poor than subsequent generations, in agreement with our findings.

Finally, we investigate the average mass-weighted metallicities [Z/H] in three different radial bins corresponding to the inner galaxy, outer galaxy and stellar halo. Both galaxies with and without shells in our sample seem to have a similar average [Z/H] within 2 R_{eff} of the galaxy center. We find, however, a bias towards a higher average [Z/H] for stars in the outer stellar halos of shell galaxies.

The results presented in this paper indicate that several of the shell galaxies in Illustris have outer shells more metal-rich than the average metallicity in the stellar halo at similar radii. We predict that multiple shell-forming progenitors, as well as different generations of shells, could leave specific signatures in the metallicity distribution of stars in shells. In addition to having the potential to find previously undetected shells, future metallicity observations could unveil information about the formation processes of shell galaxies, such as distinguishing between minor/major mergers producing shells or the number of different satellites responsible for the observed structure.

Acknowledgments: A.R.P. wishes to thank the referees for helpful, constructive comments that improved this paper. A.R.P. would like to thank Ben Cook for useful discussions on stellar metallicities in Illustris, as well as Paul Torrey, Charlie Conroy, and Jieun Choi for answering questions related to metallicities. A.R.P. also thanks Vicente Rodriguez-Gomez for the merger trees in Illustris, and Jane Huang, John Lewis and Luke Kelley for useful comments on the draft. Special thanks go to the organizers of the conference "On the Origin (and Evolution) of Baryonic Halos".

Author Contributions: A.R.P., A.P., N.C.A., and L.H. designed the research; A.R.P, A.P., and N.C.A. visually identified shell galaxies using stellar surface density maps; A.R.P. developed stellar history catalogs and post-processing tools, identified shell-forming progenitors and performed the numerical analysis; A.R.P. wrote the paper.

Conflicts of Interest: The authors declare no conflict of interest.

References

1. McConnachie, A.W.; Irwin, M.J.; Ibata, R.A.; Dubinski, J.; Widrow, L.M.; Martin, N.F.; Côté, P.; Dotter, A.L.; Navarro, J.F.; Ferguson, A.M.N.; et al. The remnants of galaxy formation from a panoramic survey of the region around M31. *Nature* **2009**, *461*, 66–69.

2. Martínez-Delgado, D.; Gabany, R.J.; Crawford, K.; Zibetti, S.; Majewski, S.R.; Rix, H.W.; Fliri, J.; Carballo-Bello, J.A.; Bardalez-Gagliuffi, D.C.; Peñarrubia, J.; et al. Stellar tidal streams in spiral galaxies of the local volume: A pilot survey with modest aperture telescopes. *Astron. J.* **2010**, *140*, 962–967.

3. Arp, H. Atlas of Peculiar Galaxies. *Astrophys. J. Suppl. Ser.* **1966**, *14*, 1.

4. Malin, D.F.; Carter, D. A catalog of elliptical galaxies with shells. *Astron. J.* **1983**, *274*, 534–540.

5. Schweizer, F. Observational evidence for mergers. In *Internal Kinematics and Dynamics of Galaxies*; Athanassoula, E., Ed.; Springer: Houten, The Netherlands, 1983; Volume 100, pp. 319–326.

6. Schweizer, F.; Ford, W.K., Jr. Fine Structure in Elliptical Galaxies. In *New Aspects of Galaxy Photometry*; Nieto, J.L., Ed.; Springer: Berlin, Germany, 1985; Volume 232, p. 145.

7. Tal, T.; van Dokkum, P.G.; Nelan, J.; Bezanson, R. The Frequency of Tidal Features Associated with Nearby Luminous Elliptical Galaxies from a Statistically Complete Sample. *Astron. J.* **2009**, *138*, 1417–1427.

8. Atkinson, A.M.; Abraham, R.G.; Ferguson, A.M.N. Faint Tidal Features in Galaxies within the Canada-France-Hawaii Telescope Legacy Survey Wide Fields. *Astron. J.* **2013**, *765*, 28.

9. Krajnović, D.; Emsellem, E.; Cappellari, M.; Alatalo, K.; Blitz, L.; Bois, M.; Bournaud, F.; Bureau, M.; Davies, R.L.; Davis, T.A.; et al. The ATLAS^{3D} project—II. Morphologies, kinematic features and alignment between photometric and kinematic axes of early-type galaxies. *Mon. Not. R. Astron. Soc.* **2011**, *414*, 2923–2949.

10. Duc, P.A. Using deep images and simulations to trace collisional debris around massive galaxies. *arXiv* **2016**, arXiv:1604.08364.

11. Arnaboldi, M.; Ventimiglia, G.; Iodice, E.; Gerhard, O.; Coccato, L. A tale of two tails and an off-centered envelope: Diffuse light around the cD galaxy NGC 3311 in the Hydra I cluster. *Astron. Astrophys.* **2012**, *545*, A37.

12. Martínez-Delgado, D.; Romanowsky, A.J.; Gabany, R.J.; Annibali, F.; Arnold, J.A.; Fliri, J.; Zibetti, S.; van der Marel, R.P.; Rix, H.W.; Chonis, T.S.; et al. Dwarfs Gobbling Dwarfs: A Stellar Tidal Stream around NGC 4449 and Hierarchical Galaxy Formation on Small Scales. *Astrophys. J. Lett.* **2012**, *748*, L24.

13. Romanowsky, A.J.; Strader, J.; Brodie, J.P.; Mihos, J.C.; Spitler, L.R.; Forbes, D.A.; Foster, C.; Arnold, J.A. The Ongoing Assembly of a Central Cluster Galaxy: Phase-space Substructures in the Halo of M87. *Astrophys. J.* **2012**, *748*, 29.

14. Foster, C.; Lux, H.; Romanowsky, A.J.; Martínez-Delgado, D.; Zibetti, S.; Arnold, J.A.; Brodie, J.P.; Ciardullo, R.; GaBany, R.J.; Merrifield, M.R.; et al. Kinematics and simulations of the stellar stream in the halo of the Umbrella Galaxy. *Mon. Not. R. Astron. Soc.* **2014**, *442*, 3544–3564.

15. Longobardi, A.; Arnaboldi, M.; Gerhard, O.; Mihos, J.C. The build-up of the cD halo of M 87: Evidence for accretion in the last Gyr. *Astron. Astrophys.* **2015**, *579*, L3.

16. Amorisco, N.C.; Martinez-Delgado, D.; Schedler, J. A dwarf galaxy's transformation and a massive galaxy's edge: Autopsy of kill and killer in NGC 1097. *arXiv* **2015**, arXiv:1504.03697.

17. Pop, A.R.; Pillepich, A.; Amorisco, N.C.; Hernquist, L. Formation and Incidence of Shell Galaxies in the Illustris Simulation. *arXiv* **2017**, arXiv:1706.06102.

18. Fabian, A.C.; Nulsen, P.E.J.; Stewart, G.C. Star formation in a galactic wind. *Nature* **1980**, *287*, 613–614.

19. Bertschinger, E. Self-similar secondary infall and accretion in an Einstein-de Sitter universe. *Astrophys. J. Suppl. Ser.* **1985**, *58*, 39–65.

20. Williams, R.E.; Christiansen, W.A. Blast wave formation of the extended stellar shells surrounding elliptical galaxies. *Astrophys. J.* **1985**, *291*, 80–87.

21. Loewenstein, M.; Fabian, A.C.; Nulsen, P.E.J. Formation of shells in elliptical galaxies from interstellar gas. *Mon. Not. R. Astron. Soc.* **1987**, *229*, 129–141.

22. Thomson, R.C.; Wright, A.E. A Weak Interaction Model for Shell Galaxies. *Mon. Not. R. Astron. Soc.* **1990**, *247*, 122.

23. Thomson, R.C. Shell formation in elliptical galaxies. *Mon. Not. R. Astron. Soc.* **1991**, *253*, 256–278.

24. Quinn, P.J. On the formation and dynamics of shells around elliptical galaxies. *Astrophys. J.* **1984**, *279*, 596–609.

25. Dupraz, C.; Combes, F. Shells around galaxies—Testing the mass distribution and the 3-D shape of ellipticals. *Astron. Astrophys.* **1986**, *166*, 53–74.

26. Hernquist, L.; Quinn, P.J. Formation of shell galaxies. I—Spherical potentials. *Astrophys. J.* **1988**, *331*, 682–698.

27. Hernquist, L.; Quinn, P.J. Formation of shell galaxies. II—Nonspherical potentials. *Astrophys. J.* **1989**, *342*, 1–16.

28. Sanderson, R.E.; Bertschinger, E. Seen and Unseen Tidal Caustics in the Andromeda Galaxy. *Astrophys. J.* **2010**, *725*, 1652–1675.

29. Sanderson, R.E.; Helmi, A. An analytical phase-space model for tidal caustics. *Mon. Not. R. Astron. Soc.* **2013**, *435*, 378–399.

30. Hernquist, L.; Quinn, P.J. Shell galaxies and alternatives to the dark matter hypothesis. *Astrophys. J.* **1987**, *312*, 17–21.

31. Seguin, P.; Dupraz, C. Dynamical friction in head-on galaxy collisions. II. N-body simulations of radial and non-radial encounters. *Astron. Astrophys.* **1996**, *310*, 757–770.

32. Bartošková, K.; Jungwiert, B.; Ebrová, I.; Jílková, L.; Křížek, M. Simulations of Shell Galaxies with GADGET-2: Multi-Generation Shell Systems. *Astrophys. Space Sci. Proc.* **2011**, *27*, 195.

33. Ebrová, I.; Jílková, L.; Jungwiert, B.; Křížek, M.; Bílek, M.; Bartošková, K.; Skalická, T.; Stoklasová, I. Quadruple-peaked spectral line profiles as a tool to constrain gravitational potential of shell galaxies. *Astron. Astrophys.* **2012**, *545*, A33.

34. Hernquist, L.; Spergel, D.N. Formation of shells in major mergers. *Astrophys. J.* **1992**, *399*, L117–L120.

35. González-García, A.C.; Balcells, M. Elliptical galaxies from mergers of discs. *Mon. Not. R. Astron. Soc.* **2005**, *357*, 753–772.

36. González-García, A.C.; van Albada, T.S. Encounters between spherical galaxies—I. Systems without a dark halo. *Mon. Not. R. Astron. Soc.* **2005**, *361*, 1030–1042.

37. González-García, A.C.; van Albada, T.S. Encounters between spherical galaxies—II. Systems with a dark halo. *Mon. Not. R. Astron. Soc.* **2005**, *361*, 1043–1054.

38. Hirschmann, M.; Naab, T.; Ostriker, J.P.; Forbes, D.A.; Duc, P.-A.; Davé, R.; Oser, L.; Karabal, E. The stellar accretion origin of stellar population gradients in massive galaxies at large radii. *Mon. Not. R. Astron. Soc.* **2015**, *449*, 528–550.

39. Kobayashi, C. GRAPE-SPH chemodynamical simulation of elliptical galaxies—I. Evolution of metallicity gradients. *Mon. Not. R. Astron. Soc.* **2004**, *347*, 740–758.

40. Carlberg, R.G. Dissipative formation of an elliptical galaxy. *Astrophys. J.* **1984**, *286*, 403–415.

41. Bekki, K.; Shioya, Y. Stellar Populations in Gas-rich Galaxy Mergers. II. Feedback Effects of Type IA and Type II Supernovae. *Astrophys. J.* **1999**, *513*, 108–127.

42. Di Matteo, P.; Pipino, A.; Lehnert, M.D.; Combes, F.; Semelin, B. On the survival of metallicity gradients to major dry-mergers. *Astron. Astrophys.* **2009**, *499*, 427–437.

43. Font, A.S.; McCarthy, I.G.; Crain, R.A.; Theuns, T.; Schaye, J.; Wiersma, R.P.C.; Dalla Vecchia, C. Cosmological simulations of the formation of the stellar haloes around disc galaxies. *Mon. Not. R. Astron. Soc.* **2011**, *416*, 2802–2820.

44. Cook, B.A.; Conroy, C.; Pillepich, A.; Rodriguez-Gomez, V.; Hernquist, L. The Information Content of Stellar Halos: Stellar Population Gradients and Accretion Histories in Early-type Illustris Galaxies. *Astrophys. J.* **2016**, *833*, 158.

45. Longhetti, M.; Bressan, A.; Chiosi, C.; Rampazzo, R. Star formation history of early-type galaxies in low density environments. IV. What do we learn from nuclear line-strength indices? *Astron. Astrophys.* **2000**, *353*, 917–929.

46. Carlsten, S.; Hau, G.K.T.; Zenteno, A. Stellar Populations of Shell Galaxies. *arXiv* **2016**, arXiv:1611.05437.

47. Bender, R.; Burstein, D.; Faber, S.M. Dynamically hot galaxies. II—Global stellar populations. *Astrophys. J.* **1993**, *411*, 153–169.

48. Johnston, K.V.; Bullock, J.S.; Sharma, S.; Font, A.; Robertson, B.E.; Leitner, S.N. Tracing Galaxy Formation with Stellar Halos. II. Relating Substructure in Phase and Abundance Space to Accretion Histories. *Astrophys. J.* **2008**, *689*, 936–957,

49. Rejkuba, M.; Greggio, L.; Harris, W.E.; Harris, G.L.H.; Peng, E.W. Deep ACS Imaging of the Halo of NGC 5128: Reaching the Horizontal Branch. *Astrophys. J.* **2005**, *631*, 262–279.

50. Rejkuba, M.; Harris, W.E.; Greggio, L.; Harris, G.L.H. How old are the stars in the halo of NGC 5128 (Centaurus A)? *Astron. Astrophys.* **2011**, *526*, A123.

51. Rejkuba, M.; Harris, W.E.; Greggio, L.; Harris, G.L.H.; Jerjen, H.; Gonzalez, O.A. Tracing the Outer Halo in a Giant Elliptical to 25 R_{eff}. *Astrophy. J. Lett.* **2014**, *791*, L2.

52. Mihos, J.C.; Harding, P.; Rudick, C.S.; Feldmeier, J.J. Stellar Populations in the Outer Halo of the Massive Elliptical M49. *Astrophy. J. Lett.* **2013**, *764*, L20.

53. Sánchez-Blázquez, P.; Forbes, D.A.; Strader, J.; Brodie, J.; Proctor, R. Spatially resolved spectroscopy of early-type galaxies over a range in mass. *Mon. Not. R. Astron. Soc.* **2007**, *377*, 759–786.

54. Foster, C.; Proctor, R.N.; Forbes, D.A.; Spolaor, M.; Hopkins, P.F.; Brodie, J.P. Metallicity gradients at large galactocentric radii using the near-infrared Calcium triplet. *Mon. Not. R. Astron. Soc.* **2009**, *400*, 2135–2146.

55. Spolaor, M.; Kobayashi, C.; Forbes, D.A.; Couch, W.J.; Hau, G.K.T. Early-type galaxies at large galactocentric radii—II. Metallicity gradients and the [Z/H]-mass, [α/Fe]-mass relations. *Mon. Not. R. Astron. Soc.* **2010**, *408*, 272–292.

56. Coccato, L.; Gerhard, O.; Arnaboldi, M. Distinct core and halo stellar populations and the formation history of the bright Coma cluster early-type galaxy NGC 4889. *Mon. Not. R. Astron. Soc.* **2010**, *407*, L26–L30.

57. Coccato, L.; Gerhard, O.; Arnaboldi, M.; Ventimiglia, G. Stellar population and the origin of intra-cluster stars around brightest cluster galaxies: The case of NGC 3311. *Astron. Astrophys.* **2011**, *533*, A138.

58. Kuntschner, H.; Emsellem, E.; Bacon, R.; Cappellari, M.; Davies, R.L.; de Zeeuw, P.T.; Falcón-Barroso, J.; Krajnović, D.; McDermid, R.M.; Peletier, R.F.; et al. The SAURON project—XVII. Stellar population analysis of the absorption line strength maps of 48 early-type galaxies. *Mon. Not. R. Astron. Soc.* **2010**, *408*, 97–132.

59. Weijmans, A.M.; Cappellari, M.; Bacon, R.; de Zeeuw, P.T.; Emsellem, E.; Falcón-Barroso, J.; Kuntschner, H.; McDermid, R.M.; van den Bosch, R.C.E.; van de Ven, G. Stellar velocity profiles and line strengths out to four effective radii in the early-type galaxies NGC3379 and 821. *Mon. Not. R. Astron. Soc.* **2009**, *398*, 561–574.

60. Greene, J.E.; Janish, R.; Ma, C.P.; McConnell, N.J.; Blakeslee, J.P.; Thomas, J.; Murphy, J.D. The MASSIVE Survey. II. Stellar Population Trends Out to Large Radius in Massive Early-type Galaxies. *Astrophy. J.* **2015**, *807*, 11,

61. González Delgado, R.M.; García-Benito, R.; Pérez, E.; Cid Fernandes, R.; de Amorim, A.L.; Cortijo-Ferrero, C.; Lacerda, E.A.D.; López Fernández, R.; Vale-Asari, N.; Sánchez, S.F.; et al. The CALIFA survey across the Hubble sequence. Spatially resolved stellar population properties in galaxies. *Astron. Astrophys.* **2015**, *581*, A103.

62. Wilkinson, D.M.; Maraston, C.; Thomas, D.; Coccato, L.; Tojeiro, R.; Cappellari, M.; Belfiore, F.; Bershady, M.; Blanton, M.; Bundy, K.; et al. P-MaNGA: Full spectral fitting and stellar population maps from prototype observations. *Mon. Not. R. Astron. Soc.* **2015**, *449*, 328–360.

63. Harris, W.E.; Harris, G.L.H.; Layden, A.C.; Wehner, E.M.H. The Leo Elliptical NGC 3379: A Metal-Poor Halo Emerges. *ApJ* **2007**, *666*, 903–918.

64. Lee, M.G.; Jang, I.S. Dual Stellar Halos in the Standard Elliptical Galaxy M105 and Formation of Massive Early-type Galaxies. *Astrophy. J.* **2016**, *822*, 70.

65. Pastorello, N.; Forbes, D.A.; Foster, C.; Brodie, J.P.; Usher, C.; Romanowsky, A.J.; Strader, J.; Arnold, J.A. The SLUGGS survey: Exploring the metallicity gradients of nearby early-type galaxies to large radii. *Mon. Not. R. Astron. Soc.* **2014**, *442*, 1003–1039.

66. Vogelsberger, M.; Genel, S.; Springel, V.; Torrey, P.; Sijacki, D.; Xu, D.; Snyder, G.; Bird, S.; Nelson, D.; Hernquist, L. Properties of galaxies reproduced by a hydrodynamic simulation. *Nature* **2014**, *509*, 177–182.

67. Vogelsberger, M.; Genel, S.; Springel, V.; Torrey, P.; Sijacki, D.; Xu, D.; Snyder, G.; Nelson, D.; Hernquist, L. Introducing the Illustris Project: Simulating the coevolution of dark and visible matter in the Universe. *Mon. Not. R. Astron. Soc.* **2014**, *444*, 1518–1547.

68. Genel, S.; Vogelsberger, M.; Springel, V.; Sijacki, D.; Nelson, D.; Snyder, G.; Rodriguez-Gomez, V.; Torrey, P.; Hernquist, L. Introducing the Illustris project: The evolution of galaxy populations across cosmic time. *Mon. Not. R. Astron. Soc.* **2014**, *445*, 175–200.

69. Springel, V. E pur si muove: Galilean-invariant cosmological hydrodynamical simulations on a moving mesh. *Mon. Not. R. Astron. Soc.* **2010**, *401*, 791–851.

70. Vogelsberger, M.; Genel, S.; Sijacki, D.; Torrey, P.; Springel, V.; Hernquist, L. A model for cosmological simulations of galaxy formation physics. *Mon. Not. R. Astron. Soc.* **2013**, *436*, 3031–3067.

71. Torrey, P.; Vogelsberger, M.; Genel, S.; Sijacki, D.; Springel, V.; Hernquist, L. A model for cosmological simulations of galaxy formation physics: Multi-epoch validation. *Mon. Not. R. Astron. Soc.* **2014**, *438*, 1985–2004.

72. Davis, M.; Efstathiou, G.; Frenk, C.S.; White, S.D.M. The evolution of large-scale structure in a universe dominated by cold dark matter. *Astrophy. J.* **1985**, *292*, 371–394.

73. Springel, V.; Yoshida, N.; White, S.D.M. GADGET: A code for collisionless and gasdynamical cosmological simulations. *New Astron.* **2001**, *6*, 79–117.

74. Dolag, K.; Borgani, S.; Murante, G.; Springel, V. Substructures in hydrodynamical cluster simulations. *Mon. Not. R. Astron. Soc.* **2009**, *399*, 497–514.

75. Amorisco, N.C. On feathers, bifurcations and shells: The dynamics of tidal streams across the mass scale. *Mon. Not. R. Astron. Soc.* **2015**, *450*, 575–591.

76. Hendel, D.; Johnston, K.V. Tidal debris morphology and the orbits of satellite galaxies. *Mon. Not. R. Astron. Soc.* **2015**, *454*, 2472–2485.

77. Salmon, J.; Quinn, P.J.; Warren, M., Using parallel computers for very large N-body simulations: Shell formation using 180 K particles. In *Dynamics and Interactions of Galaxies*; Wielen, R., Ed.; Springer: Heidelberg, Germany, 1990; pp. 216–218.

78. Cooper, A.P.; Martínez-Delgado, D.; Helly, J.; Frenk, C.; Cole, S.; Crawford, K.; Zibetti, S.; Carballo-Bello, J.A.; GaBany, R.J. The Formation of Shell Galaxies Similar to NGC 7600 in the Cold Dark Matter Cosmogony. *Astrophy. J. Lett.* **2011**, *743*, L21.

79. Boylan-Kolchin, M.; Ma, C.P.; Quataert, E. Dynamical friction and galaxy merging time-scales. *Mon. Not. R. Astron. Soc.* **2008**, *383*, 93–101.

80. Amorisco, N.C. Contributions to the accreted stellar halo: An atlas of stellar deposition. *Mon. Not. R. Astron. Soc.* **2017**, *464*, 2882–2895.

81. Rodriguez-Gomez, V.; Pillepich, A.; Sales, L.V.; Genel, S.; Vogelsberger, M.; Zhu, Q.; Wellons, S.; Nelson, D.; Torrey, P.; Springel, V.; et al. The stellar mass assembly of galaxies in the Illustris simulation: Growth by mergers and the spatial distribution of accreted stars. *Mon. Not. R. Astron. Soc.* **2016**, *458*, 2371–2390.

82. Dupraz, C.; Combes, F. Dynamical friction and shells around elliptical galaxies. *Astron. Astrophys.* **1987**, *185*, L1–L4.

83. Wilkinson, A.; Sparks, W.B.; Carter, D.; Malin, D.A. Two Colour CCD Photometry of Malin/Carter Shell Galaxies. In *Structure and Dynamics of Elliptical Galaxies*; de Zeeuw, P.T., Ed.; Springer: Houten, The Netherlands, 1987; Volume 127, p. 465.

galaxies

MDPI

Communication

Globular Clusters and the Halos of Dwarf Galaxies

Søren S. Larsen

Department of Astrophysics/IMAPP, Radboud University, P.O. box 9010, 6500 GL Nijmegen, The Netherlands; s.larsen@astro.ru.nl

Academic Editors: Duncan A. Forbes and Ericson D. Lopez
Received: 25 July 2017; Accepted: 14 August 2017; Published: 29 August 2017

Abstract: Many dwarf galaxies have disproportionately rich globular cluster (GC) systems for their luminosities. Moreover, the GCs tend to be preferentially associated with the most metal-poor stellar populations in their parent galaxies, making them attractive tracers of the halos of dwarf (and larger) galaxies. In this contribution, I briefly discuss some constraints on cluster disruption obtained from studies of metal-poor GCs in dwarf galaxies. I then discuss our recent work on detailed abundance analysis from integrated-light spectroscopy of GCs in Local Group dwarf galaxies.

Keywords: galaxies: abundances; galaxies: star clusters; globular clusters: general

1. Introduction

In most galaxies, stellar halos account for only a small fraction of all stars. Their low surface brightnesses make them challenging to study in integrated light, and even in more nearby galaxies, where individual stars can still be resolved, samples of old stellar population tracers (such as red giants) are often dominated by the more metal-rich populations. However, the number of globular clusters (GCs) per halo star usually increases steeply with decreasing metallicity, making GCs attractive tracers of the (metal-poor) halo populations in their parent galaxies. Furthermore, the GC specific frequency increases significantly towards low galaxy masses/luminosities [1–3]—a trend that mirrors the trend of overall mass-to-light ratio vs. galaxy mass. Indeed, it has been noted that GCs constitute a remarkably constant fraction of the total mass of galaxies (about 6×10^{-5}, [1,2,4]). Whether this implies a universal *formation* efficiency of GCs relative to total galaxy mass or differences in GC *disruption* efficiency [5] remains an open question. However, as will be discussed below, the fraction of metal-poor stars that belong to GCs can be so high in some dwarf galaxies that this puts useful constraints on the role of cluster disruption.

In this contribution, I first discuss how GCs can be used to put constraints on the role of cluster disruption/dissolution in dwarf galaxies (Section 2). In Section 3, I then proceed to discuss our recent results on chemical abundances of GCs from analysis of their integrated light.

2. Globular Clusters in Dwarf Galaxies: Implications for Cluster Disruption

As noted by [1], some dwarf galaxies can reach specific frequencies $S_N > 100$. Of course, it should be kept in mind that small number statistics can cause large fluctuations in the specific frequencies of dwarfs. However, even if this is taken into account (by looking at average S_N values), there is still a clear trend of increasing S_N with decreasing host galaxy M_V, reaching $\langle S_N \rangle \approx 20$ for galaxies with $-12 < M_V < -10$ [1].

The Fornax dwarf spheroidal galaxy has five GCs and an absolute magnitude $M_V = -13.2$, yielding a specific frequency of $S_N = 26$ [6]. Four of these five GCs have $[Fe/H] < -2$, while this is true for only 5% of the field stars. It has been estimated that about 20–25% of the metal-poor stars (those with $[Fe/H] < -2$) are currently members of GCs [6]. While this is a much higher fraction than in the Milky Way halo (where about 2% of the stellar mass is in GCs), this high fraction may not be particularly unusual among dwarf galaxies; the single GC in the Wolf–Lundmark–Melotte (WLM) dwarf contributes a similar fraction of the metal-poor stars, and the ratio may be even higher in the IKN dwarf galaxy in the Ursa Major group [7].

These high GC/field ratios constrain the fractions of stars that could have been lost from the globular clusters, and thus scenarios for dynamical evolution. In particular, most scenarios for the origin of *multiple populations* in GCs need to invoke the loss of a large number of "pristine" stars (i.e., stars with composition similar to that observed in the field) from the clusters in order to explain the observed large fractions of stars with anomalous abundance patterns. Typically, the loss of more than 90% of the initial cluster mass is required [8–10], which is in clear tension with the large present-day GC/field star ratios in dwarf galaxies like Fornax. However, there are also more general implications for cluster disruption. The mass functions of young cluster systems are generally well described by Schechter-like functions, $dN/dM \propto M^{-2} \exp(-M/M_c)$; i.e., they are approximately power-laws with an exponent of about -2 at low masses and exponentially truncated above some cut-off mass, M_c. This differs from the mass function of old GCs, which is roughly flat below a mass of $\sim 10^5 M_\odot$. However, the mass function of old GCs can be well described by an "evolved" version of the Schechter function, where clusters lose mass at an (average) constant rate of about $\Delta M = 2.5 \times 10^5 M_\odot$ per Hubble time. Clusters with (initial) masses below this value have thus dissolved completely, whereas clusters with present-day masses less than ΔM all had initial masses in a relatively small range above ΔM, leading to the flat present-day MF at low masses. In this simple view, the mass lost to the field from dissolving clusters would again exceed the current mass contained within surviving clusters by factors of >10 [11–14].

It is possible that the initial mass distribution of globular clusters was more top-heavy than observed in young cluster systems. In the extreme case, it might even be that the GCs we observe today in dwarf galaxies like Fornax are the only ones that ever formed—an idea that may find some support in observations of dwarf irregular galaxies that are currently experiencing—or have recently undergone—starburst activity. Examples are NGC 1569 and NGC 1705, both of which are dominated by 1–2 massive $((5 - 10) \times 10^5 M_\odot)$ young star clusters [15].

3. Chemical Abundances of GCs in Dwarf Galaxies

The preponderance of metal-poor GCs makes them attractive tracers of the metal-poor stellar populations in dwarf galaxies. Within the Local Group, high-quality spectra of the brighter GCs can be obtained in a few hours of 8–10 m telescope time, from which detailed chemical abundances can be measured. The next generation of 30–40 m telescopes will be able to push this type of measurements well beyond the Local Group.

Traditionally, measurements of metallicities and chemical abundances from integrated light have relied on techniques developed for relatively low-resolution spectra such as the Lick/IDS system of absorption line indices. However, with velocity dispersions of only a few km/s, integrated-light spectroscopy of GCs can benefit from much higher spectral resolution, which potentially makes it possible to employ techniques that are more closely related to those used in classical stellar abundance analysis. The challenge, of course, is that one still needs to model contributions from a mix of different types of stars within the cluster. We have recently tested our method developed for this type of analysis on a sample of seven Milky Way globular clusters, spanning a metallicity range from $[\text{Fe/H}] \simeq -2.3$ (M15, M30) to $[\text{Fe/H}] \simeq -0.5$ (NGC 6388). In essence, we compute "simple stellar population" (SSP) models at very high spectral resolution, based on theoretical model spectra for which we can adjust the abundances of individual elements until the best match to the observed spectra is obtained. The model atmospheres and synthetic spectra are mostly computed with the ATLAS9 and SYNTHE codes [16,17], except for the coolest giants for which we use MARCS atmospheres and the TurboSpectrum code [18–20]. Figure 1 shows example fits to three integrated-light GC spectra around the Na I doublets at 5683/5688 Å (left) and 6154/6161 Å (right). While the 6154/6161 Å lines are blended with Ca I and Sc I lines, we found that our full spectral fitting technique gave very consistent Na abundances for the two doublets, with an average difference of only 0.02 dex for the two sets of lines. From this type of fit, we can typically measure the abundances of a wide range of light, α-, Fe-peak, and heavy elements with an accuracy of ∼0.1 dex (based on comparison with measurements of individual stars) [21].

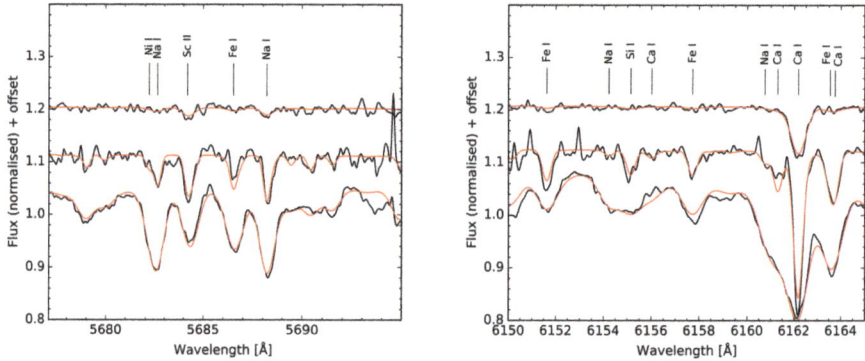

Figure 1. Fits to integrated-light spectra of the Galactic GCs NGC 7078 (**top**), NGC 6254 (**centre**), and NGC 6388 (**bottom**) for the regions near the Na I doublets at 5683/5688 Å (**left**) and 6154/6161 Å (**right**). From this type of fit, we can measure the abundances of many individual elements (e.g., Fe, Na, Mg, Ca, Sc, Ti, Cr, Mn, and Ba). From [21]. Credit: Søren S. Larsen, A&A, 601, A96, 2017, reproduced with permission ©ESO.

We have initiated an effort to carry out detailed chemical abundance analysis of GCs in dwarf galaxies in the Local Group, using the UVES and HIRES spectrographs on the VLT and Keck, respectively. Our published data for GCs in the Fornax and WLM dwarf galaxies [7,22], as well as recently obtained observations of GCs in NGC 147 and NGC 6822, indicate that the most metal-poor GCs in these dwarf galaxies (those with [Fe/H] < −1.5) generally display abundance patterns similar to those in GCs in the Milky Way, with the α-elements (Ca, Ti) being enhanced by about 0.3 dex relative to Solar-scaled composition. Na is typically enhanced compared to the composition seen in Milky Way field stars, but the integrated-light [Na/Fe] ratios are similar to the average values for individual stars in Milky Way GCs, thus providing a strong hint that multiple populations (with about half of the stars being Na-rich) are also present in these extragalactic GCs. In a few clusters, magnesium is depleted relative to other α-elements, which may indicate the presence of the Mg–Al anticorrelation. A similar effect has also been noted by other studies (e.g., in M31 GCs [23,24]). Of the GCs in dwarf galaxies that we have observed so far, only two (Fornax 4 and NGC 6822-SC7) have [Fe/H] > −1.5. Interestingly, these clusters both have approximately Solar-scaled α-element abundances, in agreement with the trends seen in field stars in dwarf galaxies [25]. Cluster SC7 also exhibits a number of other peculiarities, including a significantly sub-Solar [Sc/Fe] ratio and low [Ni/Fe] and [Na/Fe] ratios [26]—indeed, its detailed chemical abundance patterns closely match those in the galactic GC Ruprecht 106, for which an accretion origin has been suggested [27]. A few GCs in the M31 with similar abundance patterns may likewise be accreted [23,24]. This illustrates that detailed chemical abundance analysis of GCs may provide important clues to the enrichment and accretion histories of galactic halos. Finally, we note that efforts are also on-going to extend this type of analysis to younger star clusters in star-forming galaxies beyond the Local Group, where they provide a welcome alternative to more traditional methods such as measurements of strong emission lines from H II regions [28].

4. Conclusions

Globular clusters were evidently an important site of star formation in the early Universe, and this is particularly true for some dwarf galaxies. The high specific frequencies of GCs relative to metal-poor stars in galaxies such as the Fornax dSph constrain the amount of mass that could have been lost from individual clusters and from the cluster population as a whole.

With efficient and highly multiplexed spectrographs on future 30–40 m telescopes, it will be possible to apply the techniques developed for integrated-light abundance analysis to GC systems well beyond the Local Group. In combination with constraints on stellar populations from resolved imaging of individual stars, this will likely lead to a much more detailed picture of the assembly histories of galactic halos.

Conflicts of Interest: The author declares no conflict of interest.

References

1. Georgiev, I.Y.; Puzia, T.H.; Goudfrooij, P.; Hilker, M. Globular cluster systems in nearby dwarf galaxies—III. Formation efficiencies of old globular clusters. *Mon. Not. R. Astron. Soc.* **2010**, *406*, 1967–1984.
2. Harris, W.E.; Harris, G.L.H.; Alessi, M. A catalog of globular cluster systems: What determines the size of a galaxy's globular cluster population? *Astrophys. J.* **2013**, *772*, 82.
3. Miller, B.W.; Lotz, J.M. The globular cluster luminosity function and specific frequency in dwarf elliptical galaxies. *Astrophys. J.* **2007**, *670*, 1074–1089.
4. Spitler, L.R.; Forbes, D.A. A new method for estimating dark matter halo masses using globular cluster systems. *Mon. Not. R. Astron. Soc.* **2009**, *398*, L1–L5.
5. Mieske, S.; Kuepper, A.; Brockamp, M. How tidal erosion has shaped the relation between globular cluster specific frequency and galaxy luminosity. *Astron. Astrophys.* **2014**, *565*, L6.
6. Larsen, S.S.; Strader, J.; Brodie, J.P. Constraints on mass loss and self-enrichment scenarios for the globular clusters of the Fornax dSph. *Astron. Astrophys.* **2012**, *544*, L14.
7. Larsen, S.S.; Brodie, J.P.; Forbes, D.A.; Strader, J. Chemical composition and constraints on mass loss for globular clusters in dwarf galaxies: WLM and IKN. *Astron. Astrophys.* **2014**, *565*, 16.
8. Bekki, K. Secondary star formation within massive star clusters: Origin of multiple stellar populations in globular clusters. *Mon. Not. R. Astron. Soc.* **2011**, *412*, 2241–2259.
9. D'Ercole, A.; Vesperini, E.; D'Antona, E.; McMillan, S.L.W.; Recchi, S. Formation and dynamical evolution of multiple stellar generations in globular clusters. *Mon. Not. R. Astron. Soc.* **2008**, *391*, 825–843.
10. Schaerer, D.; Charbonnel, C. A new perspective on globular clusters, their initial mass function and their contribution to the stellar halo and the cosmic reionization. *Mon. Not. R. Astron. Soc.* **2011**, *413*, 2297–2304.
11. Fall, S.M.; Zhang, Q. Dynamical Evolution of the Mass Function of Globular Star Clusters. *Astrophys. J.* **2001**, *561*, 751–765.
12. Jordán, A.; McLaughlin, D.E.; Côté, P.; Ferrarese, L.; Peng, E.W.; Mei, S.; Villegas, D.; Merritt, D.; Tonry, J.L.; West, M.J. The ACS Virgo Cluster Survey. XII. The Luminosity Function of Globular Clusters in Early-Type Galaxies. *Astrophys. J. Suppl. Ser.* **2007**, *171*, 101–145.
13. Kruijssen, J.M.D.; Portegies Zwart, S.F. On the interpretation of the globular cluster luminosity function. *Astrophys. J.* **2009**, *698*, L158–L162.
14. Vesperini, E. Evolution of globular cluster systems in elliptical galaxies. II. Power-law initial mass function. *Mon. Not. R. Astron. Soc.* **2000**, *322*, 247–256.
15. O'Connell, R.W.; Gallagher, J.S., III; Hunter, D.A. Hubble Space Telescope imaging of super-star clusters in NGC 1569 and NGC 1705. *Astrophys. J.* **1994**, *433*, 65–79.
16. Kurucz, R.L. ATLAS12, SYNTHE, ATLAS9, WIDTH9, et cetera. *Mem. Soc. Astron. Ital. Suppl.* **2005**, *8*, 14.
17. Sbordone, L.; Bonifacio, P.; Castelli, F.; Kurucz, R.L. ATLAS and SYNTHE under Linux. *Mem. Soc. Astron. Ital. Suppl.* **2004**, *5*, 93.
18. Alvarez, R.; Plez, B. Near-infrared narrow-band photometry of M-giant and Mira stars: Models meet observations. *Astron. Astrophys.* **1998**, *330*, 1109–1119.

19. Gustafsson, B.; Edvardsson, B.; Eriksson, K.; Jørgensen, U.J.; Nordlund, Å.; Plez, B. A grid of MARCS model atmospheres for late-type stars. *Astron. Astrophys.* **2008**, *486*, 951–970.

20. Plez, B. TurboSpectrum: Code for Spectral Synthesis. Available online: http://ascl.net/1205.004 (accessed on 11 August 2017).

21. Larsen, S.S.; Brodie, J.P.; Strader, J. Detailed abundances from integrated-light spectroscopy: Milky Way globular clusters. *Astron. Astrophys.* **2017**, *601*, A96.

22. Larsen, S.S.; Brodie, J.P.; Strader, J. Detailed abundance analysis from integrated high-dispersion spectroscopy: Globular clusters in the Fornax dwarf spheroidal. *Astron. Astrophys.* **2012**, *546*, A53.

23. Colucci, J.E.; Bernstein, R.A.; Cohen, J.G. The detailed chemical properties of M31 star clusters. I. Fe, alpha and light elements. *Astrophys. J.* **2014**, *797*, 116.

24. Sakari, C.M.; Venn, K.A.; Mackey, D.; Shetrone, M.D.; Dotter, A.; Ferguson, A.M.N.; Huxor, A. Integrated light chemical tagging analyses of seven M31 outer halo globular clusters from the Pan-Andromeda Archaeological Survey. *Mon. Not. R. Astron. Soc.* **2015**, *448*, 1314–1334.

25. Tolstoy, E.; Hill, V.; Tosi, M. Star-Formation Histories, Abundances, and Kinematics of Dwarf Galaxies in the Local Group. *Ann. Rev. Astron. Astrophys.* **2009**, *47*, 371–425.

26. Larsen, S.S.; Brodie, J.P.; Wasserman, A.; Strader, J. Detailed abundance analysis of globular clusters in the Local Group. NGC 147, NGC 6822, and Messier 33. **2017**, in prep.

27. Villanova, S.; Geisler, D.; Carraro, G.; Moni Bidin, C.; Muñoz, C. Ruprecht 106: The first single population Globular Cluster? *Astrophys. J.* **2013**, *778*, 186.

28. Hernandez, S.; Larsen, S.; Trager, S.; Groot, P.; Kaper, L. Chemical Abundances of Two Extragalactic Young Massive Clusters. *Astron. Astrophys.* **2017**, in press.

galaxies

MDPI

Article

Hot Gaseous Halos in Early Type Galaxies

Dong-Woo Kim

Smithsonian Astrophysical Observatory, Cambridge, MA 02138, USA; kim@cfa.harvard.edu

Academic Editors: Duncan A. Forbes and Ericson D. Lopez
Received: 31 July 2017; Accepted: 27 September 2017; Published: 4 October 2017

Abstract: The hot gas in early type galaxies (ETGs) plays a crucial role in their formation and evolution. As the hot gas is often extended to the outskirts beyond the optical size, the large scale structural features identified by Chandra (including cavities, cold fronts, filaments, and tails) point to key evolutionary mechanisms, e.g., AGN feedback, merging history, accretion/stripping, as well as star formation and quenching. We systematically analyze the archival Chandra data of ETGs to study the hot ISM. Using uniformly derived data products with spatially resolved spectral information, we revisit the X-ray scaling relations of ETGs and address their implications by comparing them with those of groups/clusters and simulations.

Keywords: galaxies: elliptical and lenticular; cD–X-rays: galaxies

1. Introduction

The hot gaseous halos in early type galaxies (ETGs) are closely connected to most important physical processes throughout a galaxy's evolution. These include the AGN feedback (e.g., preventing hot gas from cooling, producing X-ray cavities/bubbles related to radio jets), stellar feedback (stellar mass loss and SN ejecta being the main source of hot gas and chemical enrichment), interactions among galaxies (e.g., as revealed by cold fronts and sloshing) and with hotter ICM (e.g., ram pressure stripping), and the assembly of dark matter (being the main force that holds up the hot gas). With the sub-arcsecond resolution of the Chandra X-ray Observatory, we are now able to accurately measure global hot gas properties by effectively excluding LMXBs (low-mass X-ray binaries) and AGNs. Studying the physical properties of the hot halos allows us to address those important aspects of galaxy formation and evolution. In this conference proceedings paper, we review the previous studies and further discuss the implications by comparing the results of groups and clusters as well as the simulations.

2. X-ray Scaling Relations

X-ray scaling relations have been widely used to investigate the origin and evolution of the hot ISM of ETGs [1–4]. Recent numerical simulation studies [5–8] have further attempted to reproduce the observed scaling relations, in order to constrain the physical mechanisms that shape the hot ISM.

2.1. $L_{X,GAS}$–L_K Relation

Chandra observations have given us the ability to separate the different X-ray emission components (e.g., nuclear, LMXB, AB + CV, hot gas) of ETGs to extract accurate measurements of $L_{X,GAS}$ [1]. This work has shown that the well-known factor of 100 spread in the previous ETG scaling relation, $L_{X,TOTAL}$–L_{OPT}, increases to a factor of 1000 when $L_{X,GAS}$ is used instead of $L_{X,TOTAL}$.

We plot the X-ray luminosities against the *K*-band luminosity in Figure 1 where different components are marked by different symbols. The contribution from ABs (active binaries) and CVs (cataclysmic variables), $L_{X,(AB + CB)}$, is marked by a linear diagonal line. The contribution from

LMXBs (blue squares), $L_{X,LMXB}$, is proportional to L_K, but with a non-negligible scatter. The LMXB integrated luminosity is about 10 times larger than that of ABs + CVs. The nuclear emission (green triangles) spans more than 2 orders of magnitude and does not seem to scale with L_K (see also [9]). The X-ray luminosity of the hot gas, $L_{X,GAS}$, after excluding all other components, ranges from a few $\times\ 10^{37}$ erg s^{-1} to a few $\times\ 10^{41}$ erg s^{-1}. We note that $L_{X,GAS}$ is often lower than $L_{X,LMXB}$, except for those gas rich ETGs with $L_{X,GAS} > 10^{40}$ erg s^{-1}. The $L_{X,GAS}$–L_K relation is still correlated, but with a large scatter, e.g., the ratio of $L_{X,GAS}/L_K$ spanning 3 orders of magnitude for a given L_K.

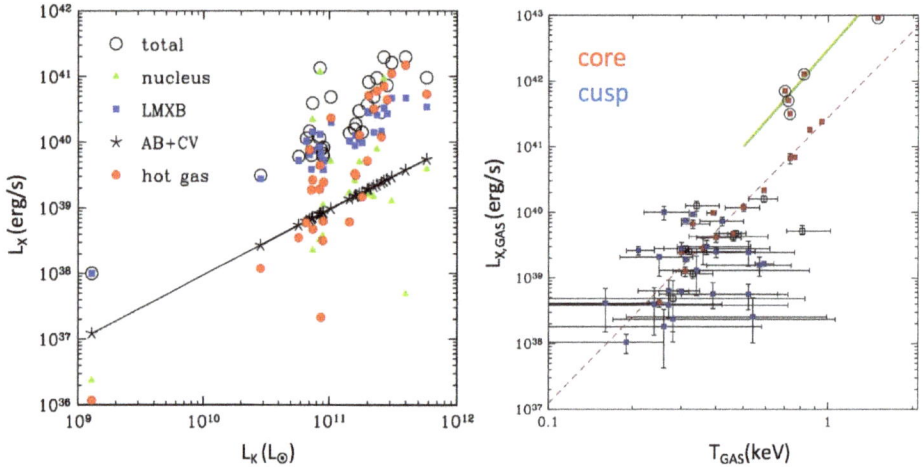

Figure 1. (**a**) X-ray luminosities of individual components are plotted against the *K*-band luminosity. The total X-ray luminosity is denoted by open black circles, nuclei by filled green triangles, LMXBs by filled blue squares, and hot gas by filled red circles. Reproduced from Reference [1]; (**b**) X-ray luminosity and temperature of hot gas plotted separately for core and cusp galaxies. Those with big green circles are cDs. Reproduced from Reference [3].

2.2. $L_{X,GAS}$–T_{GAS} Relation

With Chandra, we have also been able to accurately measure the temperature of the hot gas by effectively removing point sources to derive the $L_{X,GAS}$–T_{GAS} scaling relation for ETGs. This relation is tighter than the $L_{X,GAS}$–L_K relation and may indicate virialization of the hot gas in the dark matter halos. The best fit relation is $L_{X,GAS} \sim T_{GAS}^{4.5}$ [1,3,4] when the cD galaxies are excluded. The relation is particularly tight for the normal (non-cD) core galaxies with a scatter of only 0.2 dex rms. This tight relation holds for a range of $kT_{GAS} = 0.3$–1 keV and $L_{X,GAS} = $ a few $\times\ 10^{38}$—a few $\times\ 10^{41}$ erg s^{-1}. In contrast, no correlation exists for coreless (or cusp) galaxies (and spiral galaxies). This may be understood due to the presence of several factors in coreless galaxies (as in spiral galaxies) that may affect the retention and temperature of the hot gas. They include global rotation, flattened galaxy figures possibly with embedded disks, and recent star formation. See Figure 1a for a comparison of core and cusp galaxies.

In Figure 2, we show a schematic diagram comparing the $L_{X,GAS}$–T_{GAS} relations in the ETG samples with those reported for groups [10] and clusters of galaxies [11]. To the first approximation, the bigger the system, the hotter and more luminous the gas. The details are more complex, however. Starting from the bottom left corner, the coreless ETGs and spiral galaxies show no clear correlation. On the other hand, pure ellipticals (Es) show a tight correlation with a slope of 4.5. Furthermore, this tight relation is well-reproduced among σ-supported, hot gas rich elliptical galaxies by high-resolution simulations [5]. The groups follow a similar trend as the core Es, but are shifted toward higher $L_{X,GAS}$ values for a given T_{GAS}. The larger dark matter halo may be responsible for the retention of larger hot gaseous halos. (larger L_X). The clusters at the top right corner have a strong, but flatter relation ($L_{X,GAS} \sim T_{GAS}^3$). As the exact difference between groups and clusters is somewhat subtle, the distinction may be ambiguous among the big groups and small clusters (e.g., see [12]). The relation expected by the self-similar case (where gravity dominates) has a slope of 2 (dashed lines in Figure 2). The steep slope (3) in clusters indicates that baryonic physics is already important, even in the largest scale. In galaxies, the slope is even steeper (4.5), further indicating the increased importance of non-gravitational effects (including SF, AGN, and their feedback).

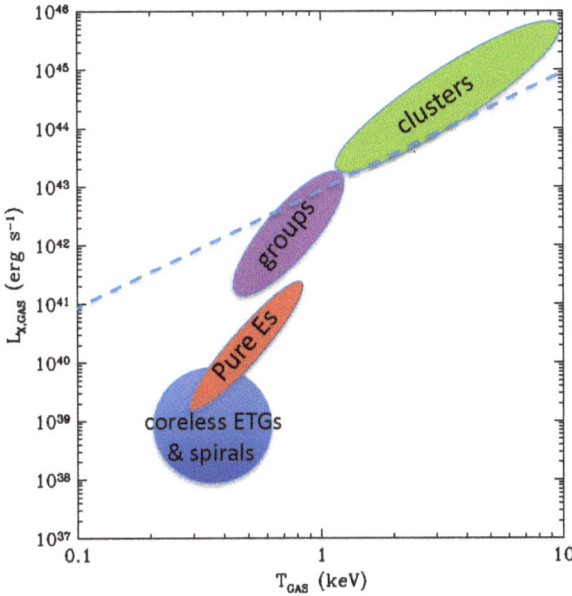

Figure 2. Comparison of the $L_{X,GAS}$–T_{GAS} relations in various samples. From the bottom left, the coreless ETGs and spirals have no correlation, while the normal (non-cD) core E galaxies have a very tight correlation ($L_{X,GAS} \sim T_{GAS}^{4.5}$). The groups have a similar trend as the core Es, but they are shifted toward higher $L_{X,GAS}$. The clusters at the top right corner have a flatter relation ($L_{X,GAS} \sim T_{GAS}^3$) compared to other sub-samples. For a reference, the self-similar expectation ($L_X \sim T^2$) is shown in dashed lines. Reproduced from Reference [3].

2.3. $L_{X,GAS}$–M_{TOTAL} Relation

The optical luminosity (L_K) is a good proxy for the integrated stellar mass of the galaxy, $M\star$; however, it does not measure the amount of dark matter (DM) mass, which may be prevalent, especially at large radii. The total mass (stellar + DM), out to radii comparable to the total extent of the hot halos of gas-rich ETGs, is the physical quantity we must know in order to explore the importance of gravitational confinement for the hot gas retention [2,13]. A number of dynamical mass measurements at large radii (within 5 R_e) have recently become available from the analysis of the kinematics of hundreds of globular clusters (GC) and planetary nebulae (PN) in individual galaxies [14,15]. With these improved X-ray and mass measurements, we have investigated the $L_{X,GAS}$–M_{TOTAL} relation [2,13] and found a tight correlation between these physically motivated quantities (even if the total mass is harder to measure than e.g., L_B or L_K). In a simple power-law form, the best fit relation is ($L_{X,GAS}/10^{40}$ erg s^{-1}) = ($M_{TOTAL}/3.2 \times 10^{11}$ M$_\odot$)^3 with an rms deviation of a factor of 3 in $L_{X,GAS} = 10^{38}$–10^{43} erg s^{-1} or M_{TOTAL} = a few $\times 10^{10}$—a few $\times 10^{12}$ M$_\odot$. More strikingly, this relation becomes even tighter with an rms deviation of a factor of 1.3 among the gas-rich galaxies (with $L_{X,GAS} > 10^{40}$ erg s^{-1}). Our results [2,13] indicate that the total mass of an ETG is the primary factor in regulating the amount of hot gas. Since the gas temperature reflects the energy input and the depth of the potential well, the $L_{X,GAS}$–T_{GAS} relation provides a complementary scaling relation to $L_{X,GAS}$–M_{TOTAL}. Interestingly, the functional form of this relation (power-law with a slope of ~3) is consistent to what would be expected from the steep $L_{X,GAS}$–T_{GAS} relation found in the above, given that $M_{TOTAL} \sim T_{GAS}^{3/2}$ (virial theorem).

3. Discussion

3.1. Comparison with Clusters

The X-ray scaling relations are well established among clusters of galaxies where the larger amount of hotter (a few–10 keV) gas is retained inside the deeper cluster potential well. In Figure 3, we show three relations (L_X–T_X, L_X–M_{500}, M_{500}–T_X) reproduced from Reference [16]. All three relations are tightly correlated among clusters of galaxies. In Figure 3, we schematically overlay those of pure Es (NGC 3379 being the smallest and NGC 4649 being the largest among the E sample). The comparisons reveal that (1) L_X of the E sample is considerably lower than that extrapolated from the cluster sample and (2) the relations (L_X against T_X or M) of the E sample are significantly steeper, while the M–T_X relation follows each other. These distinctions imply that Es lost a large portion of hot gas [17], likely by AGN/stellar feedback, while clusters have retained most of their hot gas in a closed box. Both feedback mechanisms play a significant role in increasing the importance of non-gravitational effects and making the slope deviate more from self-similar behaviors.

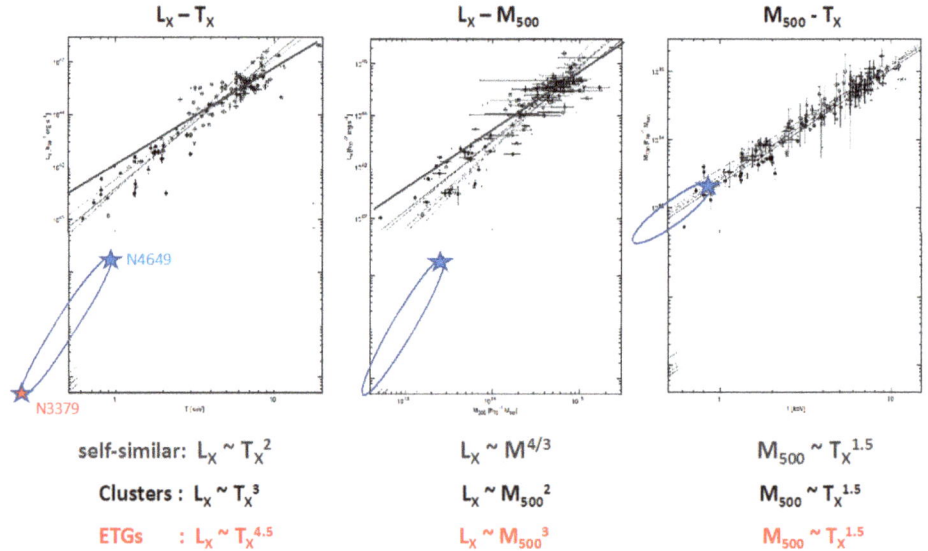

$$\text{self-similar: } L_X \sim T_X^2 \qquad L_X \sim M^{4/3} \qquad M_{500} \sim T_X^{1.5}$$

$$\text{Clusters : } L_X \sim T_X^3 \qquad L_X \sim M_{500}^2 \qquad M_{500} \sim T_X^{1.5}$$

$$\text{ETGs : } L_X \sim T_X^{4.5} \qquad L_X \sim M_{500}^3 \qquad M_{500} \sim T_X^{1.5}$$

Figure 3. Comparisons of X-ray scaling relations of ETGs and clusters. From the left, $L_{X,GAS}$–T_{GAS}, $L_{X,GAS}$–M_{500}, and M_{X500}–T_{GAS} relations are shown. The black data points and black thin lines for clusters are reproduced from Reference [16], and the red regions indicate the core ETGs. The smallest (NGC 3379) and largest (NGC 4649) among our ETG samples are marked. The self-similar predictions ($L \sim T^2$ and $L \sim M^{4/3}$) are also marked by green thick lines.

3.2. Comparison with Simulations

Recent numerical simulations have been significantly improved. They can reproduce the observational results and compare the X-ray properties in detail [5–8]. References [5,6] applied 2D high-resolution hydro simulations and considered stellar feedback, dynamical (rotation), and structural (flattening) parameters as well as AGN feedback, and References [7,8] applied 3D cosmological simulations. While the differences among different simulations and different parameters/recipes are to yet be understood, in general they can reproduce the observed relations among $L_{X,GAS}$, T_{GAS}, M_{TOTAL} (or M within 5 Re). In particular, these studies suggested that the stellar and AGN feedback are important in the early epoch to globally reduce the amount of hot gas (or $L_{X,GAS}$) and the total mass to retain the hot gas afterward. Quantitative comparisons with observables and predictions in a specific sample of galaxies are to be performed in the future.

Conflicts of Interest: The authors declare no conflict of interest.

References

1. Boroson, B.; Kim, D.-W.; Fabbiano, G. Revisiting with Chandra the Scaling Relations of the X-ray Emission Components (Binaries, Nuclei, and Hot Gas) of Early-type Galaxies. *Astrophys. J.* **2011**, *729*, 12. [CrossRef]
2. Kim, D.-W.; Fabbiano, G. X-Ray Scaling Relation in Early-Type Galaxies Dark Matter as a Primary Factor in Retaining Hot Gas. *Astrophys. J.* **2013**, *776*, 116. [CrossRef]
3. Kim, D.-W.; Fabbiano, G. X-ray Scaling Relations of core and coreless E and S0 Galaxies. *Astrophys. J.* **2015**, *812*, 127. [CrossRef]
4. Goulding, A.D.; Greene, J.E.; Ma, C.-P.; Veale, M.; Bogdan, A.; Nyland, K.; Blakeslee, J.P.; McConnell, N.J.; Thomas, J. The MASSIVE Survey. IV. The X-ray Halos of the Most Massive Early-type Galaxies in the Nearby Universe. *Astrophys. J.* **2016**, *826*, 167. [CrossRef]
5. Negri, A.; Posacki, S.; Pellegrini, S.; Ciotti, L. The effects of galaxy shape and rotation on the X-ray haloes of ETGs—II. Numerical simulations. *Mon. Not. R. Astron. Soc.* **2014**, *445*, 1351. [CrossRef]
6. Ciotti, L.; Pellegrini, S.; Negri, A.; Ostriker, J.P. The Effect of the AGN Feedback on the ISM of ETGs: 2D Simulations of the Low-Rotation Case. *Astrophys. J.* **2017**, *835*, 15. [CrossRef]
7. Choi, E.; Ostriker, J.P.; Naab, T.; Oser, L. Physics of Galactic Metals: Evolutionary Effects due to Production, Distribution, Feedback, and Interaction with Black Holes. *Astrophys. J.* **2017**, *844*, 31. [CrossRef]
8. Eisenreich, M.; Naab, T.; Choi, E.; Ostriker, J.P.; Emsellem, E. AGN feedback, quiescence and CGM metal enrichment in early-type galaxies. *Mon. Not. R. Astron. Soc.* **2017**. [CrossRef]
9. Pellegrini, S. The Nuclear X-ray Emission of Nearby Early-type Galaxies. *Astrophys. J.* **2010**, *717*, 640. [CrossRef]
10. Helsdon, S.F.; Ponman, T.J. X-ray bright groups and their galaxies. *Mon. Not. R. Astron. Soc.* **2003**, *340*, 485. [CrossRef]
11. Pratt, G.W.; Croston, J.H.; Arnaud, M.; Boehringer, H. Galaxy cluster X-ray luminosity scaling relations from a representative local sample (REXCESS). *Astron. Astrophys.* **2009**, *498*, 361. [CrossRef]
12. Sun, M.; Voit, G.M.; Donahue, M.; Jones, C.; Forman, W.; Vikhlinin, A. Chandra Studies of the X-Ray Gas Properties of Galaxy Groups. *Astrophys. J.* **2009**, *693*, 1142. [CrossRef]
13. Forbes, D.A.; Alabi, A.; Romanowsky, A.J.; Kim, D.-W.; Brodie, J.P.; Fabbiano, G. The SLUGGS survey: Revisiting the correlation between X-ray luminosity and total mass of massive early-type galaxies. *Mon. Not. R. Astron. Soc.* **2017**, *464*, L26. [CrossRef]
14. Deason, A.J.; Belokurov, V.; Evans, N.W.; McCarthy, I.G. Elliptical Galaxy Masses out to Five Effective Radii the Realm of Dark Matter. *Astrophys. J.* **2012**, *748*, 2. [CrossRef]
15. Alabi, A.; Forbes, D.A.; Romanowsky, A.J.; Brodie, J.P.; Strader, J.; Janz, J.; Pota, V.; Pastorello, N.; Usher, C.; Spitler, L.R.; et al. The SLUGGS Survey the mass distribution in early-type galaxies within five effective radii and beyond. *Mon. Not. R. Astron. Soc.* **2016**, *460*, 3838. [CrossRef]
16. Eckmiller, H.J.; Hudson, D.S.; Reiprich, T.H. Testing the low-mass end of X-ray scaling relations with a sample of Chandra galaxy groups. *Astron. Astrophys.* **2011**, *535*, A105–A143. [CrossRef]
17. Kim, D.-W.; Pellegrini, S. *"Hot Interstellar Matter in Elliptical Galaxies" Astrophysics and Space Science Library*; Springer: Berlin, Germany, 2012; Volume 378.

galaxies

Communication

How Clumpy Star Formation Affects Globular Cluster Systems

Jeremy Bailin

Department of Physics and Astronomy, University of Alabama, Box 870324, Tuscaloosa, AL 35487-0324, USA; jbailin@ua.edu

Academic Editors: Duncan A. Forbes and Ericson D. Lopez
Received: 7 June 2017; Accepted: 2 August 2017; Published: 4 August 2017

Abstract: There is now clear evidence the metallicities of globular clusters are not simple tracers of the elemental abundances in their protocluster clouds; some of the heavy elements were formed subsequently within the cluster itself. It is also manifestly clear that star formation is a clumpy process. We present a brief overview of a theoretical model for how self-enrichment by supernova ejecta proceeds in a protocluster undergoing clumpy star formation, and show that it predicts internal abundance spreads in surprisingly good agreement with those in observed Milky Way clusters.

Keywords: globular clusters; star formation; abundances; stellar halos

1. Introduction

Globular clusters (GCs) are critical tracers of galactic stellar halos. It is only very recently that heroic observations have been able to detect diffuse stellar components around galaxies beyond the Local Group (see, for example, the excellent work done by many other authors in this special edition); for most galaxies, GCs are the only easily-identifiable halo components.

GCs are almost uniformly old, and trace the early stages of galaxy formation. Observationally, they mainly differ from each other by their luminosity and colour, which more-or-less correspond to stellar mass and total metallicity. Metallicity is particularly valuable, because heavy elements were formed by previous generations of stars, implying that the metallicity of a GC traces the history of the gas cloud from which it formed.

However, this assumes that the only contribution to the metallicity of present-day GC stars is from the metallicity of the protocluster cloud. Two major pieces of evidence point towards a more complicated picture: (1) the "blue tilt", a tendency for the most massive metal-poor GCs around a large galaxy to be more metal-rich than the less-massive GCs (e.g., [1]); and (2) internal abundance spreads within GCs, especially of intermediate elements like oxygen and sodium, but sometimes also including iron (e.g., [2]). Both pieces of evidence suggest that some of the metals in GCs come from *self enrichment* (i.e., they were formed by stars within the GC itself).

In [3] (BH09), we presented a model in which protocluster clouds are able to gravitationally hold onto a fraction of core collapse supernova ejecta that depends on the balance between supernova kinetic energy and the depth of the protocluster potential well. The metals from this ejecta were then assumed to mix evenly among the low-mass slow-forming stars, increasing the total metallicity of the GC. The BH09 model reproduced the blue tilt qualitatively, and provided a decent quantitative match with reasonable modifications to the model's free parameters [4,5].

However, the BH09 model was deficient in one major way—it assumed that star formation and metals were well-mixed. This flies in the face of patently clumpy star formation regions that are observed [6] and cannot reproduce the *internal abundance spread* that is one of the main motivations for considering self-enrichment. In this contribution, I give an overview of a new extension to the BH09 model that is explicitly clumpy.

2. Materials and Methods

In brief, the model assumes that each protocluster cloud begins with a pre-enriched level of metallicity due to its history up to that point. During star formation, the cloud fragments into clumps, which undergo individual star formation events spread out over time. Parameters of the number and structure of the clumps have been calibrated using the Bolometric Galactic Plane Survey [7,8]. Supernovae from each clump can pollute later-forming clumps with metals to the degree that the gravitational potential of the entire cloud can contain them, meaning that each clump has an individual metallicity and there can be a metallicity spread within the final cluster. Full details of the model are presented in [9].

3. Results

More massive GCs both fragment into more pieces, and also are able to hold onto a larger fraction of their supernova ejecta, resulting in larger internal metallicity spreads. This is shown by the black dots in Figure 1, which shows the spread in internal iron abundances as a function of cluster mass. Observations of Milky Way GCs from [10] are overplotted. Although GCs are generally considered to have "no" iron spread (in contrast to dwarf galaxies) the measured spreads are small but non-zero. Note that the observations have been corrected to the estimated initial cluster mass using [11], in order to make them directly comparable to the model GC masses. Given that there was *no fine tuning* of the model parameters, the agreement is remarkable (in fact, the goodness of fit is probably to some degree a coincidence).

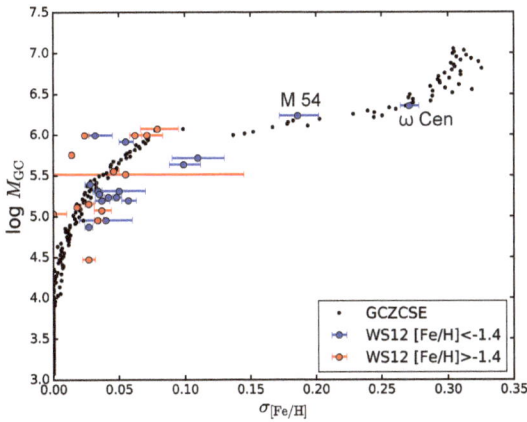

Figure 1. Internal iron abundance spread as a function of globular cluster (GC) mass. Black points denote model GCs, while blue and red data points indicate observed Milky Way metal-poor and metal-rich GCs, respectively, from [10]. Observed GCs are plotted using the estimated initial mass of [11].

4. Discussion

The agreement between the new clumpy self-enrichment model that we have presented and observations of Milky Way GCs suggests that we are indeed capturing an important facet of GC formation. Since the parameters of the model were calibrated entirely on local star formation regions, this implies that the high-intensity star formation that occurred at high redshift—when these GCs formed—was not qualitatively different from local star formation, but simply acted as a scaled-up version of processes we observe locally.

The model predicts that sufficiently massive GCs should have substantial iron spreads, and in particular matches observations of ω Cen and M 54—two objects that have often been speculated to be stripped dwarf galaxies rather than true GCs. One of the pieces of evidence that is often cited for such an identification is the iron spread, but our model predicts that GCs with these masses ought to have precisely so large of an iron spread. Therefore, the iron spread cannot be used as a piece of evidence that these high-mass GCs are not true GCs.[1]

Finally, we note that the same self-enrichment that causes internal abundance spreads also increases the total metallicity of the cluster (hence the blue tilt). If we want to use GCs as probes of the history of their natal gas cloud, we want to know the unpolluted initial metallicity of the cloud before self-enrichment occurred. The magnitude of internal abundance spread may give us a calibration of how much self enrichment has occurred, allowing us to correct GC metallicities and use them as better galaxy formation probes. This is an avenue of current research we are actively pursuing.

Acknowledgments: Support for program HST-AR-13908.001-A was provided by NASA through a grant from the Space Telescope Science Institute, which is operated by the Association for Research in Astronomy, Inc., under NASA contract NAS 5-26555. We thank the organizers and participants of the conference for a wealth of productive and enjoyable conversations.

Conflicts of Interest: The authors declare no conflict of interest.

Abbreviations

The following abbreviations are used in this manuscript:

GC   Globular Cluster

References

1. Harris, W.E.; Whitmore, B.C.; Karakla, D.; Okon, W.; Baum, W.A.; Hanes, D.A.; Kavelaars, J.J. Globular Cluster Systems in Brightest Cluster Galaxies: Bimodal Metallicity Distributions and the Nature of the High-Luminosity Clusters. *Astrophys. J.* **2006**, *636*, 90.
2. Piotto, G.; Bedin, L.R.; Anderson, J.; King, I.R.; Cassisi, S.; Milone, A.P.; Villanova, S.; Pietrinferni, A.; Renzini, A. A Triple Main Sequence in the Globular Cluster NGC 2808. *Astrophys. J.* **2007**, *661*, 53–56.
3. Bailin, J.; Harris, W.E. Stochastic Self-Enrichment, Pre-Enrichment, and the Formation of Globular Clusters. *Astrophys. J.* **2009**, *695*, 1082–1093.
4. Mieske, S.; Jordan, A.; Cote, P.; Peng, E.; Ferrarese, L.; Blakeslee, J.; Mei, S.; Baumgardt, H.; Tonry, J.; Infante, L.; et al. The ACS Fornax Cluster Survey. IX. The Color-Magnitude Relation of Globular Cluster Systems. *Astrophys. J.* **2010**, *710*, 1672.
5. Goudfrooij, P.; Kruijssen, J.M.D. Color-Magnitude Relations within Globular Cluster Systems of Giant Elliptical Galaxies: The Effects of Globular Cluster Mass Loss and the Stellar Initial Mass Function. *Astrophys. J.* **2014**, *780*, 43.
6. Larson, R.B. Turbulence and star formation in molecular clouds. *Mon. Not. R. Astron. Soc.* **1981**, *194*, 809.
7. Battisti, A.J.; Heyer, M.H. The Dense Gas Mass Fraction of Molecular Clouds in the Milky Way. *Astrophys. J.* **2014**, *780*, 173.
8. Ellsworth-Bowers, T.P.; Glenn, J.; Riley, A.; Rosolowsky, E.; Ginsburg, A.; Evans, N.J., II; Bally, J.; Battersby, C.; Shirley, Y.L.; Merello, M. The Bolocam Galactic Plane Survey. XIII. Physical Properties and Mass Functions of Dense Molecular Cloud Structures. *Astrophys. J.* **2015**, *805*, 157.
9. Bailin, J. A Model for Clumpy Self-Enrichment in Globular Clusters. **2017**, in preparation.

---

[1]   We are not arguing that these objects are necessarily GCs, simply that other evidence must be used to make the argument.

10. Willman, B.; Strader, J. "Galaxy," Defined. *Astron. J.* **2012**, *144*, 76.
11. Balbinot, E.; Gieles, M. The devil is in the tails: the role of globular cluster mass evolution on stream properties. *arXiv* **2017**, arXiv:1702.02543.

galaxies

MDPI

Article

Upper Limits to Magnetic Fields in the Outskirts of Galaxies

Ericson López *, Jairo Armijos-Abendaño, Mario Llerena and Franklin Aldás

Observatorio Astronómico de Quito, Escuela Politécnica Nacional, Av. Gran Colombia S/N, Quito 170403, Ecuador; jairo.armijos@epn.edu.ec (J.A.-A.); mario.llerena01@epn.edu.ec (M.L.); franklin.aldas@epn.edu.ec (F.A.)
* Correspondence: ericsson.lopez@epn.edu.ec

Academic Editor: Tomotsugu Goto
Received: 31 July 2017; Accepted: 18 September 2017; Published: 19 September 2017

Abstract: Based on CO(2-1) public data, we study the monoxide oxygen gas excitation conditions and the magnetic field strength of four spiral galaxies. For the galaxy outskirts, we found kinetic temperatures in the range of $\lesssim$35–38 K, CO column densities $\lesssim 10^{15}$–10^{16} cm^{-2}, and H$_2$ masses $\lesssim 4 \times 10^6$–6×10^8 M$_\odot$. An H$_2$ density $\lesssim 10^3$ cm^{-3} is suitable to explain the 2σ upper limits of the CO(2-1) line intensity. We constrain the magnetic field strength for our sample of spiral galaxies and their outskirts by using their masses and H$_2$ densities to evaluate a simplified magneto-hydrodynamic equation. Our estimations provide values for the magnetic field strength on the order of $\lesssim$6–31 µG.

Keywords: nearby galaxies; magnetic fields; molecules

1. Introduction

In this paper, we focus our attention on the study of the magnetic field strength in the outskirts of four spiral galaxies, following a different approach to those commonly based on Faraday rotation, dust polarization, synchrotron emission, etc. To constrain the magnetic field strength of spiral galaxies, we will follow the Dotson method, described in Section 4.4 of [1]; i.e., approaching the magneto-hydrodynamic force equation to derive a simple expression to estimate the upper limit of the magnetic field.

Magnetic Field Constraint

As mentioned above, we use the Dotson method [1] to constrain the magnetic field, which is mainly based on the following relation:

$$B < 3.23 \times 10^{-8} \left(\frac{R}{\text{pc}} \right)^{0.5} \left(\frac{n}{\text{cm}^{-3}} \right)^{0.5} \left(\frac{M}{\text{M}_\odot} \right)^{0.5} \left(\frac{r}{\text{pc}} \right)^{-1} \tag{1}$$

where R is the radius of the magnetic field lines, n is the molecular hydrogen gas density, M and r are the total mass and radius of the source, respectively. So, the magneto-hydrodynamic force can be used to estimate the magnetic field strength (Equation (1)) when there is available information about the shape of the field lines (see [1]). On the other hand, to estimate the density n and mass M of the source, we use the carbon monoxide emission as a tracer of the molecular gas H$_2$ [2]. This is because H$_2$ is invisible in the cold interstellar medium (around 10–20 K), so its distribution and motion must be inferred from observations of minor constituents of the clouds, such as carbon monoxide and dust [2]. Carbon monoxide is the most abundant molecule after H$_2$, CO is easily excited, and the emission of CO(1-0) at 2.6 mm is ubiquitous in the Galaxy [2]. So, CO it is a good tracer for molecular hydrogen.

2. Carbon Monoxide Data

To carry out this study, we used public CO(2-1) data, first published by [3], data obtained with the IRAM 30 m telescope[1] located in Spain. At the CO(2-1) transition frequency (230.538 GHz), the IRAM telescope has a spatial resolution of 13 arcsec. From the available data we selected a sample of four nearby spiral galaxies, whose morphology and positions are listed in Table 1. On the other hand, in Figure 1, CO(2-1) integrated intensity maps of each galaxy in our sample are shown.

Table 1. Galaxy sample morphology and positions.

Galaxy Name	RA [1] (hh:mm:ss.s)	DEC [1] (hh:mm:ss.s)	Morphology [1]	Distance [1] (Mpc)
NGC 2841	09:22:02.7	+50:58:35.3	SAa C	14.6
NGC 3077	10:03:19.1	+68:44:02.2	S0 C	3.8
NGC 3184	10:18:17.0	+41:25:27.8	SAc C	11.3
NGC 3351	10:43:57.7	+11:42:13.0	SBb C	10.5

[1] Information taken from the SIMBAD Astronomical Database.

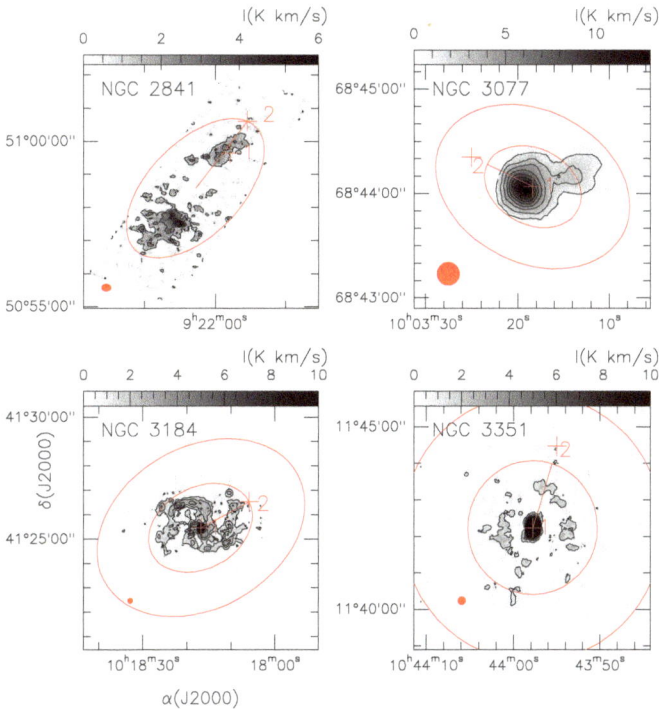

Figure 1. CO(2-1) integrated intensity maps of our galaxy sample. The red crosses show positions used to extract the spectra shown in Figure 2. The red ellipses indicate the regions used to measure CO(2-1) luminosities (see text). The tiny filled red ellipse represents the IRAM telescope beam (13 arcsec at the CO(2-1) frequency transition of 230.538 GHz). For NGC 2841, we did not measure the CO(2-1) luminosity in the galaxy outskirts, as in this particular case was not observed.

---

[1] http://www.iram-institute.org/EN/30-meter-telescope.php

3. Data Treatment

3.1. Spectra Selection

As was mentioned in the introduction, in order to constrain the magnetic field strengths for the galaxies in our sample, we use the method applied by [1], which is based on Equation (1). Therefore, for every source, the parameters R, n, M, and r should be determined or assumed. We estimate n, M, and r from CO(2-1) observations (see Section 4), while R is assumed based on the results of previous works [4–7].

To derive n and M from the CO(2-1) data we have selected the spectra by choosing two positions; one located in the nucleus of the galaxy (hereinafter the position 1) and the second located on the outskirts of the galaxy (hereinafter the position 2). CO(2-1) line emission is not detected towards the NGC 2841 nucleus; therefore, as an exception, its position 1 spectrum corresponds to a position displaced 79 arcsec from the galaxy nucleus. The inner ellipse is taken as the one that holds as much as possible the integrated intensity of CO (2-1) radiation emitted by the galaxy. The center of this ellipse defines the nucleus (position 1). Starting from the nuclei, in steps of 20 arcsec, we take CO(2-1) spectrum along the major axis of the inner ellipse. Position 2 is defined as the more contiguous positions to the galaxy nucleus, placed along the galaxy major-axis, where the CO(2-1) line emission is no longer detected above 2σ. We assume that position 2 is representative of the lowest boundary of the galactic outskirts which are well traced by the HI emission [8]. In Figure 1, positions 1 and 2 are indicated by red crosses. For all galaxies in our sample, the spectra extracted for both positions, within the 13 arcsec resolution of the IRAM telescope, are shown in Figure 2. In our study, the choice of position 2 depends on two aspects: (1) the IRAM telescope beam size and (2) the step of 20 arcsec used to find one of the nearest position to the galaxy nucleus, along the major-axis, where CO(2-1) emission is no longer detected.

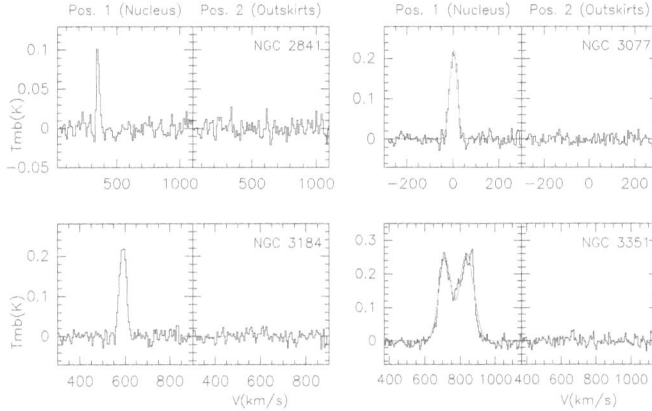

Figure 2. Spectra extracted from positions 1 and 2 indicated in Figure 1. As an example, for NGC 3077 and NGC 3351 galaxies, we show the Gaussian fits (indicated with red curves) applied to the CO(2-1) lines.

3.2. Gaussian Fitting and CO Luminosities

Gaussian fits to the CO(2-1) lines (indicated in Figure 2) have been performed. Then, the large velocity gradient (LVG) modeling [9] is employed to estimate the gas density (n), based on the CO(2-1) line width (ΔV) and the line intensity (I). For position 1 (nucleus) we derived the values of these parameters, which are presented in Table 2. For position 2 (outskirts), we estimated 2σ upper limits for the CO(2-1) line intensity, which are also given in Table 2. This table includes the radial velocity (V) and the average $\overline{\Delta V}$ obtained from spectra extracted along the major-axis of the galaxy disk.

To estimate CO integrated luminosities (L_{CO}) for the galaxies in our sample, we used the inner ellipse for the nuclear region (the L_{CO} will be used later to estimate H_2 masses). As seen in Figure 1, and as it was indicated before, this ellipse is defined to enclose almost all the CO(2-1) emission arising from the galaxy disk. In turn, to estimate a 2σ upper limit on the L_{CO} for the galaxy outskirts, we use a ringlike region with axes equal to twice the sizes of the inner ellipse. The derived L_{CO} values for both regions of each galaxy are given in Table 3.

Table 2. CO(2-1) line properties derived for our galaxy sample.

Galaxy Name	Region	$\Delta V(\sigma)$	$V(\sigma)$	I [2]	$\overline{\Delta V}$ [3]
		km s^{-1}	km s^{-1}	K	km s^{-1}
NGC 2841	Position displaced 79 arcsec from nucleus [1]	25.2 (2.0)	352.0	0.1	23.4
	outskirts	...	...	$\lesssim$0.03	...
NGC 3077	nucleus	38.2 (1.1)	1.8 (0.5)	0.2	40.2
	outskirts	...	...	$\lesssim$0.02	...
NGC 3184	nucleus	35.4 (1.0)	590.3 (0.5)	0.2	25.6
	outskirts	...	...	$\lesssim$0.03	...
NGC 3351	nucleus	80.9 (2.3)	713.1 (0.9)	0.3	45.9
	outskirts	...	...	$\lesssim$0.03	...

[1] As CO(2-1) line emission is not detected at the galaxy nucleus, this position is displaced 79 arcsec from the galaxy nucleus (see Figure 1); [2] For galaxy outskirts, this parameter is a 2σ upper limit based on the rms noise; [3] Average value obtained from several spectra extracted along the major-axis of the galaxy disk.

Table 3. Parameters derived for the galaxy sample.

Galaxy Name	Region	L_{CO} [2]	r [3]	M_{H_2} [4]	$(M_{H_2} + M_{HI})$ [5]	B
		$\times10^6$ K km s^{-1} pc^2	kpc	$\times10^7$ M$_\odot$	$\times10^8$ M$_\odot$	μG
NGC 2841	disk	2.0	8.5	103.0	89.6	$\lesssim$31
	outskirts [1]	...	...	...	...	...
NGC 3077	disk	2.3	0.4	1.3	0.4	$\lesssim$6
	outskirts	$\lesssim$0.7	0.8	$\lesssim$0.4	$\lesssim$0.1	$\lesssim$7
NGC 3184	disk	188.0	6.6	103.0	36.8	$\lesssim$14
	outskirts	$\lesssim$81.1	13.2	$\lesssim$44.6	$\lesssim$15.9	$\lesssim$19
NGC 3351	disk	226.0	7.1	124.0	32.1	$\lesssim$11
	outskirts	$\lesssim$109.0	14.2	$\lesssim$59.8	$\lesssim$15.5	$\lesssim$15

[1] For NGC 2841 we were not able to estimate L_{CO} luminosity and other relative parameters in the outskirts, as this region was not observed; [2],[4],[5] For galaxy outskirts, these parameters are a 2σ upper limit; [3] This radius traces the boundary of the galaxy disk region, where the CO(2-1) line emission is still detectable above 2σ (see Section 3.1).

4. Results

4.1. Galaxy Mass

The molecular hydrogen mass (M_{H_2}) for the galaxy disk and its outskirts is derived using the the equation given by [3]:

$$\frac{M_{H_2}}{M_\odot} = 5.5 \frac{R_{21}}{0.8} \left(\frac{L_{CO}}{\text{K km s}^{-1} \text{pc}^2} \right) \tag{2}$$

where R_{21} is the CO(2-1)/CO(1-0) line intensity ratio (equal to 0.8) and L_{CO} is the CO luminosity. In this equation, a CO(1-0)/H_2 conversion factor of 2×10^{20} cm^{-1} (K km s^{-1})$^{-1}$ is adopted [3]. In Table 3, the M_{H_2} values are listed for both galaxy regions. We found that the H_2 outskirts masses derived for our galaxies sample vary within $\lesssim 4 \times 10^6$–1×10^9 $M_\odot$. The total mass ($M_{H_2} + M_{HI}$) is also presented in the same table. The atomic hydrogen mass has been estimated by using the M_{H_2}/M_{HI} ratio given in [3].

4.2. H_2 Density, CO Column Density, and Kinetic Temperature

As was mentioned above, the H_2 density is derived for the nucleus and outskirts positions in the galaxies of our sample by using the LVG approach. We fit the CO(2-1) line intensity or its limit and the average line width $\overline{\Delta V}$, while the CO column density (N_{CO}), H_2 molecular density (n_{H_2}), and the kinetic temperature (T_{kin}) are considered as free parameters. The CO(1-0) line intensity is known based on the CO(2-1)/CO(1-0) line intensity ratio of about 0.8 derived by [3]. Note that at position 2 (i.e., outskirt positions) , we do not detect CO(2-1) emission and we have derived 2σ upper limits for these line intensities. We found that a n_{H_2} density of about 10^3 cm^{-3} is suitable to fit the CO(2-1) line intensity for the four galaxies in our sample. For positions 1 and 2, the N_{CO} and T_{kin} that provide the best fits to the CO(2-1) line intensity (or its limit) are given in Table 4. For the galaxy outskirts, we found $T_{kin} \lesssim 35$–38 K and $N_{CO} \lesssim 10^{15}$–10^{16} cm^{-2}.These findings tell us that the molecular gas in galaxy outskirts is relatively cold. For the nucleus positions, the T_{kin} are found within 35–38 K and the N_{CO} within $\sim 10^{15}$–10^{16} cm^{-2}. So, the column density N_{CO} exhibits the greatest changes, whereas the kinetic temperature T_{kin} is relatively constant. The physical parameters presented in Table 4 seem to be similar for both nuclei and outskirts, but in the case of the nuclei the given values are accurate, while for the outskirts they correspond to the 2σ upper limits.

Table 4. Physical parameters derived for our galaxy sample.

Component	n_{H_2}	T_{kin}	N_{CO}
outskirts	$\lesssim 10^3$ cm^{-3}	$\lesssim 35$–38 K	$\lesssim 10^{15}$–10^{16} cm^{-2}
nucleus	10^3	35–38 K	$\sim 10^{15}$–10^{16} cm^{-2}

4.3. Magnetic Fields in Galaxies and Their Outskirts

As mentioned previously, in order to constrain the magnetic field strength for a given galaxy and its outskirts, we use Equation (1) given by [1]. This equation includes the n_{H_2}, the mass M, and radius r of the galaxy, and the radius of curvature R of the magnetic field lines. n_{H_2} and M were derived in the previous sections, but the mass that we use in Equation (1) refers to the dust mass, which is obtained by the relation ($M_{H_2} + M_{HI}$)/100; i.e., assuming the typical dust-to-gas mass ratio of 0.01. In this section, we assume that R is equal to the radius r of the studied regions. This assumption is crude, but it is based on several studies of spiral galaxies such as NGC 4736, M51, NGC 1097, and NGC 1365, which reveal magnetic field lines that extend as far as their galactic arms [4–6]. The derived values for the magnetic field strength are listed in Table 3. For the galaxy outskirts, we have considered the upper limits of M and n as fixed values in order to calculate the magnetic field strength. For the NGC 2841 galaxy, no value was estimated for the magnetic field in its outskirts, because this region was not observed in this object. We found magnetic field magnitudes on the order of $\lesssim 6$–31 µG for the galaxies in our sample and their outskirts. These limits agree with those values of ~ 20–60 µG found in spiral galaxies [4,6,7].

5. Conclusions

In the present contribution, we have estimated the magnetic field strength in the galaxy nuclei and in the outskirts of a sample of four spiral galaxies. For that, we have used an approximate expression of the magnetohydrodynamics to find an upper limit for the magnetic field magnitudes. The magnetic field strength lies within the range of $\lesssim 6$–31 µG, which is in good agreement with the values provided by other authors for spiral galaxies. A first good idea about the strength of the magnetic field can be obtained directly from the estimation of molecular hydrogen mass and radio of the source, without the necessity of a magnetohydrodynamical model or the use of a traditional technique like Faraday rotation or Zeeman line broadening. This is a rough estimation that works if the gas pressure is uniform and the viscosity is neglected. A better approach can be obtained by keeping more terms in the magnetohydrodynamics force equation [1] to impede gravitational collapse.

Moreover, instead of using the total hydrogen mass $(M_{H_2} + M_{HI})$ in our magnetic field calculations, the mass of the dust has been considered as enough to get values in good agreement with the ~20–60 µG observed in spiral galaxies [4,6,7]. This fact suggests that the dust is the main component that influences the magnetic field strength better than the molecular gas.

On the other hand, for the galaxy outskirts we found a kinetic temperature T_{kin} of about $\lesssim$35–38 K and a column density $N_{CO} \lesssim 10^{15}$–10^{16} cm^{-2}. These findings tell us that the molecular gas in galaxy outskirts is relatively cold. Moreover, the M_{H_2} masses on the outskirts of the galaxy are in the range of $\lesssim 4 \times 10^6$–6×10^8 M$_\odot$, and a n_{H_2} density of $\lesssim 10^3$ cm^{-3} is suitable to explain the 2σ upper limits to the CO(2-1) line intensity. Then, it seems that if the densities and temperatures are low in the outskirts it would result in a higher M_{H_2} mass at a given CO(2-1) luminosity if the outskirts volume is increasing. It is interesting that the values of both H$_2$ density and kinetic temperature are relatively similar in the nuclear region of the studied galaxies, but not the CO column density that varies by a factor of 10 along our sample.

In future research, we plan to go deeper in understanding the magnetic field structure in galaxy halos, studying more spiral galaxies and employing other molecular hydrogen tracers. Additionally, understanding the variations of the field direction associated with the column density changes is part of our future work.

Author Contributions: E. López, J. Armijos-Abendaño, M. Llerena and F. Aldás performed the data analysis. E. López and J. Armijos-Abendaño wrote the manuscript. All authors contributed to the discussion and interpretation of the results.

Conflicts of Interest: The authors declare no conflict of interest.

References

1. Dotson, J.L. Polarization of the Far-Infrared Emission from M17. *Astrophys. J.* **1996**, *470*, 566–576.

2. Neininger, N.; Guélin, M.; Ungerechts, H.; Lucas, R.; Wielebinski, R. Carbon monoxide emission as a precise tracer of molecular gas in the Andromeda galaxy. *Nature* **1998**, *395*, 871–873.

3. Leroy, A.K.; Walter, F.; Bigiel, F.; Usero, A.; Weiss, A.; Brinks, E.; de Blok, W.J.G.; Kennicutt, R.C.; Schuster, K.F.; Kramer, C.; et al. Heracles: The HERA CO Line Extragalactic Survey. *Astrophys. J.* **2009**, *137*, 4670–4696.

4. Beck, R.; Fletcher, A.; Shukurov, A.; Snodin, A.; Sokoloff, D.D.; Ehle, M.; Moss, D.; Shoutenkov, V. Magnetic fields in barred galaxies. IV. NGC 1097 and NGC 1365. *Astron. Astrophys.* **2005**, *444*, 739–765.

5. Chyzy, K.; Buta, R. Discovery of a Strong Spiral Magnetic Field Crossing the Inner Pseudoring of NGC 4736. *Astrophys. J. Lett.* **2008**, *677*, L17.

6. Fletcher, A.; Beck, R.; Shukurov, A.; Berkhuijsen, E.M.; Horellou, C. Magnetic fields and spiral arms in the galaxy M51. *Mon. Not. R. Astron. Soc.* **2011**, *412*, 2396–2416.

7. Knapik, J.; Soida, M.; Dettmar, R.; Beck, R.; Urbanik, M. Detection of spiral magnetic fields in two flocculent galaxies. *Astron. Astrophys.* **2000**, *362*, 910–920.

8. Sofue, Y. The most completely sampled rotation curves for galaxies. *Astrophys. J.* **1996**, *458*, 120–131.

9. Van der Tak, F.F.S.; Black, J.H.; Schöier, F.L.; Jansen, D.J.; van Dishoeck, E.F. A computer program for fast non-LTE analysis of interstellar line spectra. With diagnostic plots to interpret observed line intensity ratios. *Astron. Astrophys.* **2007**, *468*, 627–635.

galaxies

[MDPI]

Conference Report

Interstellar Reddening Effect on the Age Dating of Population II Stars

Sergio Ortolani [1,2,*], Santi Cassisi [3] and Maurizio Salaris [4]

[1] Dipartimento di Fisica e Astronomia, Universitá di Padova, 35122 Padova, Italy

[2] INAF—Osservatorio Astronomico di Padova, 35122 Padova, Italy

[3] INAF—Osservatorio Astronomico di Teramo, via M. Maggini, sn, 64100 Teramo, Italy; cassisi@oa-teramo.inaf.it

[4] Astrophysics Research Institute, Liverpool John Moores University, Liverpool L3 5RF, UK; M.Salaris@ljmu.ac.uk

* Correspondence: sergio.ortolani@unipd.it

Academic Editors: Duncan A. Forbes and Ericson D. López
Received: 27 April 2017; Accepted: 18 June 2017; Published: 22 June 2017

Abstract: The age measurement of the stellar halo component of the Galaxy is based mainly on the comparison of the main sequence turn-off luminosity of the globular cluster (GC) stars with theoretical isochrones. The standard procedure includes a vertical shift, in order to account for the distance and extinction to the cluster, and a horizontal one, to compensate the reddening. However, the photometry is typically performed with broad-band filters where the shape of the stellar spectra introduces a shift of the effective wavelength response of the system, dependent on the effective temperature (or color index) of the star. The result is an increasing distortion—actually a rotation and a progressive compression with the temperature—of the color-magnitude diagrams relatively to the standard unreddened isochrones, with increasing reddening. This effect is usually negligible for reddening $E(B-V)$ on the order of or smaller than 0.15, but it can be quite relevant at larger extinction values. While the ratio of the absorption to the reddening is widely discussed in the literature, the importance of the latter effect is often overlooked. In this contribution, we present isochron simulations and discuss the expected effects on age dating of high-reddening globular clusters.

Keywords: interstellar matter; stellar evolution; galactic halo

1. Introduction

Interstellar reddening causes a reduction of the flux received from the stars following the well-known Whitford law [1,2]. The relations between the absorption in different bands and the color excess are discussed widely in the literature and are well established (e.g., [3,4]). However, in practical cases of wide-band photometry, the interstellar extinction convolution with the spectra of stars having different T_{eff} also causes a shift in the effective wavelength of the bands, whose effects are not much considered in the literature.

The variation of the $R_V = A_V/E(B-V)$ ratio as a function of the intrinsic $(B-V)$ color was calculated by [5]. More recently, [6] published detailed results on the selective extinctions for stars with temperatures in the range $3500 \leq T_{eff}(K) \leq 40,000$, metallicities equal to $[Fe/H] = -2.0, -1.0$, and $+0.5$, and luminosity classes I, III, and V, based on Kurucz synthetic spectra and Scheffer interstellar extinction. They found that R_V changes by about 15% over the full temperature range. One of the main results of the paper is an increase of the R_V ratio for the spectral type of the stars most used in the extinction determination for globular clusters (GCs), and a consequent shortening of the distance scale in the Galaxy, but the result is biased by a relatively high choice of the normalization of R_V to a value of 3.346 at 17,000 K.

The large variation of R_V with spectral type has been confirmed by [7], which used the [8] library of spectra and a normalization $R_V = 3.07$ for Vega ($(B - V) = 0$)—considerably lower than Grebel and Roberts' assumption. The range of his results is roughly consistent with the original Schmidt–Kaler findings, with an increase of R_V up to 3.6 for the coolest stars.

Recent examples of extinction temperature-dependent corrections implemented in theoretical isochrones are from [9,10]. Although the basis of the stellar temperature dependence is well established, none of these papers clearly addressed the issue of the systematic effects of the reddening on GC ages derived from isochrone fitting in color-magnitude diagrams (CMDs).

In this contribution, we show how even a relatively moderate extinction $A_V = 1$ can produce a significant bias when the age is derived from main sequence (MS) turn off (TO) fitting using isochrones not corrected for temperature-dependent extinction. Our simulations include the Johnson BVI and ACS@HST F606W/F775W/F814W photometric bands.

2. Methodological Approach

Numerical simulations were perfomed with the BaSTI evolutionary models ([11,12], available at the following URL: http://www.oa-teramo.inaf.it/BASTI). We selected isochrones with age in the range 12.0–13.0 Gyr, for an α−enhanced mixture [α/Fe] = +0.40, metallicity Z = 0.004, and initial helium abundance Y = 0.256, corresponding to [Fe/H] = −1.0.

A temperature-dependent extinction with $A_V = 1$ and $A_V = 2$ has been applied to the isochrones (hereinafter *simulated isochrones*). To account for this effect, we used the web interface at http://stev.oapd.inaf.it/cgi-bin/cmd, which implements the results by [9] to determine the extinctions in the selected Johnson–Cousins (UBVI) and ACS@HST F606W, F775W, F814W photometric filters, covering the full range of effective temperature of our isochrones, for our selected A_V values. These extinctions are calculated assuming the extinction law by [3] with a reference $R_V = 3.1$. Finally, the simulated isochrones have been shifted and matched to the unreddened ones at two points: one on the MS at $M_V = 5.5$, and another one on the rising sub-giant branch (SGB).

Figure 1 shows the results in the $(V, B - V)$ CMD. The MS turn-off of the 12.0 Gyr reddened isochrone matches very well with that of the 12.5 Gyr one, if not corrected for temperature-dependent effects. This means that unreddened isochrones matched to the simulated one give an age older by about 0.5 Gyr at $A_V = 1$. With an extinction $A_V = 2$, the distortion of the shape of the simulated isochrone with respect to the unreddened ones causes a 12.0 Gyr reddened isochrone to overlap almost perfectly with the 13 Gyr unreddened one. In both cases, the derived older age is a consequence of the shorter color range spanned by the SGB. This is a consequence of the compression in color of the reddened isochrone, due to the relatively higher reddening of the hotter stars compared to the cooler ones. The effect of a lower MS TO luminosity—due to the higher extinction in the blue—is also present but less important.

As shown in Figure 2, this effect is reduced by about a factor of two in the $(V, V - I)$ CMD, due to the longer wavelengths of V/I bands compared to B/V, and the consequent smaller shift of the effective wavelengths with temperature. The conclusion is that a bias of $\approx$0.5–0.7 Gyr on the age estimate is produced only for an absorption $A_V = 2.0$ or larger.

The case of the ACS@HST photometric bands is shown in Figures 3 and 4. For the F775W/F606W CMD, the effect of the age bias is already very strong—about 1.5 Gyr—at $A_V = 1$, so we have not investigated the case $A_V = 2$. This effect is reduced in the F606W/F814W CMD. The higher sensitivity of extinction in the ACS@HST bands with stellar effective temperature of the ACS@HST is due to the wider wavelength range of the ACS filters compared to the Johnson counterparts. Whilst B and V have a typical FWHM around 100 nm, the ACS filters have a width ranging from 150 (F775W) to 250 nm (F606W and F804W).

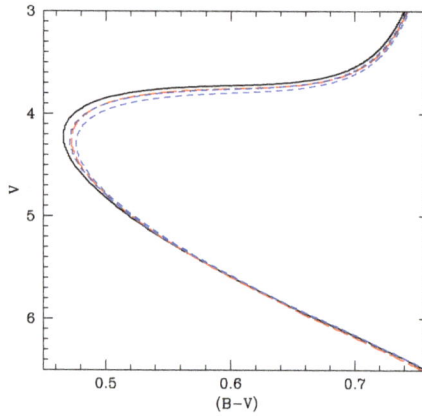

Figure 1. $(V, B - V)$ color-magnitude diagram (CMD) of a t = 12.0 Gyr simulated isochrones for $A_V = 1$ (red dashed line), compared to a set of unreddened isochrones with t = 12.0 (solid black line), 12.5, and 13.0 Gyr (dashed blue lines) according to the procedure discussed in the text.

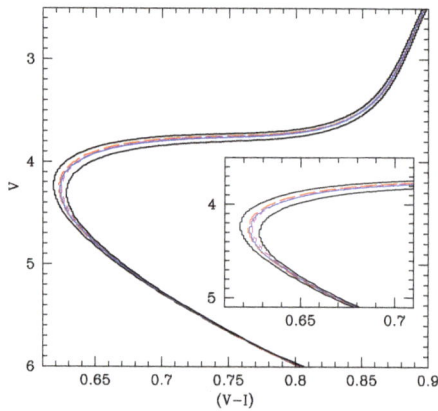

Figure 2. As Figure 1 but for the $(V, V - I)$ CMD, for two assumptions about A_V.

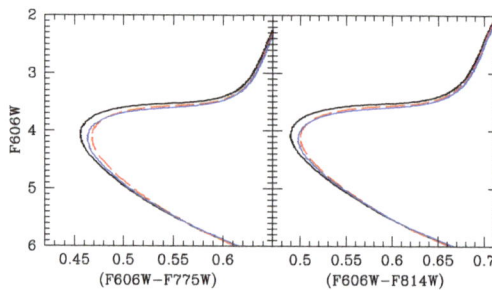

Figure 3. Left panel: as Figure 1, but for the $(F606W, F606W - F775W)$ ACS@HST CMD. **Right panel**: as left panel, but in the $(F606W, F606W - F814W)$ CMD.

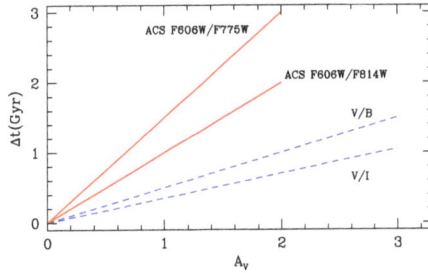

Figure 4. Bias on ages derived using isochrones accounting for T_{eff}-dependent extinction, compared to the case of neglecting this effect, for various A_V values and band combinations.

3. Final Remarks

In an age of Galactic GC surveys based on isochrone fitting without correction for T_{eff}-dependent extinction, systematic age trends can be introduced, because reddening is a function of Galactic latitude. Given the relation between galactocentric distance and galactic extinction, neglecting T_{eff}-dependent extinction can bias estimated ages depending on the GC position in the Galaxy. The effect is generally very small at large galactocentric distances, but in the inner halo/bulge, at galactocentric distances shorter than 5 Kpc, most of the GCs have visual extinctions larger than $A_V = 1.0$. In these cases, neglecting T_{eff}-dependent extinction causes a bias towards ages older by more than 0.5 Gyr when using the Johnson B/V bands, this figure increasing by a factor of two in optical ACS bands. In a forthcoming paper, we plan to investigate the impact of using appropriate T_{eff}-dependent extinction corrections on GC age dating, based on the use of the standard horizontal and vertical methods (see [13] and references therein).

Author Contributions: All authors have equally contributed to the analysis of the data and the writing of the paper.

Conflicts of Interest: The authors declare no conflict of interest.

References

1. Whitford, A.E. An Extension of the Interstellar Absorption-Curve. *Astrophys. J.* **1948**, *107*, 102.
2. Whitford, A.E. The law of interstellar reddening. *Astron. J.* **1958**, *63*, 201–207.
3. Cardelli, J.A.; Clayton, G.C.; Mathis, J.S. The relationship between infrared, optical, and ultraviolet extinction. *Astrophys. J.* **1989**, *345*, 245–256.
4. Fitzpatrick, E.L. Correcting for the Effects of Interstellar Extinction. *Publ. Astron. Soc. Pac.* **1999**, *111*, 63–75.
5. Schmidt-Kaler, T. Die Verfärbung als Funktion der interstellaren Absorption und der Energieverteilung des kontinuierlichen Sternspektrums. *Astron. Nachr.* **1961**, *286*, 113.
6. Grebel, E.K.; Roberts, W.J. Heterochromatic extinction. I. Dependence of interstellar extinction on stellar temperature, surface gravity, and metallicity. *Astron. Astrophys. Suppl. Ser.* **1995**, *109*, 293–312.
7. McCall, M.L. On Determining Extinction from Reddening. *Astron. J.* **2004**, *128*, 2144–2169.
8. Pickles, A.J. A Stellar Spectral Flux Library: 1150–25000 Å. *Publ. Astron. Soc. Pac.* **1998**, *110*, 863–878.
9. Girardi, L.; Dalcanton, J.; Williams, B.; de Jong, R.; Gallart, C.; Monelli, M.; Groenewegen, M.A.T.; Holtzman, J.A.; Olsen, K.A.G.; Seth, A.C.; et al. Revised Bolometric Corrections and Interstellar Extinction Coefficients for the ACS and WFPC2 Photometric Systems. *Publ. Astron. Soc. Pac.* **2008**, *120*, 583.
10. Casagrande, L.; VandenBerg, D.A. Synthetic stellar photometry—I. General considerations and new transformations for broad-band systems. *Mon. Not. R. Astron. Soc.* **2014**, *444*, 392–419.
11. Pietrinferni, A.; Cassisi, S.; Salaris, M.; Castelli, F. A Large Stellar Evolution Database for Population Synthesis Studies. I. Scaled Solar Models and Isochrones. *Astrophys. J.* **2004**, *612*, 168–190.

12. Pietrinferni, A.; Cassisi, S.; Salaris, M.; Castelli, F. A Large Stellar Evolution Database for Population Synthesis Studies. II. Stellar Models and Isochrones for an α-enhanced Metal Distribution. *Astrophys. J.* **2006**, *642*, 797–812.

13. Cassisi, S.; Salaris, M. *Old Stellar Populations: How to Study the Fossil Record of Galaxy Formation*; John Wiley & Sons: New York, NY, USA, 2013.

galaxies

MDPI

Article

On the Kinematics, Stability and Lifetime of Kinematically Distinct Cores: A Case Study

Felix Schulze [1,2,*], Rhea-Silvia Remus [1] and Klaus Dolag [1,3]

[1] Faculty of Physics, Universitäts-Sternwarte München, Ludwig-Maximilians-Universität, Scheinerstr. 1, D-81679 München, Germany; rhea@usm.uni-muenchen.de (R.-S.R.); dolag@usm.uni-muenchen.de (K.D.)
[2] Max Planck Institute for Extraterrestrial Physics, Giessenbachstrasse 1, D-85740 Garching, Germany
[3] Max Planck Institute for Astrophysics, Karl-Schwarzschild-Str. 1, D-85748 Garching, Germany
* Correspondence: fschulze@usm.lmu.de

Academic Editors: Duncan A. Forbes and Ericson D. Lopeze
Received: 6 July 2017; Accepted: 11 August 2017; Published: 17 August 2017

Abstract: We present a case study of a early-type galaxy (ETG) hosting a kinematically distinct core (KDC) formed in a binary high resolution 1:1 spiral galaxy merger simulation. The runtime of the simulation is pushed up to 10 Gyr to follow the complete evolution of various physical properties. To investigate the origin of the KDC, the stellar component residing within the KDC is dissected, revealing that the rotational signal is purely generated by stars that belong to the KDC for at least 0.5 Gyr and are newly formed during the merging process. Following the orientation of the total stellar angular momentum of the KDC, we show that it performs a motion comparable to the precession of a gyroscope in a gravitational potential. We draw the conclusion that the motion of the KDC is a superposition of an intrinsic rotation and a global precession that gets gradually damped over cosmic time. Finally, the stability of the KDC over the complete runtime of the simulation is investigated by tracing the evolution of the widely used λ_R parameter and the misalignment angle distribution. We find that the KDC is stable for about 3 Gyr after the merger and subsequently disperses completely on a timescale of $\approx$1.5 Gyr.

Keywords: kinematically distinct cores; galaxy merger; numerical simulation; galaxy formation

1. Introduction

The stellar kinematics of galaxies represent a meaningful benchmark for modern models of galaxy formation. Recent advances in integral field spectroscopy revealed a rich variety of kinematical features in the line-of-sight velocity distribution (LOSVD), especially in early-type galaxies (Emsellem et al., 2004 [1]; Krajnović et al., 2011 [2]). These features embody the final state of a complex assembly history and evolution shaping the dynamical and kinematical appearance of galaxies. A particularly interesting class of kinematic appearances are galaxies that exhibit kinematically distinct cores (KDCs), which are kinematically decoupled from their host galaxy, often visible in an inclined net rotation of the central core component. Providing full two-dimensional observations of the stellar kinematics of statistically meaningful samples, the ATLAS^{3D} (Cappellari et al., 2011 [3]) and CALIFA (Sánchez et al., 2012 [4]) surveys unveiled a significant fraction of ETGs exhibiting KDCs.

Results from McDermid et al., 2006 [5], who investigated the central region of ETGs using the OASIS spectrograph, suggest two fundamental types of KDCs: the first type are KDCs that exhibit an old stellar population (>8 Gyr) contemporary to the surrounding host galaxy. This indicates that those KDCs are not a result of recent merging and were more likely to be formed through accreted material or merging at earlier times. KDCs of this type are typically extended to kpc scale while residing in non-rotating galaxies. The second type of KDCs are comprised of a more compact younger stellar population extending characteristically out to a few 100 pc. These KDCs characteristically exist in fast rotating galaxies emphasising the different formation histories of the two types.

A diagnostically conclusive method to probe the general formation pathways of KDCs is to utilise binary merger simulations of spiral galaxies. Within this framework, it is possible to follow the evolution of galaxy properties in great detail. Early theoretical studies suggest that young KDCs can arise from a binary galaxy merger on a retrograde orbit via in situ star formation (Balcells et al., 1990 [6]; Hernquist et al., 1991 [7]). In a more recent study, Tsatsi et al., 2015 [8] showed that old KDCs can also originate from a initially prograde merger through a reversal of the orbital spin induced by reactive forces due to substantial mass loss. Furthermore, Hoffman et al., 2010 [9] showed that the initial gas fraction (f_{gas}) of the progenitors has a substantial impact on the existence of a KDC in the center of the merger remnant. Analysing a sample of 56 1:1 binary spiral mergers with varying orbital parameters and f_{gas}, they show that, for $f_{gas} < 10\%$, the remnants do not host a KDC, while, for $10\% < f_{gas} < 40\%$, the fraction of remnants hosting a KDC increases. However, while the origin of young KDCs seem to be well understood by now, their stability and lifetime has not been studied in great detail.

In this work we present a case-study for the evolution of a second-type KDC formed in a major merger event, demonstrating that these structures might only be visible for a short timespan.

2. Simulation

We perform a case study of a single binary merger simulation selected from a sample of 10 high resolution simulations outlined in Schauer et al., 2014 [10]. From the 10 simulations, only two have a KDC: none of the 3:1 mergers show any sign of a KDC, while the presence of a bulge does not seem to influence the existence of a KDC. In fact, the two simulations that exhibit a KDC are identical in all configurations except for the inclusion of a bulge in the progenitor spiral galaxies. Additionally, albeit our sample is small, we also see a coupling between the initial gas fraction f_{gas} and the appearance of a KDC, similar to the results presented by Hoffman et al., 2010 [9]. From the different orbital configurations in this sample, only one leads to a KDC. This suggests that the orbital configuration of the initial merger setup is important for the formation of a KDC; however, this is not studied in this work.

Using the TreeSPH-code GADGET2 (Springel et al., 2005 [11]), all simulations implement various physical processes like star formation, supernova feedback (Springel et al., 2003 [12]) and black hole feedback (Springel et al., 2005 [13]). We use the 1:1 spiral-spiral merger that manifests the following orbital configuration: inclinations $i_1 = -109°$ and $i_2 = 180°$, pericenter arguments $\omega_1 = 60°$ and $\omega_2 = 0°$ according to [14] (for further information about the simulation, see Johansson et al., 2009 [15], Johansson et al., 2009 [16] and Remus et al., 2013 [17]). The two spiral progenitors are identical clones with an initial gas fraction of 20% hosting a stellar bulge as well as a central black hole. In order to investigate the stability and kinematics of the KDC, we trace several properties like LOSVD, stellar and gaseous angular momentum, mass and star formation rates (SFR) of the KDC in time as well as those of the hosting galaxy. The simulation is run up to 10.2 Gyr. The simulation allows to subdivide the total stellar population into the two subpopulations of initial stars already present in the progenitor galaxies, and stars formed during the simulation runtime. The gravitational softening length for stars formed from the gas during the simulation is $\epsilon = 0.1$ kpc/h, while the softening for the initial stellar component is set to $\epsilon = 0.2$ kpc/h. This choice ensures that the maximum gravitational force exerted from a particle does not depend on its mass (Johansson et al., 2009 [15]).

The selection of this particular merger for this study is based on a visual inspection of the LOSVD at a simulation time of 2.7 Gyr (see Figure 1), revealing the presence of a counter rotating kinematically distinct core in the centre. Throughout this study, LOSVD maps are Voronoi-binned using the method outlined in Cappellari & Copin, 2003 [18], ensuring a minimum of 100 particles per cell to reduce the statistical uncertainty of the mean velocity.

The binary merger proceeds as follows: subsequent to the approaching phase lasting until $t = 0.66$ Gyr, a rapid merging phase with two encounters follows. The first encounter takes place at

$t_1 = 0.66$ Gyr, while the second and final encounter happens at $t_2 = 1.3$ Gyr. Afterwards, the remnant relaxes under the influence of dynamical friction and violent relaxation.

3. Results

3.1. Global Properties

Before investigating the kinematics of the KDC in detail, we have to determine the spatial extent of the KDC. Figure 1 shows the LOSVD of the merger remnant projection onto the three coordinate planes after a simulation runtime of 2.7 Gyr that is 1.5 Gyr after the second encounter. As can clearly be seen, the merger remnant hosts a distinct central rotating feature that is present in all three projections (i.e., never seen face-on) and is clearly misaligned to the rotation of its surroundings. From the visual appearance of the LOSVD map, we conservatively estimate the radius of the KDC to be 1.5 kpc, as shown by the solid black circle in Figure 1. This corresponds roughly to a third of the stellar half-mass radius ($r_{1/2} = 5.2$ kpc) of the remnant, which is illustrated by the black dashed line, and therefore much larger than the resolution limit. The half-mass radius is determined to be the radius of a three-dimensional sphere containing half of the total stellar mass centred on the galaxy center of mass.

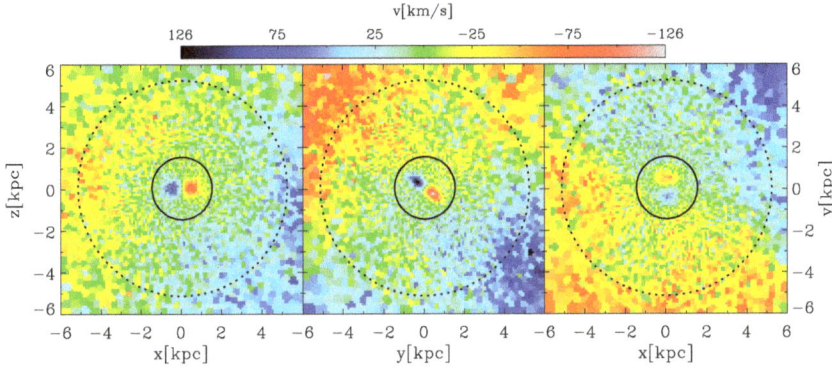

Figure 1. Each panel displays a LOSVD map of the central region of the merger remnant hosting the KDC in different projections at $t = 2.7$ Gyr. From left to right: x–z plane, y–z plane and x–y plane. The color bar on the top indicates the velocity scaling adapted individually for each panel. The black solid circle marks the estimated KDC radius of 1.5 kpc, while the dashed circle indicates the half-mass radius.

Figure 2 displays the temporal evolution of the stellar (green) and gaseous (blue) mass within the core as well as the total SFR of the galaxy in black. The starting point in time is chosen to be directly before the second encounter to also capture the starburst triggered by the second encounter. This starburst lasts for approximately 0.7 Gyr, with a maximum SFR of 28 $M_\odot$/yr.

During this period, the stellar mass inside the KDC increases until it plateaus at $\approx 3 \times 10^{10}$ $M_\odot$. Afterwards, M_{STAR} stays constant, excluding significant infall of stars. Consistently with a starburst, the amount of gas in the KDC decreases drastically by roughly one order of magnitude within this timeframe. The subsequent modest decrease of M_{GAS} suggest ongoing star formation activity within the KDC, however at much lower rates.

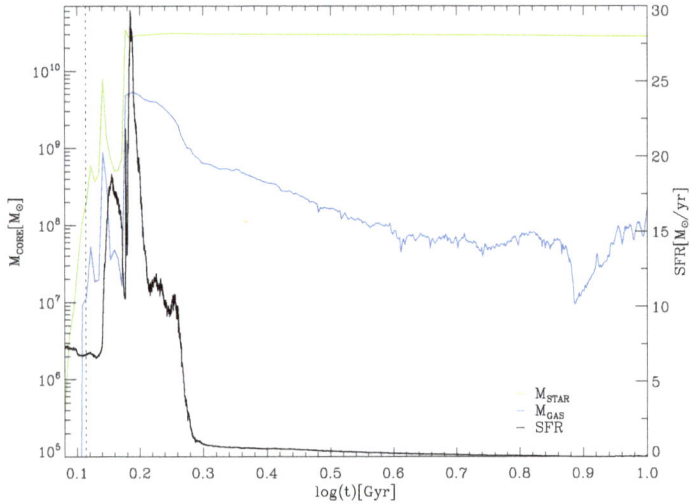

Figure 2. Temporal evolution of global core properties and the total star formation rate (SFR) of the complete galaxy. The black line indicates the SFR of the entire galaxy, while the green and blue curves represent the total stellar and gaseous mass within the KDC, respectively. The vertical dashed line marks the moment of the second encounter.

3.2. Dissecting the KDC

To investigate the origin of the KDC signal in the LOSVD maps, we explore the contributions of stellar populations with different properties to the signal. We distinguish between stars that are set up in the initial conditions and stars that are produced during the runtime of the simulation, denoting them *'Initial Stars (IS)'* and *'Newly Formed Stars (NFS)'*. The initial gas from which the NFS are formed is expected to form a disc in the centre during the merger due to its hydrodynamical nature and therefore generates a orbital configuration characteristic for discs. In contrast, the IS are only affected by violent relaxation and dynamical friction. Of course, the interplay between those processes and their efficiency is complex and depends on the parameters of the merger and hence cannot be predicted easily.

In addition, we split the stars inside the core into two groups, independently of whether they are IS or NFS: stellar particles that are permanently located within the KDC are classified as *'Permanent Core Stars (PCS)'*, and stars which are localised only temporarily within the core denoted *'Temporary Core Stars (TCS)'*. TCS are expected to move on highly elliptical orbits with apocenters beyond the KDC radius. For this purpose, we use a algorithm that iteratively removes particles that move outside of the KDC. At a runtime of 2.2 Gyr, where the KDC is fully evolved, the algorithm selects all stellar particles within the KDC radius, traces this parent sample of stars to the consecutive snapshot, and removes the particles that leave the KDC radius from the sample. This procedure is reiterated for 24 subsequent snapshots until a runtime of 2.7 Gyr. In this manner, the algorithm creates a sample of stellar particles that reside within the KDC for at least 0.5 Gyr. This corresponds to approximately one characteristic orbital period within the KDC. By construction, this algorithm also disregards stellar particles falling into the KDC during the application of the procedure from the sample.

Figure 3 directly compares the visual contribution of the PCS and TCS to the rotational signature in the LOSVD map at $t = 2.7$ Gyr.

The left panel displays a zoom-in onto the KDC including all stellar particles in the line-of-sight. Velocity maps for the TCS and PCS are shown in the central and right panel, respectively. As can clearly be seen, the rotational signal is the net rotation induced by the PCS component overlaying the TCS component, which is dominated by random motion. The fact that the maximum velocities reached by the PCS is higher than those of the full stellar sample infers that the random moving component diminishes the signal as expected.

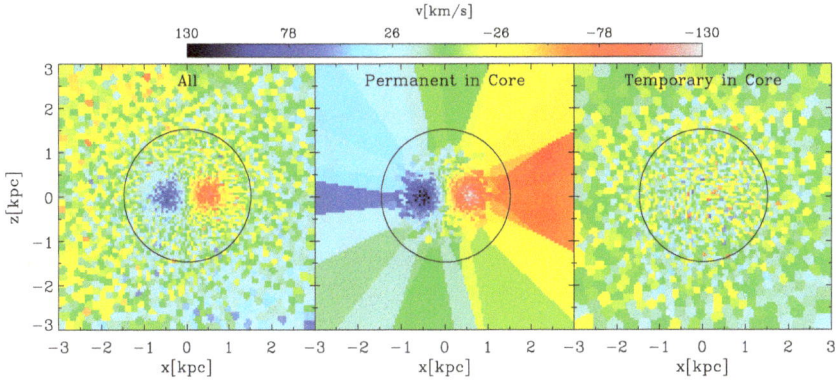

Figure 3. Left panel: LOSVD of the central 3 kpc in the *x*–*z* projection considering all stars in the line-of sight; **central panel**: LOSVD map of the central 3 kpc in the *x*–*z* projection taking into account stars that are determined to belong permanently to the KDC; **right panel**: LOSVD map of the central 3 kpc in the *x*–*z* projection including stars that only temporarily reside in the core. The panels share a common velocity scaling given in the color bar. The solid black circle indicates the core radius of $r_{CORE} = 1.5$ kpc.

As we are interested in the origin of the KDC, we further separate the PCS into the two populations of IS and NFS. The result is displayed in Figure 4: the left velocity map replicates the central panel of Figure 3, while the central and right panels show the subdivision into IS and NFS, respectively. Comparing the central and right panels demonstrates that the KDC is dominated by the newly formed stars in agreement with the results from Hoffman et al., 2010 [9]. The NFS component exhibits a peak velocity that is a roughly a factor of two higher than the peak velocity of the IS component. We reason that the minor rotation in the IS component might be generated by the drag caused by the fast rotating particles.

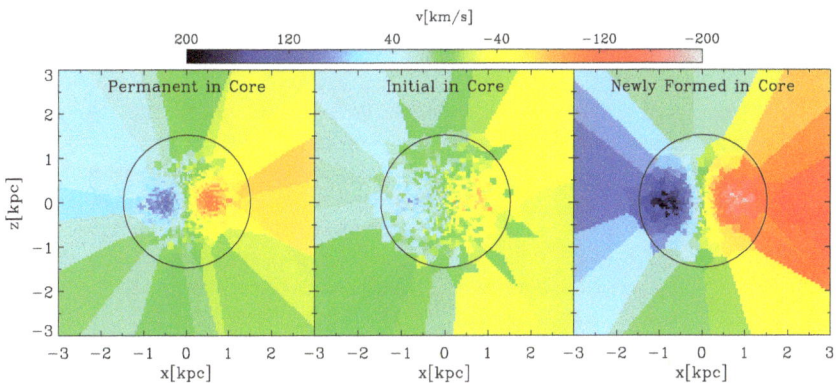

Figure 4. LOSVD maps of the central 3 kpc in the *x*–*z* projection. **Left panel**: all permanent core stars; **central panel**: permanent core stars that are initially in one of the progenitors; **right panel**: permanent core stars that are formed in situ during the simulation. The panels share a common velocity scaling given in the color bar. The solid black circle indicates the core radius of $r_{CORE} = 1.5$ kpc.

From our analysis we can conclude that the KDC signal in the merger remnant is generated mainly by stellar particles that permanently reside within the core and are formed during the merging process in situ. Therefore, the KDC seems to be in agreement with the younger more compact KDCs found observationally by McDermid et al., 2006 [5].

3.3. Evolution of Stellar KDC Kinematics

To understand the evolution of the KDC, we investigate its kinematical behaviour over time with respect to the kinematical behaviour of the surrounding host galaxy. The previous section revealed that the actual KDC is composed of stellar particles formed in situ during the merger and hence might retain a fraction of the orbital angular momentum. To measure the orientation of the KDC, we calculate the total angular momentum of all stars located inside the KDC ($J_{STAR,CORE}$) for each snapshot, and calculate its angle to the three coordinate axes. In addition, the same angles are determined for the host galaxy to exclude a global mutual motion of the KDC and the galaxy within the simulated box. The temporal evolution of the angles is shown in the main panel of Figure 5. The red, black and green curves represent the angles between the KDC and the coordinate axes, while the orange line marks the angle between the angular momentum of the host galaxy and the z-axis.

The KDC angular momentum follows a completely unexpected behaviour: subsequent to a violent phase of relaxation lasting until $t = 1.7$ Gyr, the angles with respect to the coordinate axes reveal an oscillation of the KDC, which gets gradually dampened in the further evolution. The imprint of the oscillation is visible up to $t \approx 7$ Gyr, and the period of the oscillation is nearly equal for each axis and is determined to be $\delta t \approx 0.5$ Gyr. As a two-dimensional visualisation, the upper and lower rows of Figure 5 show the velocity field in the x–z plane projection at six points in times equally distributed over one oscillation period: from an initial alignment with the x-axis at 2.7 Gyr, the KDC turns upwards by 90° at $t \approx 2.93$ Gyr. In the following snapshot, it gets more diffuse and almost disappears until it reverts to its initial configuration at $t = 3.27$ Gyr. The last panel shows the initiation of a new oscillation cycle indicated by the slight upturn.

Of course, the projected appearance in the velocity maps is strongly influenced by the rotation of the KDC around the other two coordinate axes. Furthermore, it is difficult to deduce the actual movement in three dimensions from the angles and the velocity maps. Therefore, Figure 6 illustrates the direction of the angular momentum vector of the KDC in three-dimensional space at three points in time. The curves drawn on the coordinate planes visualise the projected track of the vector's until its current position. In all three projections, the tracks describe slightly distorted circles that get smaller due to the damping of the oscillation until the angular momentum stabilises in direction.

From our analysis, we conclude that the total stellar angular momentum of the KDC performs a three-dimensional motion comparable to the precession of a gyroscope in a gravitational potential superimposed with a general tilt of the precession axis. From this, we infer that the kinematics of the KDC can be subdivided into an intrinsic rotation and a superimposed global figure rotation.

This result raises the questions of which mechanism generates this periodic, well-defined motion of the KDC and how it is damped. In the previous section we showed that the rotating component of the KDC comprises stars formed during the merger and that the intrinsic rotation is the result of the process, whereby gas condenses into a disc. We speculate that the precession is a residue of the orbital angular momentum of the merger event retained by the gaseous component.

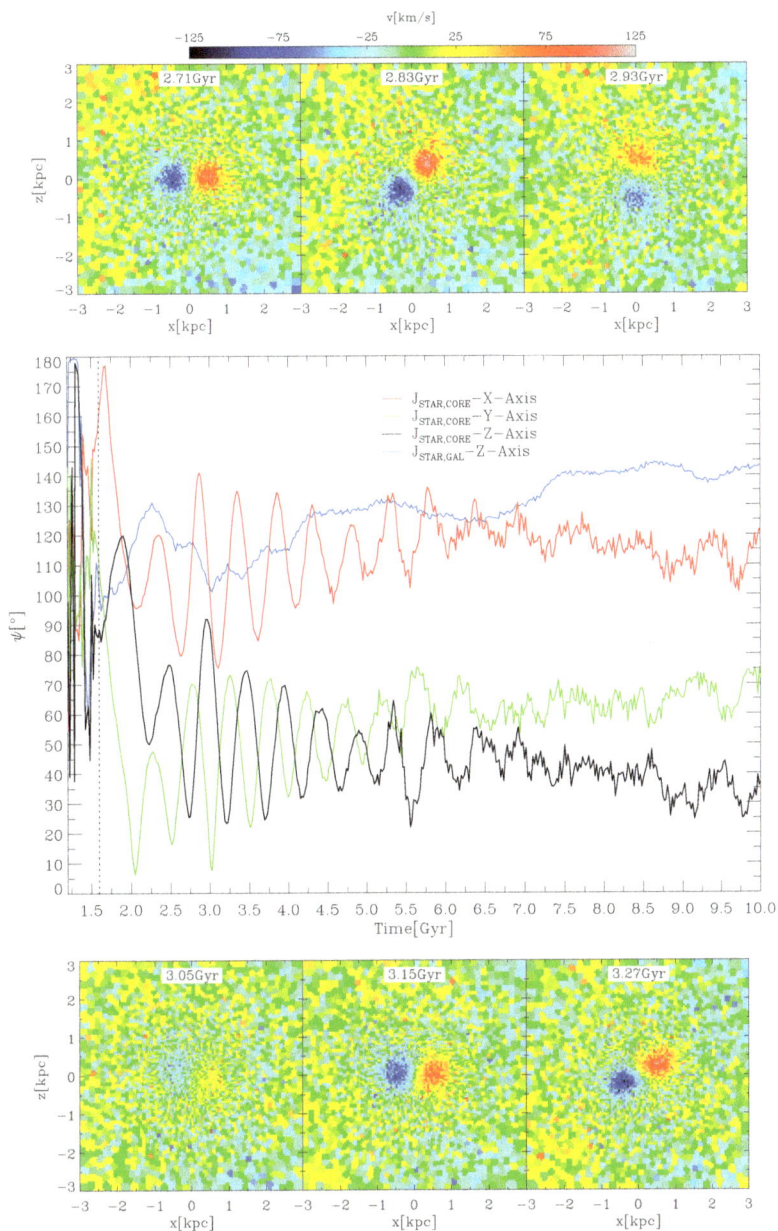

Figure 5. The main panel shows the temporal evolution of the angle between the total stellar angular momentum of the KDC and the three coordinate axes as given in the legend. The blue line for comparison shows the evolution of the angle between the angular momentum of the surrounding galaxy and the z-axis. The dashed vertical line marks the time at which the stellar mass accretion of the KDC is finished (see Figure 5). The upper and lower rows display a temporal sequence of the LOSVD map of the central 3 kpc in the x–z projection.

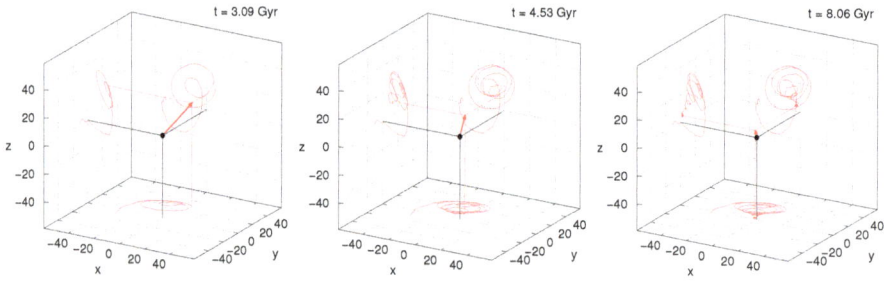

Figure 6. Illustration of the direction of the total stellar angular momentum in three-dimensional space as a red vector at three different time steps, of the simulation as given in the label. The red curves drawn on the coordinate planes trace the projected paths of the vector's endpoint until its current position.

3.4. Fading of the KDC

After revealing the origin and kinematics of the KDC in the previous sections, we now examine the temporal stability of the KDC with regard to its kinematical properties. The timescale on which the young compact KDCs are stable is still widely unknown since only numerical simulations with sufficient resolution can trace their evolution. In contrast, classical old KDCs are expected to be stable for up to 10 Gyr since their stellar population is indistinguishable from their host galaxy.

One of the few investigations of KDC stability was conducted by McDermid et al., 2006 [19] using a simulation based on SSP model spectra. They find that the KDC is fading due to the stellar evolution of the young KDC stars, as their luminosity-weighted contribution is diminished due to a increasing mass-to-light ratio. Following this argument, the KDC is still present, however forced to the background of the velocity field.

As shown in the previous section, our analysis provides a reasonable dissection of the core into its rotating and dispersion dominated component by differentiating the IS and NFS components confirmed by their visual appearance in the LOSVD maps.

A sequence of LOSVD maps in various decisive steps of the KDC evolution considering all stars in the line-of-sight from $t = 3.6$ Gyr to $t = 5.2$ Gyr are shown in the upper and lower rows of Figure 7. The projection plane is held constant in all panels. After a runtime of 3.63 Gyr, the KDC is still clearly visible performing a precession in the two following snapshots. Between $t = 4.27$ Gyr and $t = 5.33$ Gyr, the KDC signal gradually gets weaker until it vanishes almost completely. By testing different projections, we confirmed that this fading is real and not just due to a precession of the core, as in the previous time steps where a weakening of the KDC signal could be detected in one projection, while it strengthened in another projection. Up to the full time of the simulation, the KDC does not build up again in any projection, clearly indicating that the KDC was dissolved. Hence, we conclude that the KDC is fading on a timescale of 1.6 Gyr.

In order to quantify the rotational support of the stellar populations of the KDC, we calculate the λ_R parameter for the core, which is given by

$$\lambda_R = \frac{\langle R\,|V|\rangle}{\langle R\,\sqrt{V^2 + \sigma^2}\rangle} = \frac{\sum_{i=1}^{N_p} F_i\,R_i\,|\overline{V}_i|}{\sum_{i=1}^{N_p} F_i\,R_i\,\sqrt{\overline{V}_i^2 + \sigma_i^2}}, \tag{1}$$

where the summation runs over each pixel within the chosen aperture. F_i, R_i, $|\overline{V}_i|$ and σ_i are the flux, projected distance to the galaxy centre, mean stellar velocity and velocity dispersion of the ith pixel, respectively (Emsellem et al., 2007 [20]). For simulations, the fluxes are replaced by stellar masses as we do not have luminosities, while assuming a constant mass-to-light ratio within each galaxy. λ_R is a measure of the ratio of random to ordered motion. Furthermore, λ_R is an observationally accessible quantity, albeit sensitive to projection effects.

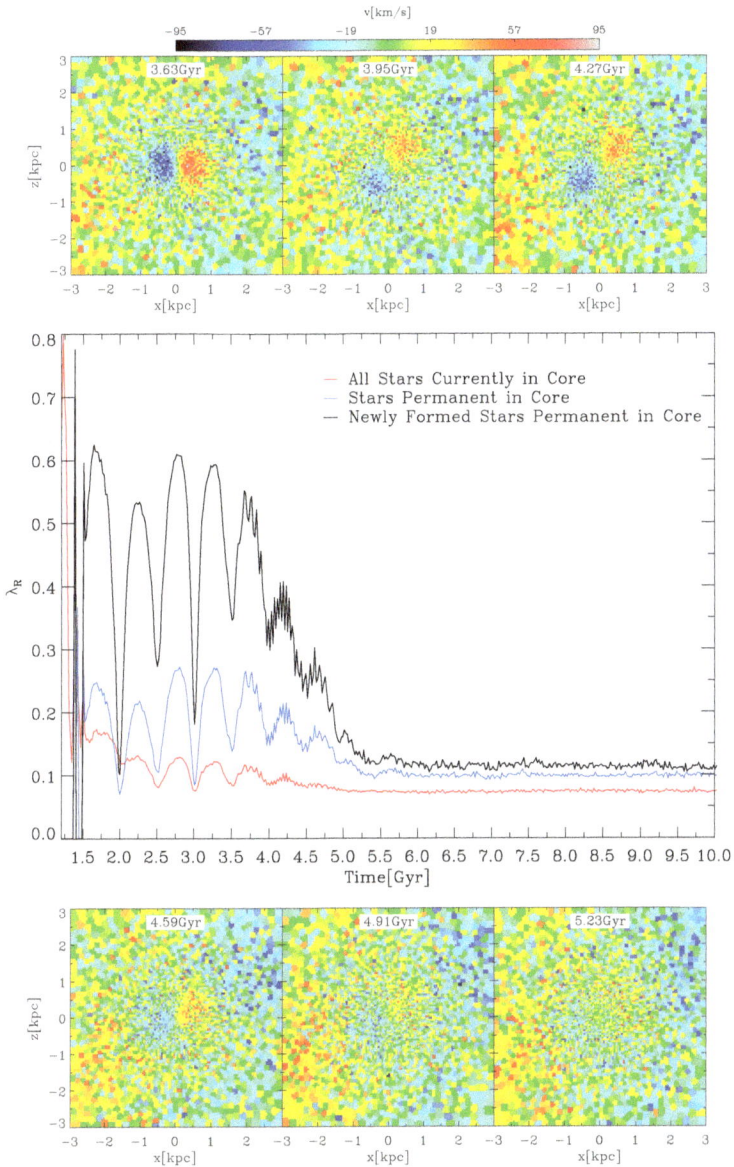

Figure 7. The upper and lower rows display a temporal sequence of the LOSVD maps of the central 3 kpc in the x–z projection visualising the gradual fade of the KDC. The main panel shows the temporal evolution of λ_R in the x–z projection split up into three stellar populations: All stars that reside inside the core at a given time are represented by the red curve, while the blue and black curve trace the evolution for the permanent core stars and the newly formed permanent core stars, respectively.

The temporal evolution of λ_R in the core region, measured in the x–z plane and considering different populations of stars, is displayed in the main panel of Figure 7. The red curve shows λ_R for all stellar particles within the KDC at each time step, while blue and black curves represent the PCS and newly formed PCS population, respectively. A common feature of all three populations is an imprinted oscillation, reflecting the projection effects induced by the precession of the KDC revealed in the previous section.

Comparing the mean level of λ_R for the three populations, the red and blue curves exhibit significantly less rotational support than the black curve. For the total population, this is due to the included, highly dispersive TCS component. In the case of the PCS, we showed that the IS subcomponent gets partly dragged into rotation, however, at a factor of two slower than the NFS subcomponent. Thus, the dispersion within the PCS is still rather large and explains the lower λ_R values of the blue curve.

The newly formed PCS component shows λ_R-values as high as $\lambda_R \approx 0.6$, which is a clearly rotation-dominated signal. However, due to the precession of the core, it can reach values as low as $\lambda_R \approx 0.1$ when the core precesses out of the x–z-plane, extinguishing any rotational signal. Between $t = 3.75$ Gyr and $t = 5.25$ Gyr, we find a gradual decrease to $\lambda_R \approx 0.13$, which is clearly in the slow rotating regime populated by non-rotating early-type galaxies, on a timescale of 1.5 Gyr. This is in agreement with the fading timescale inferred from the inspection of the LOSVD maps.

To further constrain the fading of the KDC, we take full advantage of the three-dimensional information from the simulation. A useful method to determine the degree of ordered rotation within a system of particles is to calculate the distribution of the angles between the angular momentum of each particle and the total angular momentum vector of the complete system. For a rotating system, the distribution of these angles is expected to feature a peak at small angles, while a completely dispersion-dominated system is expected to show a random distribution. The distribution of these angles at two points in time for the PCS is displayed in Figure 8 separated for the IS (blue) and NFS (red) populations. The highlighted times are chosen such that, at the earlier point ($t = 2.76$ Gyr), the KDC is fully developed, while, at the later time ($t = 10$ Gyr), the KDC is completely dispersed.

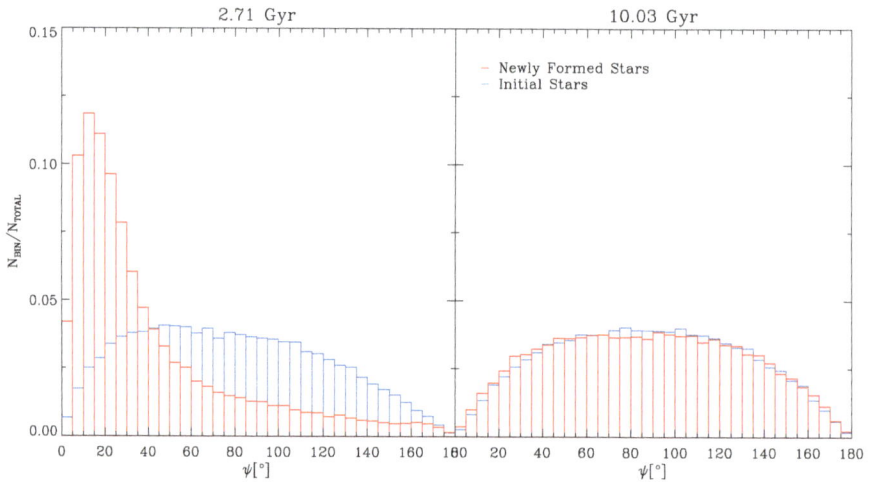

Figure 8. Distribution of the misalignment angles between the total stellar angular momentum vector of the KDC after and the angular momentum vectors of the individual stars inside the KDC. The newly formed stars are represented by the red histograms, while the blue histograms illustrate the initial stars. **left:** $t = 2.71$ Gyr; **right:** $t = 10$ Gyr.

The angle distributions shown in Figure 8 conclusively confirm the kinematical difference between the IS and NFS population already found in Section 3.2. Already at $t = 2.7$ Gyr, the IS component shows a nearly randomised distribution of angles, indicating a dispersion dominated system. In contrast, the distribution for the NFS population completely changes its shape over the period of 7.3 Gyr. At $t = 2.7$ Gyr, the distribution features a clear peak at low angles, indicating ordered rotation in the system. At $t = 10$ Gyr, the alignment disappears entirely, evolving into a purely random distribution. Therefore, we conclude that the population of NFS generating the KDC signal at earlier times disperses in its further evolution. This is most likely caused by the interaction between the PCS and the dispersion dominated TCS component. We conclude that, depending on the mass ratio between the two populations, gravitational mixing drives the combined system towards a homogeneous distribution in phase space, dissolving the KDC.

4. Conclusions

Results from highly resolved IFU observations suggest a dichotomy of KDCs found in the centres of galaxies. We present an extensive case study of a KDC formed in a high resolved 1:1 binary disc galaxy merger simulation. We identify the KDC to be consistent with the class of young and compact KDCs, where the rotational signal is mainly generated by stars that are formed in situ during the merger event. Furthermore, the dissection of the KDC reveals a significant dispersion dominated population of stars following highly elliptical orbits permeating the KDC.

We trace the kinematical properties of the KDC over cosmic time, revealing a global gyroscopic precession motion of the KDC. From this result, we infer a superimposed motion of the KDC consisting of an intrinsic rotation and a global precession of the complete system. We suspect the precession to be induced by the orbital angular momentum of the merger retained in the gaseous component.

Furthermore, we demonstrate that the KDC is stable for about 3 Gyr after the merger event. We show that the amount of ordered motion within the KDC subsequently drops significantly on a timescale of 1.5 Gyr, leading to a dispersion of the KDC. This dispersion happens on a similar timescale on which the global precession of the KDC fades. We suspect that the effect of gravitational mixing between the rotational component and the intruding, permeating dispersion-dominated population causes the KDC to disperse. Therefore, we conclude that the visibility of a young KDC in the cores of early-type galaxies indicates a (recent) dominant merger event; however, a more statistical approach is required for a generalised statement regarding this matter.

Acknowledgments: Felix Schulze gratefully thanks the complete Cast Group at the University Observatory Munich for helpful discussions. The authors thank Peter Johansson for running the simulation and providing the output data. Felix Schulze also acknowledges the Max Planck Institute for Extraterrestrial Physics for providing the funding to attend the conference *On the Origin and Evolution of Baryonic Galaxy Halos*.

Author Contributions: F.S. and R.-S.R. analysed the data and wrote the paper. K.D. helped to physically interpret the results and provided useful input during the project.

Conflicts of Interest: The authors declare no conflict of interest.

References

1. Emsellem, E.; Cappellari, M.; Peletier, R.F.; McDermid, R.M.; Bacon, R.; Bureau, M.; Copin, Y.; Davies, R.L.; Krajnović, D.; Kuntschner, H.; et al. The SAURON project-III. Integral-field absorption-line kinematics of 48 elliptical and lenticular galaxies. *Mon. Not. Roy. Astron. Soc.* **2004**, *352*, 721–743.
2. Krajnović, D.; Emsellem, E.; Cappellari, M.; Alatalo, K.; Blitz, L.; Bois, M.; Bournaud, F.; Bureau, M.; Davies, R.L.; Davis, T.A.; et al. The ATLAS^{3D} project-II. Morphologies, kinemetric features and alignment between photometric and kinematic axes of early-type galaxies. *Mon. Not. Roy. Astron. Soc.* **2011**, *414*, 2923–2949.
3. Cappellari, M.; Emsellem, E.; Krajnović, D.; McDermid, R.M.; Scott, N.; Verdoes Kleijn, G.A.; Young, L.M.; Alatalo, K.; Bacon, R.; Blitz, L.; et al. The ATLAS^{3D} project-I. A volume-limited sample of 260 nearby early-type galaxies: Science goals and selection criteria. *Mon. Not. Roy. Astron. Soc.* **2011**, *413*, 813–836.
4. Sánchez, S.F.; Kennicutt, R.C.; Gil de Paz, A.; van de Ven, G.; Vílchez, J.M.; Wisotzki, L.; Walcher, C.J.; Mast, D.; Aguerri, J.A.L.; Albiol-Pérez, S.; et al. CALIFA, the Calar Alto Legacy Integral Field Area survey. I. Survey presentation. *Astron. Astrophys.* **2012**, *538*, A8.

5. McDermid, R.M.; Bacon, R.; Kuntschner, H.; Emsellem, E.; Shapiro, K.L.; Bureau, M.; Cappellari, M.; Davies, R.L.; Falcón-Barroso, J.; Krajnović, D.; et al. Stellar kinematics and populations of early-type galaxies with the SAURON and OASIS integral-field spectrographs. *New Astron. Rev.* **2006**, *49*, 521–535.

6. Balcells, M.; Quinn, P.J. The formation of counterrotating cores in elliptical galaxies. *Astrophys. J.* **1990**, *361*, 381–393.

7. Hernquist, L.; Barnes, J.E. Origin of kinematic subsystems in elliptical galaxies. *Nature* **1991**, *354*, 210–212.

8. Tsatsi, A.; Macciò, A.V.; van de Ven, G.; Moster, B.P. A New Channel for the Formation of Kinematically Decoupled Cores in Early-type Galaxies. *Astrophys. J. Lett.* **2015**, *802*, L3.

9. Hoffman, L.; Cox, T.J.; Dutta, S.; Hernquist, L. Orbital Structure of Merger Remnants. I. Effect of Gas Fraction in Pure Disk Mergers. *Astrophys. J.* **2010**, *723*, 818–844.

10. Schauer, A.T.P.; Remus, R.S.; Burkert, A.; Johansson, P.H. The Mystery of the σ-Bump—A New Signature for Major Mergers in Early-type Galaxies? *Astrophys. J. Lett.* **2014**, *783*, L32.

11. Springel, V. The cosmological simulation code GADGET-2. *Mon. Not. Roy. Astron. Soc.* **2005**, *364*, 1105–1134.

12. Springel, V.; Hernquist, L. Cosmological smoothed particle hydrodynamics simulations: A hybrid multiphase model for star formation. *Mon. Not. Roy. Astron. Soc.* **2003**, *339*, 289–311.

13. Springel, V.; Di Matteo, T.; Hernquist, L. Modelling feedback from stars and black holes in galaxy mergers. *Mon. Not. Roy. Astron. Soc.* **2005**, *361*, 776–794.

14. Toomre, A.; Toomre, J. Galactic Bridges and Tails. *Astrophys. J.* **1972**, *178*, 623–666.

15. Johansson, P.H.; Naab, T.; Burkert, A. Equal- and Unequal-Mass Mergers of Disk and Elliptical Galaxies with Black Holes. *Astrophys. J.* **2009**, *690*, 802–821.

16. Johansson, P.H.; Burkert, A.; Naab, T. The Evolution of Black Hole Scaling Relations in Galaxy Mergers. *Astrophys. J. Lett.* **2009**, *707*, L184–L189.

17. Remus, R.S.; Burkert, A.; Dolag, K.; Johansson, P.H.; Naab, T.; Oser, L.; Thomas, J. The Dark Halo—Spheroid Conspiracy and the Origin of Elliptical Galaxies. *Astrophys. J.* **2013**, *766*, 71.

18. Cappellari, M.; Copin, Y. Adaptive spatial binning of integral-field spectroscopic data using Voronoi tessellations. *Mon. Not. Roy. Astron. Soc.* **2003**, *342*, 345–354.

19. McDermid, R.M.; Emsellem, E.; Shapiro, K.L.; Bacon, R.; Bureau, M.; Cappellari, M.; Davies, R.L.; de Zeeuw, T.; Falcón-Barroso, J.; Krajnović, D.; et al. The SAURON project-VIII. OASIS/CFHT integral-field spectroscopy of elliptical and lenticular galaxy centres. *Mon. Not. Roy. Astron. Soc.* **2006**, *373*, 906–958.

20. Emsellem, E.; Cappellari, M.; Krajnović, D.; van de Ven, G.; Bacon, R.; Bureau, M.; Davies, R.L.; de Zeeuw, P.T.; Falcón-Barroso, J.; Kuntschner, H.; et al. The SAURON project-IX. A kinematic classification for early-type galaxies. *Mon. Not. Roy. Astron. Soc.* **2007**, *379*, 401–417.

MDPI

Letter

One Piece at a Time—Adding to the Puzzle of S0 Formation

Carlos Escudero [1,2,*,†], **Favio Faifer** [1,2,†] and **Lilia Bassino** [1,2,†]

[1] Facultad de Ciencias Astronómicas y Geofísicas, Universidad Nacional de La Plata, Paseo del Bosque s/n, La Plata B1900FWA, Argentina; favio@fcaglp.unlp.edu.ar (F.F.); lbassino@fcaglp.unlp.edu.ar (L.B.)
[2] Instituto de Astrofísica de La Plata (IALP; CCT La Plata, CONICET-UNLP), Paseo del Bosque s/n, La Plata B1900FWA, Argentina
*   Correspondence: cgescudero@fcaglp.unlp.edu.ar; Tel.: +54-221-483-7324
†   These authors contributed equally to this work.

Academic Editors: Duncan A. Forbes and Ericson D. Lopeze
Received: 29 June 2017; Accepted: 3 August 2017; Published: 7 August 2017

Abstract: Understanding the origin of galaxies remains a topic of debate in the current astronomy. In this work, we have focused on lenticular (S0) galaxies located in low-density environments, using their associated globular cluster (GC) systems as a tool. Initially, we have started the study of three S0 galaxies—NGC 2549, NGC 3414 and NGC 5838—using photometric data in several filters obtained with the GMOS camera mounted on the Gemini North telescope. The different GC systems, as well as their host galaxies, have shown particular features, such as multiple GC subpopulations and low-brightness substructures. These pieces of evidence show that the mentioned galaxies have suffered several merger/interaction events, even the accretion of satellite companions, probably causing their current morphologies.

Keywords: globular cluster; lenticular galaxies; galaxy halos

1. Introduction

One of the biggest challenges that persists in astronomy today is understanding how the galaxies we observe were formed. In this context, a fundamental aspect lies in identifying those influential factors in the formation and evolution of galaxies of a given morphological type. In this regard, the importance of globular clusters (GCs) has been recognized as tracers of the first formation stages of the galaxies, and also as a useful tool for obtaining information about different epochs, regions, and physical processes that would otherwise be inaccessible [1].

To a greater or lesser extent, GC systems reveal a bimodal colour distribution, indicating the presence of at least two subpopulations of GCs, usually referred to as "blue" and "red". This colour bimodality has usually been interpreted as a metallicity bimodality, suggesting that the galaxy has experienced two periods of intense star formation. However, the presence of trimodal colour distributions has been observed in some massive galaxies [2–4], which would indicate that new GCs have been formed through merger events, and/or have been stripped off of smaller galaxies.

As each history has a beginning, the beginning of the work presented here focused on the analysis of the GC systems associated with lenticular galaxies (S0) located in relatively low-density environments, such as groups and/or the field (Escudero et al., in preparation). To this end, we have used excellent photometric data obtained with Gemini/GMOS through the filters $g'r'i'$. This data set in itself constitutes an important contribution to the study of this type of galaxy, since it will serve as the starting point for future spectroscopic work, and will allow the different processes that govern the galaxies and their GC systems to be delineated.

2. Sample of Galaxies

In this work, we have focused on three S0 galaxies: NGC 2549, NGC 3414 and NGC 5838, essentially classified as S0, and are located in poor (NGC 2549) as well as rich groups (NGC 3414 and NGC 5838). Table 1 shows some properties thereof.

Following is a brief description of the galaxies presented here, with relevant information obtained from previous studies in the literature.

NGC 2549: This edge-on S0 galaxy is the central galaxy of a low-density group or "poor group". It presents an X-shape distribution of light in the bulge [5], and evidences a ring-like structure towards larger radii [6,7]. The spectroscopic study by Sil'chenko et al. [8] shows that this galaxy presents a young bulge with an age ∼ 2 Gyr, whereas the disk reveals an increase of the age from 6.4 to 12 Gyr.

NGC 3414: Defined as a peculiar S0, it is the central galaxy of the LGG 227 group [9]. The appearance of this galaxy has given diverse interpretations about its nature. Whitmore et al. [10] suggested that it was an edge-on galaxy with a large polar ring, whereas Chitre and Jog [11] considered it as face-on galaxy with a prominent bar. Age estimates of the bulge and disk are in ∼ 12 Gyr [8].

NGC 5838: This SA0 galaxy is located in the group of galaxies NGC 5846. However, NGC 5838 is the dominant galaxy of a smaller subgroup. It presents kinematically decoupled nuclear regions, with two rings or dust disks and young stars [12]. Its bulge contains old stellar populations with a slight colour gradient [13]. On the other hand, Michard and Marchal [14] suggest that the outer disk could show weak signs of a spiral pattern. McDermid et al. [15] estimated ages within the effective radius, obtaining 11.27 Gyr.

Table 1. Galaxy sample. Morphological type taken from the NASA/IPAC Extragalactic Database (NED), right ascension and declination (NED), V magnitudes from the RC3 catalogue, distance modulus of Tully et al. [16], Tonry et al. [17] and Theureau et al. [18] for NGC 2549, NGC 3414, and NGC 5838, respectively. Last column indicates the mean surface density of galaxies inside a cylinder of height $h = 600$ km s^{-1} centered on the galaxy which contains the 10 nearest neighbors [19].

Galaxy	Type	α_{J2000} (h:m:s)	δ_{J2000} (d:m:s)	M_V (mag)	$(m - M)_0$ (mag)	$\log \Sigma_{10}$ (Mpc^{-2})
NGC 2549	SA0^0(r)	08:18:58.3	+57:48:10.9	−19.44	30.51	−0.71
NGC 3414	S0 pec	10:51:16.2	+27:58:30.3	−21.11	32.01	−0.16
NGC 5838	SA0$^-$	15:05:26.2	+02:05:57.6	−21.24	32.01	−0.39

3. Results

We present the analysis developed on each galaxy and the preliminary results obtained. Initially, we obtained the surface brightness profile of the galaxies using the IRAF task ELLIPSE. In order to avoid the light contribution from nearby bright or extended objects, before executing ELLIPSE we masked them in the image. In each case we modelled the galaxy light allowing the centre, ellipticity and position angle of the isophotes to vary freely. Subsequently, the smooth ellipse models were subtracted from the original images.

NGC 2549: when modeling and subtracting the diffuse galaxy halo (panel (a) in Figure 1), different stellar structures are evident. In particular, some excesses of light are clearly observed at the position of the two possible rings (orange letters A and B). On the other hand, when analyzing the colour distribution of the GC system (panel (b) in Figure 1), it does not show signs of a clear bimodality, although the system is obviously not unimodal. The colour histogram is dominated by the presence of the blue GC subpopulation (modal value $(g' - i')_0 \sim 0.84$ mag), and a less conspicuous red GC subpopulation $((g' - i')_0 \sim 1.07$ mag). In addition, these red GCs have a "knee" extending to the red end in $(g' - i')_0$. When observing the spatial distribution of the latter (red circles in panel (a)), they have a slight concentration towards the galaxy (∼ 16 candidates within 4.2 kpc of galactocentric radius). These features suggest the accretion and/or merger of some lower-mass neighbours.

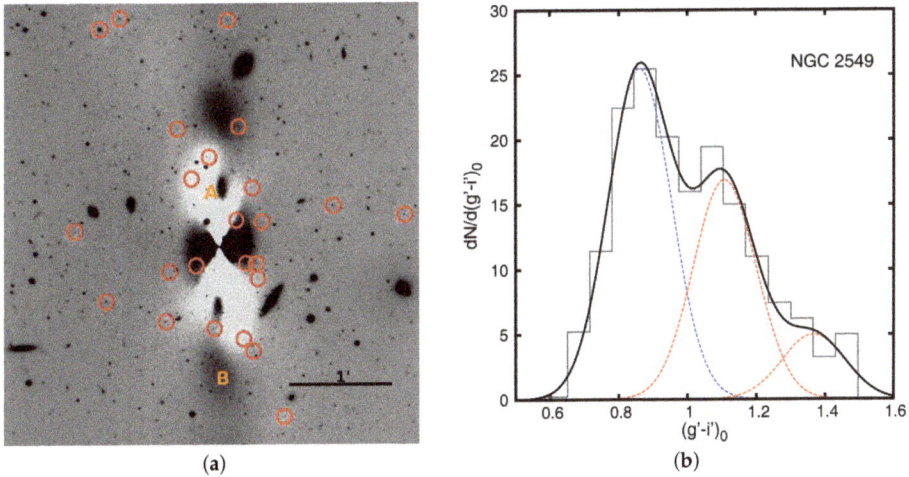

Figure 1. (a) Image after the galaxy light is removed. The letters A and B indicate the excesses of light possibly associated with two rings. Red circles show the position of the globular cluster (GC) candidates with reddest colours. (b) Colour histogram $(g' - i')_0$ for the GC candidates of NGC 2549. The dashed lines show the Gaussian fit performed for each GC subpopulation.

NGC 3414: In this case, by subtracting the model on the original image (panel (a) in Figure 2), two clear structures arise. On the one hand, the probable bar or disk perpendicular to the dust spiral, and shell structures on opposite sides, located at a distance of $R_{gal} \sim 1.7$ arcmin (orange letters A and B, respectively). Shell structures are usually interpreted as evidence that the galaxy has experienced a dry minor merger event recently. Furthermore, the asymmetric dust structure would indicate the accretion or merger of some gas-rich minor galaxy. In this case, the GC colour distribution presents a broad and flat shape (panel (b) in Figure 2). To study the different possible GC subpopulations in this galaxy, we separated our sample into several radial bins. Panel (c) shows that there are at least three peaks in the innermost region, which are found in the modal values $(g' - i')_0 \sim 0.75$, $(g' - i')_0 \sim 0.96$, and $(g' - i')_0 \sim 1.09$ mag, in addition to a possible group of GC significantly redder in $(g' - i')_0 \sim 1.24$ mag. The first and third peaks correspond to typical values for the aforementioned blue and red GC, while the intermediate and the reddest peaks may be associated with different subpopulations present in this galaxy. Towards larger galactocentric radii (panel (d)), the number of blue candidates begins to increase, clearly observing two peaks (blue and red GCs). The spatial distribution of the possible intermediate subpopulation and those GCs with reddest colours are shown in panel (a) (green crosses and red circles, respectively). The latter ones show a clear concentration towards the center of NGC 3414 which confirms the nature of GCs associated with the galaxy. The mean integrated colour shown by this group may have originated from highly enriched material during an old fusion with neighboring galaxies, although the reddening effect caused by the dust cannot be ruled out.

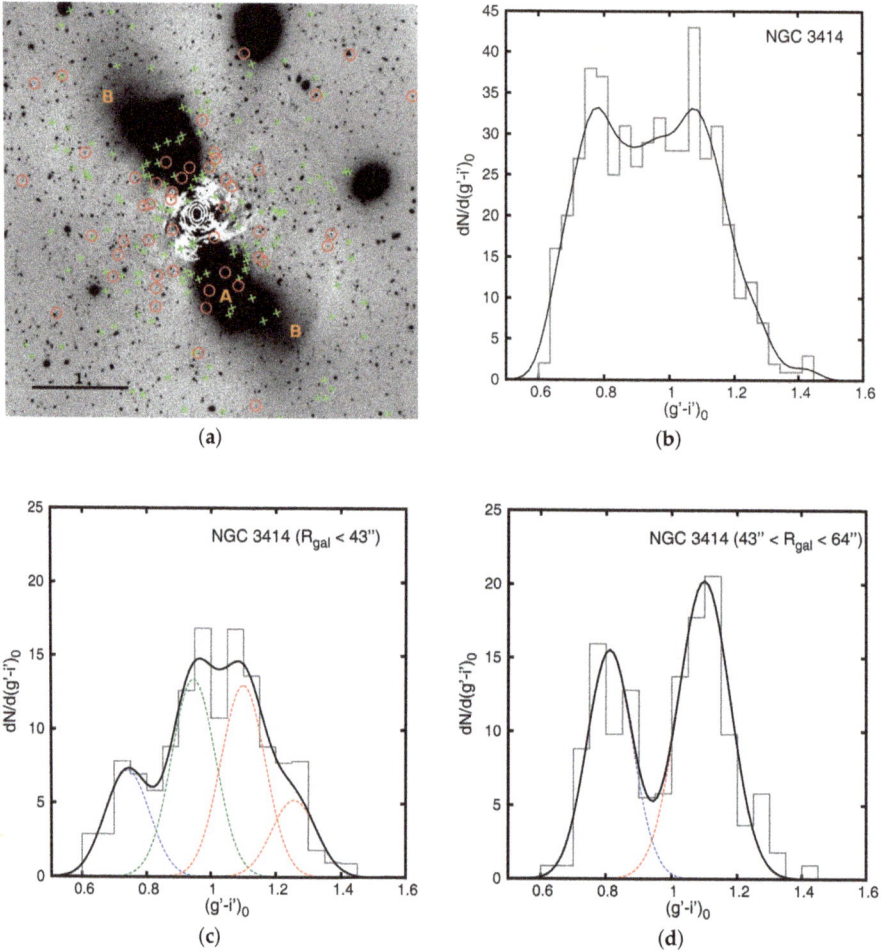

Figure 2. (**a**) Image after the galaxy light is removed. The orange letters A and B indicate the bar/disk and the shells structures, respectively. Red circles and green crosses show the position of the GC candidates with the reddest colour $(g' - i')_0$ and those with intermediate colour, respectively. (**b**) Colour histogram $(g' - i')_0$ for the GC candidates of NGC 3414. Solid black line represents the smoothed colour distribution. (**c,d**) Colour histograms with the GC candidates according to the galactocentric distance < 43 and $43 - 64$ arcsec, respectively. The dashed lines show the Gaussian fit performed for each GC subpopulation.

NGC 5838: to highlight the various complex structures present in this galaxy, we have used the unsharp masking technique. In panel (a) of Figure 3, a ring of dust (orange letter A) is observed towards the central part, whereas towards the outer zones it is possible to observe stellar rings and/or thin remains of a possible spiral structure, or the debris of a low-mass object destroyed in a fusion (B, C, D). The GC system of NGC 5838 (panel (b) in Figure 3) has a clear trimodal integrated colour distribution with peaks in the modal values $(g' - i')_0 \sim 0.79$, $(g' - i')_0 \sim 1.06$, and $(g' - i')_0 \sim 1.28$ mag. The first two are in good agreement with typical colors for the blue and red GC subpopulations, with a clear "valley" between them in $(g' - i')_0 \sim 0.90$ mag. On the other hand, when analyzing the spatial distribution of the reddest candidates (red circles in panel (a)), it is observed that they present an elongated distribution along the semi-major axis of the galaxy. As mentioned in the case of NGC 3414, the presence of GCs with these particularly red colours may be due to the effects of internal reddening coming from the galaxy, and/or the existence of a very rich metal subpopulation formed in a third GC formation.

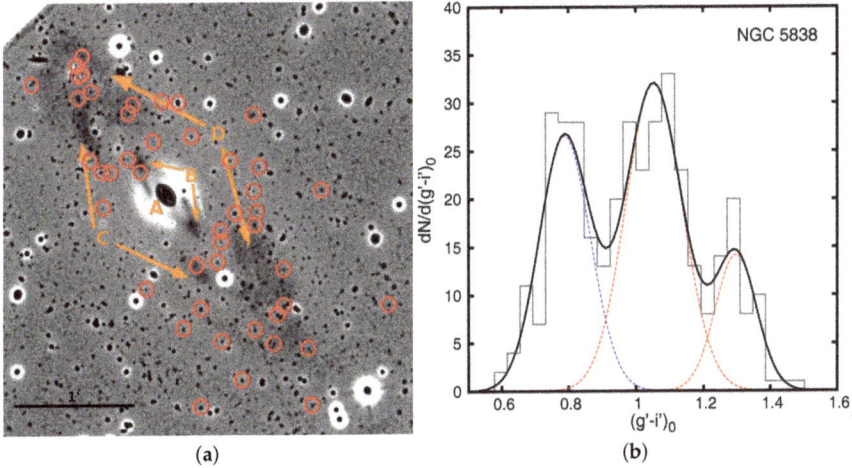

(a)   (b)

Figure 3. (**a**) Image after the galaxy light is removed. The orange letter A indicates the inner ring of dust. Letters B, C, and D show the outer stellar rings and/or the thin remains of a possible spiral structure. (**b**) Colour histogram $(g' - i')_0$ for the GC candidates of NGC 5838. The dashed lines show the Gaussian fit performed for each GC subpopulation.

4. Conclusions

We have obtained photometric data from three S0 galaxies located in relatively low density environments in order to initially characterize their GC systems and look for evidence about their formation. As shown in this work, the various and particular properties exhibited by some of these systems, such as the existence of multiple GC subpopulations and low surface-brightness substructures, indicate that most of the studied galaxies have probably undergone several merger/interaction-accretion events. The presence of these features may be related to the formation history of the GCs [20–22], mainly of those that present different colours to the typical blue and red subpopulations. These pieces of evidence leads directly to a question: is it possible that these events are responsible for their current morphologies? The answer is still open.

Acknowledgments: This research was funded with grants from Consejo Nacional de Investigaciones Científicas y Técnicas de la República Argentina, Agencia Nacional de Promoción Científica y Tecnológica, and Universidad Nacional de La Plata, Argentina.

Conflicts of Interest: The authors declare no conflict of interest.

References

1. Brodie, J.P.; Strader, J. Extragalactic Globular Clusters and Galaxy Formation. *Annu. Rev. Astron. Astrophys.* **2006**, *44*, 193–267.
2. Blom, C.; Spitler, L.R.; Forbes, D.A. Wide-field imaging of NGC 4365's globular cluster system: the third subpopulation revisited. *Mon. Not. Roy. Astron. Soc.* **2012**, *420*, 37–60.
3. Caso, J.P.; Richtler, T.; Bassino, L.P.; Salinas, R.; Lane, R.R.; Romanowsky, A. The paucity of globular clusters around the field elliptical NGC 7507. *Astron. Astrophys.* **2013**, *555*, A56.

4. Sesto, L.A.; Faifer, F.R.; Forte, J.C. The complex star cluster system of NGC 1316 (Fornax A). *Mon. Not. Roy. Astron. Soc.* **2016**, *461*, 4260–4275.

5. Laurikainen, E.; Salo, H. Observed properties of boxy/peanut/barlens bulges. In *Galactic Bulges*; Springer: Basel, Switzerland, 2016.

6. Seifert, W.; Scorza, C. Disk structure and kinematics of S0 galaxies. *Astron. Astrophys.* **1996**, *310*, 75–92.

7. Michard, R.; Poulain, P. Colour distributions in E-S0 galaxies. - V. Colour data for strongly inclined lenticulars. *Astron. Astrophys. Suppl.* **2000**, *141*, 1–22.

8. Sil'chenko, O.K.; Proshina, I.S.; Shulga, A.P.; Koposov, S.E. Ages and abundances in large-scale stellar discs of nearby S0 galaxies. *Mon. Not. Roy. Astron. Soc.* **2012**, *427*, 790–805.

9. Garcia, A.M. General study of group membership. II - Determination of nearby groups. *Astron. Astrophys. Suppl.* **1993**, *100*, 47–90.

10. Whitmore, B.C.; Lucas, R.A.; McElroy, D.B.; Steiman-Cameron, T.Y.; Sackett, P.D.; Olling, R.P. New observations and a photographic atlas of polar-ring galaxies. *Astron. J.* **1990**, *100*, 1489–1522.

11. Chitre, A.; Jog, C.J. Luminosity profiles of advanced mergers of galaxies using 2MASS data. *Astron. Astrophys.* **2002**, *388*, 407–424.

12. Peletier, R.F.; Balcells, M.; Davies, R.L.; Andredakis, Y.; Vazdekis, A.; Burkert, A.; Prada, F. Galactic bulges from Hubble Space Telescope NICMOS observations: ages and dust. *Mon. Not. Roy. Astron. Soc.* **1999**, *310*, 703–716.

13. Peletier, R.F.; Balcells, M. Near-infrared surface photometry of bulges and disks of spiral galaxies. The data. *New Astron.* **1997**, *1*, 349–362.

14. Michard, R.; Marchal, J. Quantitative morphology of E-S0 galaxies. III. Coded and parametric description of 108 galaxies in a complete sample. *Astron. Astrophys. Suppl.* **1994**, *105*.

15. McDermid, R.M.; Alatalo, K.; Blitz, L.; Bournaud, F.; Bureau, M.; Cappellari, M.; Crocker, A.F.; Davies, R.L.; Davis, T.A.; de Zeeuw, P.T.; et al. The ATLAS^{3D} Project - XXX. Star formation histories and stellar population scaling relations of early-type galaxies. *Mon. Not. Roy. Astron. Soc.* **2015**, *448*, 3484–3513.

16. Tully, R.B.; Courtois, H.M.; Dolphin, A.E.; Fisher, J.R.; Héraudeau, P.; Jacobs, B.A.; Karachentsev, I.D.; Makarov, D.; Makarova, L.; Mitronova, S.; et al. Cosmicflows-2: The Data. *Astron. J.* **2013**, *146*, 86.

17. Tonry, J.L.; Dressler, A.; Blakeslee, J.P.; Ajhar, E.A.; Fletcher, A.B.; Luppino, G.A.; Metzger, M.R.; Moore, C.B. The SBF Survey of Galaxy Distances. IV. SBF Magnitudes, Colors, and Distances. *Astrophys. J.* **2001**, *546*, 681–693.

18. Theureau, G.; Hanski, M.O.; Coudreau, N.; Hallet, N.; Martin, J.M. Kinematics of the Local Universe. XIII. 21-cm line measurements of 452 galaxies with the Nançay radiotelescope, JHK Tully-Fisher relation, and preliminary maps of the peculiar velocity field. *Astron. Astrophys.* **2007**, *465*, 71–85.

19. Cappellari, M.; Emsellem, E.; Krajnović, D.; McDermid, R.M.; Serra, P.; Alatalo, K.; Blitz, L.; Bois, M.; Bournaud, F.; Bureau, M.; et al. The ATLAS^{3D} project - VII. A new look at the morphology of nearby galaxies: the kinematic morphology-density relation. *Mon. Not. Roy. Astron. Soc.* **2011**, *416*, 1680–1696.

20. Sikkema, G.; Carter, D.; Peletier, R.F.; Balcells, M.; Del Burgo, C.; Valentijn, E.A. HST/ACS observations of shell galaxies: inner shells, shell colours and dust. *Astron. Astrophys.* **2007**, *467*, 1011–1024.

21. Blom, C.; Forbes, D.A.; Foster, C.; Romanowsky, A.J.; Brodie, J.P. The SLUGGS Survey: new evidence for a tidal interaction between the early-type galaxies NGC 4365 and NGC 4342. *Mon. Not. Roy. Astron. Soc.* **2014**, *439*, 2420–2431.

22. Bassino, L.P.; Caso, J.P. The merger remnant NGC 3610 and its globular cluster system: a large-scale study. *Mon. Not. Roy. Astron. Soc.* **2017**, *466*, 4259–4271.

![galaxies logo]

Conference Report

Resolving the Extended Stellar Halos of Nearby Galaxies: The Wide-Field PISCeS Survey

Denija Crnojević

Department of Physics & Astronomy, Texas Tech University, Box 41051, Lubbock, TX 79409-1051, USA; denija.crnojevic@ttu.edu

Academic Editors: Duncan A. Forbes and Ericson D. Lopez
Received: 15 June 2017; Accepted: 14 August 2017; Published: 17 August 2017

Abstract: The wide-field Panoramic Imaging Survey of Centaurus and Sculptor (PISCeS) investigates the resolved stellar halos of two nearby galaxies (the spiral NGC 253 and the elliptical Centaurus A, $D \sim 4$ Mpc) out to a galactocentric radius of 150 kpc. The survey to date has led to the discovery of 11 confirmed faint satellites and stunning streams/substructures in two environments substantially different from the Local Group; i.e., the loose Sculptor group of galaxies and the Centaurus A group dominated by an elliptical. The newly discovered satellites and substructures, with surface brightness limits as low as $\sim$32 mag/arcsec^2, are then followed-up with HST imaging and Keck/VLT spectroscopy to investigate their stellar populations. The PISCeS discoveries clearly testify the past and ongoing accretion processes shaping the halos of these nearby galaxies, and provide the first census of their satellite systems down to an unprecedented $M_V < -8$.

Keywords: galaxies: evolution; galaxies: groups; galaxies: halos; galaxies: photometry

---

1. The Past, Present, and Future of Near-Field Cosmology

The past decade has seen a tremendous effort to study the properties of our Milky Way (MW) and its neighbouring Local Group (LG) galaxies in great detail. In particular, the wide-field surveys of the MW-analogue M31 (PAndAS, SPLASH; [1,2]) have highlighted significant differences in the halo properties, accretion history, and satellite populations of the two LG giant spirals. This is unsurprising, as theoretical simulations predict a large halo-to-halo scatter in the properties of MW-sized halos (e.g., [3]), but it certainly underlines the need for in-depth surveys of galaxies beyond the LG.

A variety of approaches can be adopted to refine our knowledge of galaxy halos and their inhabitants: for example, deep pencil-beam surveys of a select sample of nearby (<10–15 Mpc) galaxies can be performed from the space (the Galaxy Halos, Outer disks, Substructure, Thick disks, and Star clusters, or GHOSTS, survey; e.g., [4]); or a larger number of more distant (>15–20 Mpc), unresolved galaxies can be investigated in integrated light (e.g., [5–7]). However, it remains an observational challenge to produce wide and deep resolved stellar maps such as those available for LG galaxies. With the advent of ground-based wide-field imagers, the first PAndAS-like maps for nearby MW-analogues are finally starting to be obtained (e.g., [8,9]).

2. The PISCeS Survey

The Panoramic Imaging Survey of Centaurus and Sculptor (PISCeS) was designed to help bridge the gap between our detailed knowledge of LG galaxies' properties and the integrated light information from unresolved galaxies at larger distances by targeting two resolved galaxies at $\sim$4 Mpc. To facilitate a comparison to the MW and M31, the PISCeS targets are $\sim$ MW-mass galaxies, a spiral (NGC 253, also known as Sculptor) and an elliptical (NGC 5128, or Centaurus A, Cen A). The former is part of a loose filament of galaxies, the latter of a rich group, thus giving us the opportunity to investigate halos and faint satellite populations evolving under different environmental conditions.

PISCeS ultimately aims at obtaining a deep, wide-field view of the extended halos of Sculptor and Cen A, by resolving individual stars out to a galactocentric radius of $\sim$150 kpc (i.e., comparable to the PAndAS survey). The survey is performed with the Magellan/Megacam imager [10] in the g and r filters, and its areal coverage is $\sim$95% and $\sim$80% complete for Cen A and Sculptor, respectively (the total area corresponds to $\sim$16 deg^2, or $\sim$95 Megacam pointings).

After performing point-spread-function photometry, we are able to resolve the uppermost $\sim$1.5 mag of the red giant branch (RGB) population in our targets, with a limiting magnitude of $r \sim 27$ mag. In the color–magnitude space, RGB stars have a clearly distinct sequence from those of unresolved background galaxies and foreground Galactic stars. These predominantly old populations are ideal tracers for the extended halo, and allow surface brightness values as faint as $\sim$32 mag/arcsec^2 to be reached without the complications inherent to intergrated light studies (e.g., sky subtraction, flat-fielding, Galactic cirrus, etc.).

3. Results

In Figure 1, we report the RGB stellar density map derived for Cen A (from [11]; the data do not include our latest observing run): overdensities in the number of RGB stars reveal a plethora of previously unknown faint satellites, streams, and substructures. To date, the PISCeS results can be summarized as follows:

- *satellites of Sculptor:* we search for previously unknown satellites by performing an initial visual inspection of the Magellan images, which is complemented by an identification of RGB stellar overdensities. To date, we have confirmed two new faint satellites around Sculptor ($M_V \sim -10$ and -12), of which one (Scl-MM-Dw2) is an intriguing tidally disrupting ultra-diffuse galaxy ([12,13]; i.e., a galaxy with an effective radius larger than $\sim$1.5 kpc and a low central surface brightness, in this case $\mu_{V,0} \sim 26$ mag arcsec^{-2}, following the definition by [14]). We are following-up Scl-MM-Dw2 with a novel coadded stellar spectroscopy technique ([15]) which will allow us to assess the possible presence of velocity and metallicity gradients along its extent;

- *satellites of Cen A:* within the PISCeS footprint we have discovered nine new satellites of Cen A, with luminosities in the range $M_V \sim -8/-14$, as confirmed by our HST follow-up imaging (GO-13856 and GO-14259; see also [11]). Of particular interest are: CenA-MM-Dw3, a disrupting dwarf at $\sim$90 kpc from Cen A with extended tidal tails ($\sim$2 deg), which has the faint surface brightness of an ultra-diffuse galaxy, a prominent nuclear star cluster and a strong metallicity gradient along its tails; and CenA-MM-Dw1, another ultra-diffuse dwarf with a fainter candidate satellite of its own and a globular cluster system similar to the one of the Fornax dwarf in the LG. The properties of the new satellites are consistent with those of LG dwarfs of comparable luminosities; PISCeS extends the faint end of the satellite luminosity function in the Cen A group by two magnitudes. The preliminary Cen A galaxy luminosity function is steeper than the ones of the MW and M31, but similar to the one derived for the rich M81 group by [16];

- *halo substructures in Cen A:* as seen from Figure 1, Cen A appears to have had a rather active accretion/interaction history: many of the low surface brightness features highlighted by the PISCeS map resemble those previously identified in M31's halo, even though they appear to be less numerous; these features are being followed-up with HST imaging in order to derive their star formation histories and possibly characterize their progenitors' stellar content. Once the substructures are identified, they can be decoupled from the smooth stellar halo to more robustly derive the latter's profile and shape (e.g., [17]);

- *globular cluster population around Cen A:* the wide-field photometry of Cen A's halo has allowed us to identify $\sim$1000 globular cluster candidates and several ultra-compact dwarf galaxies, identified with a two aperture photometry technique. Some of these objects are spatially correlated with stellar streams, similarly to what was found for M31 [18], and will guide our interpretation of past accretion events as well as provide valuable constraints to derive Cen A's total halo mass.

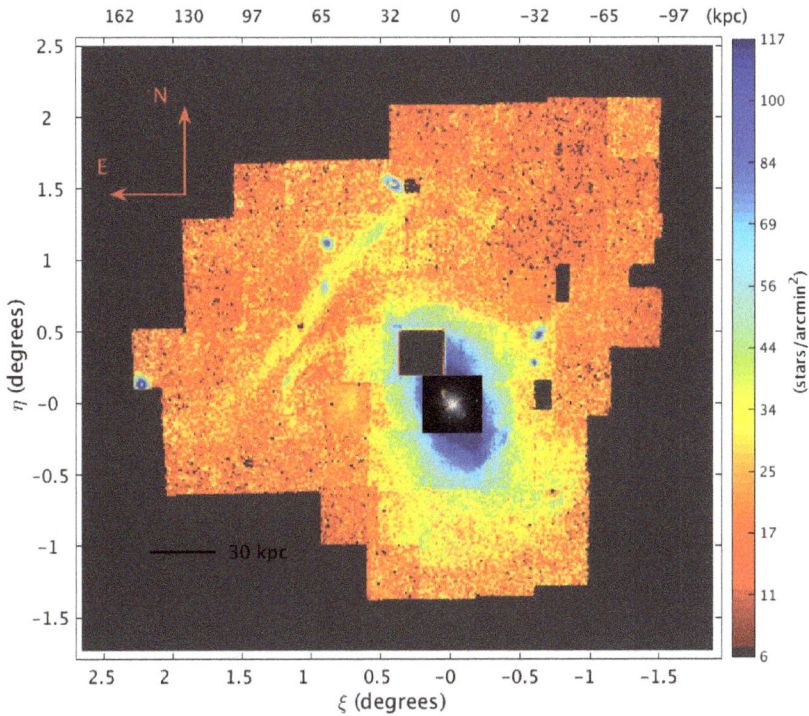

Figure 1. Stellar density map of old, metal-poor RGB stars in the extended halo of Cen A, from the Magellan/Megacam PISCeS survey (Figure 3 from [11]). PISCeS reaches significantly fainter surface brightness limits (down to $\mu_V \sim 32$ mag/arcsec^2) with respect to integrated light alone, thus allowing us to decode the past evolutionary history of Cen A. The density scale is shown on the right; the physical scale is reported on the upper axis. The central regions of the galaxy are replaced by a color image (the star-count map in this region suffers from incompleteness due to stellar crowding). The next generation of telescopes will allow us to obtain comparable maps for tens of galaxies in the Local Volume, thus revolutionizing our understanding of galaxy evolution.

4. Summary and Future Work

The PISCeS survey pushes the limits of near-field cosmology beyond the LG, and enables a comparison of external galaxies' resolved halos to the wide-field surveys of the MW companion M31. The detailed characterization of the extended stellar halos of Sculptor and Cen A will shed light onto their in situ vs. accreted halo components, their metallicity gradients, their faint satellite populations, their halo mass, and shape. These are among the first efforts to explore the properties of extended halos and faint satellite populations for a range of host galaxy morphologies and environments. Such efforts are also starting to be extended to isolated hosts with lower masses (e.g., the Magellanic Analog Dwarf Companions And Stellar Halos, MADCASH; [19]), in order to provide a much needed comparison to the recently discovered satellite system of the Large Magellanic Cloud (e.g., [20] and references therein).

In the next decade, the advent of ground-based and space-borne telescopes (e.g., LSST, JWST, Euclid, GMT, TMT, E-ELT, WFIRST) will open up a new window for near-field cosmology. The next generation of resolved wide-field surveys will be capable of reaching the horizontal branch magnitude level for $\sim$100 galaxies within $\sim$10 Mpc, thus delivering maps comparable to (and deeper than) the one shown in Figure 1 for Cen A (e.g., [21]). These upcoming efforts will ultimately provide crucial constraints to theoretical models of galaxy formation and evolution.

Acknowledgments: D.C. thanks the SOC/LOC of the conference "On the Origin (and Evolution) of Baryonic Galaxy Halos" for impeccable organization, for a very successful and stimulating conference, and for the opportunity to visit beautiful Ecuador. D.C. acknowledges the contributions of the PISCeS team to the research presented here.

Conflicts of Interest: The authors declare no conflict of interest.

References

1. McConnachie, A.W.; Irwin, M.J.; Ibata, R.A.; Dubinski, J.; Widrow, L.M.; Martin, N.F.; Côté, P.; Dotter, A.L.; Navarro, J.F.; Ferguson, A.M.N.; et al. The remnants of galaxy formation from a panoramic survey of the region around M31. *Nature* **2009**, *461*, 66–69.

2. Gilbert, K.M.; Kalirai, J.S.; Guhathakurta, P.; Beaton, R.L.; Geha, M.C.; Kirby, E.N.; Majewski, S.R.; Patterson, R.J.; Tollerud, E.J.; Bullock, J.S.; et al. Global Properties of M31's Stellar Halo from the SPLASH Survey. II. Metallicity Profile. *Astrophys. J.* **2014**, *796*, 76.

3. Pillepich, A.; Vogelsberger, M.; Deason, A.; Rodriguez-Gomez, V.; Genel, S.; Nelson, D.; Torrey, P.; Sales, L.V.; Marinacci, F.; Springel, V.; et al. Halo mass and assembly history exposed in the faint outskirts: The stellar and dark matter haloes of Illustris galaxies. *Mon. Not. R. Astron. Soc.* **2014**, *444*, 237–249.

4. Monachesi, A.; Bell, E.F.; Radburn-Smith, D.J.; Bailin, J.; de Jong, R.S.; Holwerda, B.; Streich, D.; Silverstein, G. The GHOSTS survey-II. The diversity of halo colour and metallicity profiles of massive disc galaxies. *Mon. Not. R. Astron. Soc.* **2016**, *457*, 1419–1446.

5. Martínez-Delgado, D.; Gabany, R.J.; Crawford, K.; Zibetti, S.; Majewski, S.R.; Rix, H.; Fliri, J.; Carballo-Bello, J.A.; Bardalez-Gagliuffi, D.C.; Peñarrubia, J.; et al. Stellar Tidal Streams in Spiral Galaxies of the Local Volume: A Pilot Survey with Modest Aperture Telescopes. *Astrophys. J.* **2010**, *140*, 962–967.

6. Duc, P.A.; Cuillandre, J.C.; Karabal, E.; Cappellari, M.; Alatalo, K.; Blitz, L.; Bournaud, F.; Bureau, M.; Crocker, A.F.; Davies, R.L.; et al. The ATLAS^{3D} project-XXIX. The new look of early-type galaxies and surrounding fields disclosed by extremely deep optical images. *Mon. Not. R. Astron. Soc.* **2015**, *446*, 120–143.

7. Spavone, M.; Capaccioli, M.; Napolitano, N.R.; Iodice, E.; Grado, A.; Limatola, L.; Cooper, A.P.; Cantiello, M.; Forbes, D.A.; Paolillo, M.; et al. VEGAS: A VST Early-type GAlaxy Survey. II. Photometric study of giant ellipticals and their stellar halos. *Astron. Astrophys.* **2017**, *603*, A38.

8. Mouhcine, M.; Ibata, R.; Rejkuba, M. A Panoramic View of the Milky Way Analog NGC 891. *Astrophys. J. Lett.* **2010**, *714*, L12–L15.

9. Okamoto, S.; Arimoto, N.; Ferguson, A.M.N.; Bernard, E.J.; Irwin, M.J.; Yamada, Y.; Utsumi, Y. A Hyper Suprime-Cam View of the Interacting Galaxies of the M81 Group. *Astrophys. J. Lett.* **2015**, *809*, L1.

10. McLeod, B.; Geary, J.; Conroy, M.; Fabricant, D.; Ordway, M.; Szentgyorgyi, A.; Amato, S.; Ashby, M.; Caldwell, N.; Curley, D.; et al. Megacam: A Wide-Field CCD Imager for the MMT and Magellan. *Publ. ASP* **2015**, *127*, 366–382.

11. Crnojević, D.; Sand, D.J.; Spekkens, K.; Caldwell, N.; Guhathakurta, P.; McLeod, B.; Seth, A.; Simon, J.D.; Strader, J.; Toloba, E. The Extended Halo of Centaurus A: Uncovering Satellites, Streams, and Substructures. *Astrophys. J.* **2016**, *823*, 19.

12. Sand, D.J.; Crnojević, D.; Strader, J.; Toloba, E.; Simon, J.D.; Caldwell, N.; Guhathakurta, P.; McLeod, B.; Seth, A.C. Discovery of a New Faint Dwarf Galaxy Associated with NGC 253. *Astrophys. J. Lett.* **2014**, *793*, L7.

13. Toloba, E.; Sand, D.J.; Spekkens, K.; Crnojević, D.; Simon, J.D.; Guhathakurta, P.; Strader, J.; Caldwell, N.; McLeod, B.; Seth, A.C. A Tidally Disrupting Dwarf Galaxy in the Halo of NGC 253. *Astrophys. J. Lett.* **2016**, *816*, L5.

14. Van Dokkum, P.G.; Abraham, R.; Merritt, A.; Zhang, J.; Geha, M.; Conroy, C. Forty-seven Milky Way-sized, Extremely Diffuse Galaxies in the Coma Cluster. *Astrophys. J. Lett.* **2015**, *798*, L45.

15. Toloba, E.; Sand, D.; Guhathakurta, P.; Chiboucas, K.; Crnojević, D.; Simon, J.D. Spectroscopic Confirmation of the Dwarf Spheroidal Galaxy d0994+71 as a Member of the M81 Group of Galaxies. *Astrophys. J. Lett.* **2016**, *830*, L21.

16. Chiboucas, K.; Jacobs, B.A.; Tully, R.B.; Karachentsev, I.D. Confirmation of Faint Dwarf Galaxies in the M81 Group. *Astrophys. J.* **2013**, *146*, 126.

17. Rejkuba, M.; Harris, W.E.; Greggio, L.; Harris, G.L.H.; Jerjen, H.; Gonzalez, O.A. Tracing the Outer Halo in a Giant Elliptical to 25 R$_{eff}$. *Astrophys. J. Lett.* **2014**, *791*, L2.

18. Mackey, A.D.; Huxor, A.P.; Ferguson, A.M.N.; Irwin, M.J.; Tanvir, N.R.; McConnachie, A.W.; Ibata, R.A.; Chapman, S.C.; Lewis, G.F. Evidence for an Accretion Origin for the Outer Halo Globular Cluster System of M31. *Astrophys. J. Lett.* **2010**, *717*, L11–L16.

19. Carlin, J.L.; Sand, D.J.; Price, P.; Willman, B.; Karunakaran, A.; Spekkens, K.; Bell, E.F.; Brodie, J.P.; Crnojević, D.; Forbes, D.A.; et al. First Results from the MADCASH Survey: A Faint Dwarf Galaxy Companion to the Low-mass Spiral Galaxy NGC 2403 at 3.2 Mpc. *Astrophys. J. Lett.* **2016**, *828*, L5.

20. Drlica-Wagner, A.; Bechtol, K.; Rykoff, E.S.; Luque, E.; Queiroz, A.; Mao, Y.Y.; Wechsler, R.H.; Simon, J.D.; Santiago, B.; Yanny, B.; et al. Eight Ultra-faint Galaxy Candidates Discovered in Year Two of the Dark Energy Survey. *Astrophys. J.* **2015**, *813*, 109.

21. Greggio, L.; Falomo, R.; Uslenghi, M. Studying stellar halos with future facilities. In *The General Assembly of Galaxy Halos: Structure, Origin and Evolution*; IAU Symposium; Bragaglia, A., Arnaboldi, M., Rejkuba, M., Romano, D., Eds.; Cambridge University Press: Cambridge, UK, 2016; Volume 317, pp. 209–214.

galaxies

MDPI

Conference Report

Revisiting the Globular Cluster Systems of NGC 3258 and NGC 3268

Juan Pablo Caso [1,2,]* and Lilia Bassino [1,2]

[1] Instituto de Astrofísica de La Plata (CCT La Plata—CONICET, UNLP), Facultad de Ciencias Astronómicas y Geofísicas de la Universidad Nacional de La Plata, Paseo del Bosque S/N, B1900FWA La Plata, Argentina; lbassino@fcaglp.unlp.edu.ar

[2] Consejo Nacional de Investigaciones Científicas y Técnicas, Rivadavia 1917, C1033AAJ Ciudad Autónoma de Buenos Aires, Argentina

* Correspondence: jpcaso@fcaglp.unlp.edu.ar

Academic Editors: Ericson D. Lopez and Duncan A. Forbes
Received: 27 June 2017; Accepted: 22 August 2017; Published: 31 August 2017

Abstract: We present a photometric study of NGC 3258 and NGC 3268 globular cluster systems (GCSs) with a wider spatial coverage than previous works. This allowed us to determine the extension of both GCSs, and obtain new values for their populations. In both galaxies, we found the presence of radial colour gradients in the peak of the blue globular clusters. The characteristics of both GCSs point to a large evolutionary history with a substantial accretion of satellite galaxies.

Keywords: galaxies: elliptical and lenticular, cD; galaxies: evolution; galaxies: star clusters: individual: NGC 3258 & NGC 3268

1. Introduction

The majority of the members in globular cluster (GC) populations are usually old stellar systems (e.g., [1,2]). They were formed under extreme environmental conditions, which are probably reachable only in massive star formation episodes during major mergers [3]. This implies a direct connection between the episodes that built up the globular cluster systems (GCSs) and the stellar population of the host galaxy. Hence, the study of a GCS is important to obtain a comprehensive picture of the evolutionary history of galaxies.

Our target galaxies, NGC 3258 and NGC 3268, are thought to make-up the bulk of the Antlia galaxy cluster, located in the Southern sky at a low Galactic latitude (≈19 deg). The central part seems to consist of two groups, each one dominated by one of these giant ellipticals (gEs) with similar luminosity. These two subgroups might be in a merging process, but surface brightness fluctuations distances [4–6] and radial velocities analysis [7,8] are not conclusive.

There are several studies about these GCSs [9–11] and the connection between their bright-end and ultra-compact dwarfs [12,13]. The aim of the present study is to complement previous ones by taking advantage of wider and deeper datasets. This results in a more accurate contamination estimation, and the possibility of calculating the GCSs' spatial extensions and total populations.

2. Materials and Methods

The dataset consists of two wide fields (36×36 arcmin^2) obtained at the Cerro Tololo 4-m telescope with the MOSAIC II camera in filters (C, T_1). One field contains both galaxies, while the other is located to the east (Figure 1). These double the areal coverage from Dirsch et al. [9], the more extended GCSs study of these galaxies. We also used four fields from VLT obtained with the FORS1 camera (6.8×6.8 arcmin^2) in filters (V, I) (see Figure 1). Table 1 contains the basic information of the observations. For MOSAIC data, we selected $T_{1,0} = 23.85$ as the magnitude limit; for fainter magnitudes, the completeness falls below 60%. In the case of FORS1 data, we selected $V_0 = 25.5$ as the magnitude limit, which implies a completeness limit of 60% for GCs at less than 1 arcmin from the galaxies centre and 70% for larger galactocentric distances.

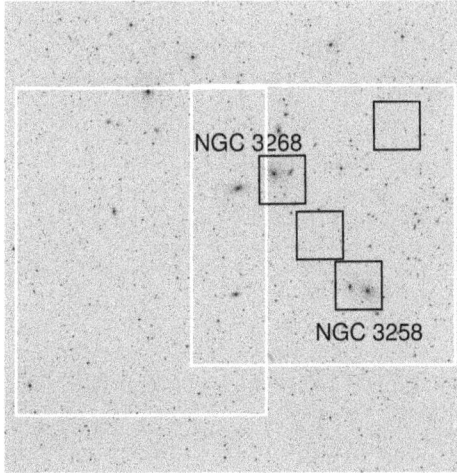

Figure 1. The MOSAIC (white regions) and FORS1 (black regions) fields are overlaid on a $70 \times 70 \, \text{arcmin}^2$ DSS image of the Antlia cluster. North is up and east is to the left.

Table 1. Basic data from observations. FWHM: full width at half maximum.

Name	Obs. Date	Exp. Time	Typical *FWHM*
		MOSAIC data	
Central Field	4/5 April 2002	$4 \times 600 \, \text{s}$ in $C - 4 \times 600 \, \text{s}$ in T_1	$1''$
East Field	24/25 May 2004	$7 \times 900 \, \text{s}$ in $C - 5 \times 600 \, \text{s}$ in T_1	$1.1''$
		FORS1 data	
	27/28 March 2003	$5 \times 300 \, \text{s}$ in $V - 5 \times 700 \, \text{s}$ in I	$0.6''$

GC candidates were selected in both datasets from point sources with colours in the usual range of GCs, i.e., $0.9 < (C - T_1)_0 < 2.3$ and $0.4 < (V - I)_0 < 1.6$, and fainter than $M_V \approx -10.5$ to avoid ultra-compact dwarfs (e.g., [14,15]), which implies $T_{1,0} > 21.6$ and $V_0 > 22.2$ at the assumed distance. We refer to Caso et al. [16] for further details on the data reduction and photometry.

3. Results

3.1. GCSs Spatial Extension

In order to complement the available datasets, taking advantage of their different properties, we derived the radial distributions of the GC candidates following a two-step procedure. First, we fitted power-laws to blue and red subpopulations, masking regions close to neighbour galaxies and bright stars. Considering that previous studies point to an overlap of both GCSs (e.g., [10]), we fitted the radial distributions iteratively, avoiding regions where the other GCS might be contributing to the observed surface density. We assumed that the outer limit of a GCS is achieved when its GC surface density is equal to 30 per cent of the background level. This criterion has been applied to several GCS studies, including those with a large field of view (FOV) (e.g., [17,18]). The background region selected to correct for contamination in the MOSAIC photometry is located in the eastern portion of our field, at more than 25 arcmin from both galaxies. It spans 489 arcmin^2, and the mean projected densities of point sources with similar colours and magnitudes than GC candidates are ≈0.45 arcmin^{-2} and ≈0.2 arcmin^{-2}, respectively, for blue and red GC candidates.

In the case of NGC 3258, the blue GCS reaches ≈17 arcmin (i.e., ≈170 kpc), while the red subpopulation appears more concentrated towards the galaxy, with an extension of ≈6 arcmin (i.e., ≈60 kpc). For NGC 3268, the extension of both subpopulations are similar, ≈14 arcmin and ≈12 arcmin for blue and red GCs, respectively.

For the innermost regions of the GCSs, we obtained the radial profiles from the FORS1 data, applying background and completeness corrections to the GC surface densities. In these cases, we fitted a modified Hubble distribution [9,19]:

$$n(r) = a \left(1 + \left(\frac{r}{r_0}\right)^2\right)^{-\beta} \tag{1}$$

to consider the usual flattening in the inner GC radial profiles [20–22]. Figure 2 shows the background and completeness corrected radial profiles from FORS1 data for blue (filled squares) and red (filled circles) subpopulations. Open symbols represent the radial profiles obtained from MOSAIC data, properly scaled to match the deeper FORS1 data. Solid curves indicate the fitted Hubble distributions, while dashed ones correspond to the scaled power-laws fitted to MOSAIC data. Table 2 shows the parameters associated with the Hubble profiles.

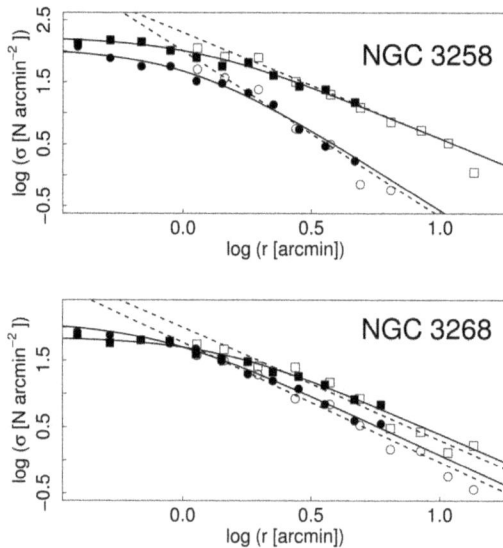

Figure 2. Background and completeness corrected radial profiles from FORS1 data for blue (filled squares) and red (filled circles) globular clusters (GCs). The open symbols shows the MOSAIC data, properly scaled. Solid curves represent the Hubble modified profile fitted to FORS1 data, and dashed ones the power-laws fitted to MOSAIC data.

The wider FOV allows us a better determination of the radial profile and the background level, which implies a more accurate estimation of the radial extension in both systems. The joint analysis of MOSAIC and FORS1 data results in a more general fit of the GCs radial distributions than previous studies.

Table 2. Parameters of the modified Hubble distribution fitted to the FORS1 radial profiles.

	NGC 3258		NGC 3268	
	Blue GCs	**Red GCs**	**Blue GCs**	**Red GCs**
a	2.23 ± 0.05	2.05 ± 0.06	1.85 ± 0.03	2.09 ± 0.06
r_0	1.11 ± 0.10	1.05 ± 0.08	1.40 ± 0.10	0.77 ± 0.08
$beta$	-3.5 ± 0.2	-5.8 ± 0.3	-3.4 ± 0.26	3.6 ± 0.3

3.2. Total Population of the GCSs

In order to calculate the population of the GCSs we obtained the background and completeness GC luminosity functions (GCLFs) from FORS1 data for both galaxies (Figure 3). It is largely documented in literature that GCLF in elliptical galaxies can be approximated by Gaussian profiles with a turn-over magnitude (TOM) $M_V = -7.4$ (e.g., [1,23]). In both galaxies, the expected TOM was close to our magnitude limit. Hence, we used the distance moduli calculated by Tully et al. [6] to determine the TOM, $mM = 32.56 \pm 0.14$ for NGC 3258 and $mM = 32.74 \pm 0.14$ for NGC 3268. The rest of the Gaussian parameters were fitted from the data. From this procedure, we established that GCs brighter than $V_0 = 25.5$ represent $\approx 62\%$ for NGC 3258 and $\approx 56\%$ in the case of NGC 3268.

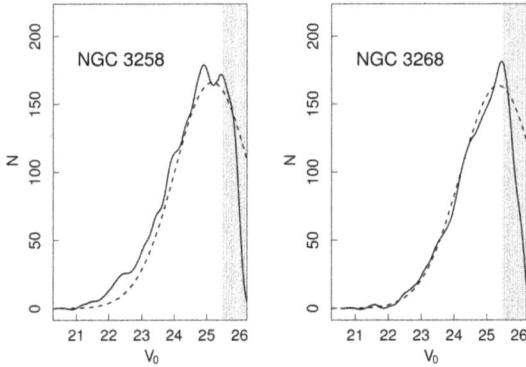

Figure 3. In each panel the smoothed GC luminosity function from FORS1 data is represented with a solid line, while the dashed one represents the Gaussian fitted to the data. The shaded regions indicate the regions with fainter magnitudes than our limit due to the completeness drop.

Then, we numerically integrated the GC radial profiles derived in the previous section along the entire radial extension of the GCSs. Finally, we corrected these values by the fraction of missing GCs due to our completeness limits. The results indicate a population of ≈ 6600 blue GCs and ≈ 1400 red ones for NGC 3258, which implies a fraction of red-to-total GCs of $f_{red} \approx 0.18$. In the case of NGC 3268, we obtained ≈ 5200 blue and ≈ 3000 red GCs, with $f_{red} \approx 0.38$.

The fraction of red GCs calculated in this work for both galaxies is similar to the values indicated by Bassino et al. [10]. Our results point to more populated GCSs, mainly due to the larger radial extension derived in this contribution. For instance, in case we integrate our radial distribution for NGC 3258 up to 10 arcmin, the number of GCs is in agreement with Bassino et al. [10] values. The same analysis for NGC 3268 results in a larger number of GCs.

3.3. Radial Gradients in the Colour Distribution

The left panels of Figure 4 show the smoothed and background corrected colour distribution for GC candidates around NGC 3258 (upper panel) and NGC 3268 (lower panel) from the MOSAIC data. The sample was split in three radial regimes. We statistically subtracted the contamination and applied the algorithm Gaussian mixture modeling (GMM) [24] to the clean samples to calculate the colour for the blue GC peak in the three ranges. After repeating the procedure 25 times in order to reduce statistical noise, the mean colours in the $(C - T_1)$ filters resulted 1.27 ± 0.01, 1.21 ± 0.01, and 1.12 ± 0.02, respectively, in the case of NGC 3258, and 1.37 ± 0.02, 1.28 ± 0.02, and 1.22 ± 0.03, respectively, for NGC 3268. These mean colours are indicated in the panels with vertical lines.

The right panels of Figure 4 are analogues for the FORS1 data. The smaller FOV and the lower sensitivity of $(V - I)$ colours with respect to $(C - T_1)$ make results noisy, but blue peaks seem to get bluer for GC candidates at galactocentric distances larger than 150″.

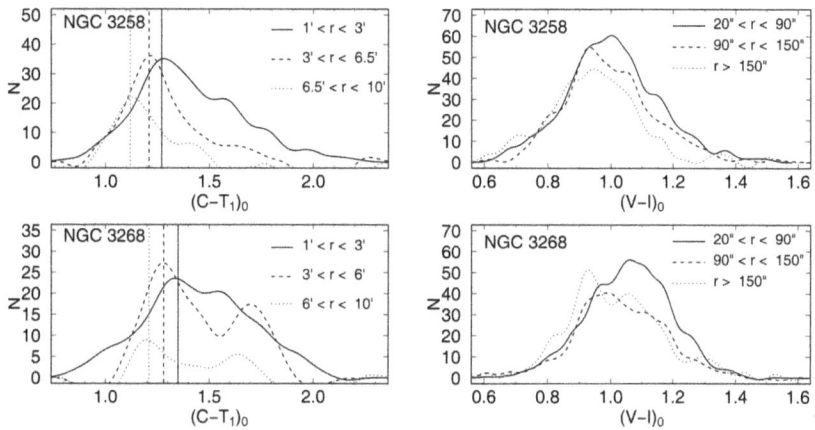

Figure 4. Left panels: smoothed colour distributions for GC candidates from MOSAIC data, split in three radial distance regimes. Vertical lines indicate mean colours for the blue subpopulation, obtained with Gaussian mixture modeling (GMM) [24]. **Right panels**: analogue figures for GC candidates from FORS1 data.

4. Summary

The rich and spatially extended GCSs in both galaxies point to rich evolutionary histories. The radial gradient in the colour peak of blue GCs found in NGC 3258 and NGC 3268 could be explained by the correlation between the colour and metallicity for blue GCs and the galaxy masses (e.g., [25,26]), and points to the relevant role of the accretion of satellite galaxies in the build-up of the outer regions of GCSs. Similar radial gradients have been found in other giant ellipticals, pointing to two phases in galaxy formation (e.g., [27]).

Acknowledgments: This research was funded with grants from Consejo Nacional de Investigaciones Científicas y Técnicas de la República Argentina (PIP 112-201101-00393), Agencia Nacional de Promoción Científica y Tecnológica (PICT-2013-0317), and Universidad Nacional de La Plata (UNLP 11-G124), Argentina.

Author Contributions: This paper was totally carried out by the authors.

Conflicts of Interest: The authors declare no conflict of interest.

References

1. Brodie, J.P.; Strader, J. Extragalactic Globular Clusters and Galaxy Formation. *Ann. Rev. Astron. Astrophys.* **2006**, *44*, 193–267.
2. Tonini, C. The Metallicity Bimodality of Globular Cluster Systems: A Test of Galaxy Assembly and of the Evolution of the Galaxy Mass-Metallicity Relation. *Astrophys. J.* **2013**, *762*, 39.
3. Kruijssen, J.M.D. Globular cluster formation in the context of galaxy formation and evolution. *Class. Quant. Grav.* **2014**, *31*, 244006.
4. Blakeslee, J.P.; Lucey, J.R.; Barris, B.J.; Hudson, M.J.; Tonry, J.L. A synthesis of data from fundamental plane and surface brightness fluctuation surveys. *Mon. Not. Roy. Astron. Soc.* **2001**, *327*, 1004–1020.
5. Cantiello, M.; Blakeslee, J.P.; Raimondo, G.; Mei, S.; Brocato, E.; Capaccioli, M. Detection of Radial Surface Brightness Fluctuations and Color Gradients in Elliptical Galaxies with the Advanced Camera for Surveys. *Astrophys. J.* **2005**, *634*, 239–257.
6. Tully, R.B.; Courtois, H.M.; Dolphin, A.E.; Fisher, J.R.; Héraudeau, P.; Jacobs, B.A.; Karachentsev, I.D.; Makarov, D.; Makarova, L.; Mitronova, S.; et al. Cosmicflows-2: The Data. *Astron. J.* **2013**, *146*, 86.
7. Hess, K.M.; Jarrett, T.H.; Carignan, C.; Passmoor, S.S.; Goedhart, S. KAT-7 Science Verification: Cold Gas, Star Formation, and Substructure in the Nearby Antlia Cluster. *Mon. Not. Roy. Astron. Soc.* **2015**, *452*, 1617–1636.
8. Caso, J.; Richtler, T. Deconstructing the Antlia cluster core. *Astron. Astrophys.* **2015**, *584*, doi:10.1051/ 0004-6361/201527136.
9. Dirsch, B.; Richtler, T.; Bassino, L.P. The globular cluster systems of NGC 3258 and NGC 3268 in the Antlia cluster*. *Astron. Astrophys.* **2003**, *408*, 929–939.
10. Bassino, L.P.; Richtler, T.; Dirsch, B. VLT photometry in the Antlia cluster: The giant ellipticals NGC 3258 and NGC 3268 and their globular cluster systems. *Mon. Not. Roy. Astron. Soc.* **2008**, *386*, 1145–1156.
11. Harris, W.E.; Whitmore, B.C.; Karakla, D.; Okoń, W.; Baum, W.A.; Hanes, D.A.; Kavelaars, J.J. Globular Cluster Systems in Brightest Cluster Galaxies: Bimodal Metallicity Distributions and the Nature of the High-Luminosity Clusters. *Astrophys. J.* **2006**, *636*, 90–114.
12. Caso, J.P.; Bassino, L.P.; Richtler, T.; Smith Castelli, A.V.; Faifer, F.R. Ultracompact dwarfs around NGC 3258 in the Antlia cluster. *Mon. Not. Roy. Astron. Soc.* **2013**, *430*, 1088–1101.
13. Caso, J.P.; Bassino, L.P.; Richtler, T.; Calderón, J.P.; Smith Castelli, A.V. Ultracompact dwarfs around NGC 3268. *Mon. Not. Roy. Astron. Soc.* **2014**, *442*, 891–899.
14. Mieske, S.; Hilker, M.; Misgeld, I. The specific frequencies of ultra-compact dwarf galaxies. *Astron. Astrophys.* **2012**, *537*, A3.
15. Norris, M.A.; Kannappan, S.J.; Forbes, D.A.; Romanowsky, A.J.; Brodie, J.P.; Faifer, F.R.; Huxor, A.; Maraston, C.; Moffett, A.J.; Penny, S.J.; et al. The AIMSS Project - I. Bridging the star cluster-galaxy divide. *Mon. Not. Roy. Astron. Soc.* **2014**, *443*, 1151–1172.
16. Caso, J.P.; Bassino, L.P.; Gómez, M. Globular cluster systems as tracers of the evolutionary history in NGC 3258 and NGC 3268. *Mon. Not. Roy. Astron. Soc.* **2017**, *470*, 3227–3238.
17. Bassino, L.P.; Faifer, F.R.; Forte, J.C.; Dirsch, B.; Richtler, T.; Geisler, D.; Schuberth, Y. Large-scale study of the NGC 1399 globular cluster system in Fornax. *Astron. Astrophys.* **2006**, *451*, 789–796.
18. Caso, J.P.; Richtler, T.; Bassino, L.P.; Salinas, R.; Lane, R.R.; Romanowsky, A. The paucity of globular clusters around the field elliptical NGC 7507. *Astron. Astrophys.* **2013**, *555*, A56.
19. Binney, J.; Tremaine, S. Galactic Dynamics. *Nature* **1987**, *326*, 219.
20. Elson, R.A.W.; Grillmair, C.J.; Forbes, D.A.; Rabban, M.; Williger, G.M.; Brodie, J.P. HST imaging of the globular clusters in the Fornax cluster-NGC 1379. *Mon. Not. Roy. Astron. Soc.* **1998**, *295*, 240.
21. Capuzzo-Dolcetta, R.; Mastrobuono-Battisti, A. Globular cluster system erosion in elliptical galaxies. *Astron. Astrophys.* **2009**, *507*, 183–193.
22. Brodie, J.P.; Romanowsky, A.J.; Strader, J.; Forbes, D.A.; Foster, C.; Jennings, Z.G.; Pastorello, N.; Pota, V.; Usher, C.; Blom, C.; et al. The SAGES Legacy Unifying Globulars and GalaxieS Survey (SLUGGS): Sample Definition, Methods, and Initial Results. *Astrophys. J.* **2014**, *796*, 52.
23. Jordán, A.; McLaughlin, D.E.; Côté, P.; Ferrarese, L.; Peng, E.W.; Mei, S.; Villegas, D.; Merritt, D.; Tonry, J.L.; West, M.J. The ACS Virgo Cluster Survey. XII. The Luminosity Function of Globular Clusters in Early-Type Galaxies. *Astrophys. J. Suppl.* **2007**, *171*, 101–145.

24. Muratov, A.L.; Gnedin, O.Y. Modeling the Metallicity Distribution of Globular Clusters. *Astrophys. J.* **2010**, *718*, 1266–1288.

25. Strader, J.; Brodie, J.P.; Forbes, D.A. Metal-Poor Globular Clusters and Galaxy Formation. *Astron. J.* **2004**, *127*, 3431–3436.

26. Peng, E.W.; Jordán, A.; Côté, P.; Blakeslee, J.P.; Ferrarese, L.; Mei, S.; West, M.J.; Merritt, D.; Milosavljević, M.; Tonry, J.L. The ACS Virgo Cluster Survey. IX. The Color Distributions of Globular Cluster Systems in Early-Type Galaxies. *Astrophys. J.* **2006**, *639*, 95–119.

27. Forbes, D.A.; Spitler, L.R.; Strader, J.; Romanowsky, A.J.; Brodie, J.P.; Foster, C. Evidence for two phases of galaxy formation from radial trends in the globular cluster system of NGC 1407. *Mon. Not. Roy. Astron. Soc.* **2011**, *413*, 2943–2949.

galaxies

MDPI

Conference Report

The Baryonic Halos of Isolated Elliptical Galaxies

Ricardo Salinas [1,*], Adebusola Alabi [2,3], Nicklas Hammar [1,4], Tom Richtler [5], Richard R. Lane [6] and Mischa Schirmer [1]

[1]   Gemini Observatory, Colina el Pino s/n, Casilla 603, La Serena 1700000, Chile; nhammar@live.ca (N.H.); mschirmer@gemini.edu (M.S.)
[2]   Centre for Astrophysics & Supercomputing, Swinburne University, Hawthorn VIC 3122, Australia; aalabi@swin.edu.au
[3]   Department of Astronomy and Astrophysics, University of California Observatories, Santa Cruz, CA 95064, USA
[4]   Department of Physics and Astronomy, University of Victoria, Victoria, BC V8W 3P2, Canada
[5]   Departamento de Astronomía, Universidad de Concepción, Concepción 3349001, Chile; tom@astroudec.cl
[6]   Instituto de Astrofísica, P. Universidad Católica de Chile, Santiago 7820436, Chile; rlane@astro.puc.cl
*   Correspondence: rsalinas@gemini.edu

Academic Editors: Duncan A. Forbes and Ericson D. López
Received: 23 June 2017; Accepted: 15 August 2017; Published: 18 August 2017

Abstract: Without the interference of a number of events, galaxies may suffer in crowded environments (e.g., stripping, harassment, strangulation); isolated elliptical galaxies provide a control sample for the study of galaxy formation. We present the study of a sample of isolated ellipticals using imaging from a variety of telescopes, focusing on their globular cluster systems as tracers of their stellar halos. Our main findings are: (a) GC color bimodality is common even in the most isolated systems; (b) the specific frequency of GCs is fairly constant with galaxy mass, without showing an increase towards high-mass systems like in the case of cluster ellipticals; (c) on the other hand, the red fraction of GCs follows the same inverted V shape trend with mass as seen in cluster ellipticals; and (d) the stellar halos show low Sérsic indices which are consistent with a major merger origin.

Keywords: galaxies: star clusters: general; galaxies: halos; galaxies: elliptical and lenticular; CD

1. Introduction

As tracers of star formation, galaxy assembly, and mass distribution, globular clusters have provided important clues to our understanding of the formation and structure of early-type galaxies. However, their study has been mostly confined to galaxy clusters (e.g., [1]), leaving the properties of the globular cluster systems (GCSs) of isolated ellipticals as a mostly uncharted territory.

Having poorer merger histories, isolated ellipticals are particularly relevant to understanding the environmental influence on the formation of a GCS. In hierarchical-merging inspired models, the properties of GCSs and dark halos of host galaxies in low-density environments should be very different from their high-density counterparts [2,3]. In particular, the slope of the outer stellar halo profile will differ depending on the accretion history of each galaxy (e.g., [4]).

In this contribution, we present results from our ongoing work on isolated ellipticals, focusing on the properties of their GCSs and giving preliminary results on deep imaging of their stellar halos.

2. Globular Clusters in Isolated Ellipticals

2.1. Observations

Targets observable from the southern hemisphere were selected from the isolated ellipticals catalogues of [5,6].

Observations were conducted using CTIO/MOSAIC-II (in the *C* and *R* filters, NGC 5812, NGC 3585, and NGC 7507), Gemini-S/GMOS (*g* and *i* filters, NGC 2271, NGC 2865, NGC 3962, NGC 4240, and IC 4889) and VLT/VIMOS (*B* and *R*; NGC 720, NGC 821, NGC 1162, NGC 7796, and ESO-194G021). In the present contribution we compile the results for nine of these ellipticals. Expanded details of these results have been given in [7–10].

Photometry of GCs was performed with DAOPHOT/ALLSTAR [11] after subtraction of a model of the parent galaxy in each image with IRAF/ellipse. We selected GCs based on their point-source appearance and colors consistent with old single stellar populations.

2.2. Results

Color bimodality is one of the main features of GCSs, and our work shows that isolated ellipticals are not an exception. Of the nine galaxies studied, six present clear bimodality, while the unclear status of the remaining three can probably be attributed to a population of intermediate age, which, based on the predictions from stellar population models, is expected to have optical colors between the classic red and blue globular clusters, hence blurring the bimodality signal.

A second diagnostic for the formation of a GCS is the red fraction; that is, the number of metal-rich over the total number of clusters (e.g., [1]). The models presented in [3] predict that the red fraction will be a function of the accretion history; while galaxies with a rich accretion history are expected to be dominated by blue clusters (and hence have a low red fraction), the opposite is expected for galaxies with a poor accretion history. Our measurement of the red fraction in our sample, compared to galaxies in low-density environments and in Virgo can be seen in Figure 1 (top panel). The red fraction of field (blue symbols) and isolated ellipticals (red symbols), cannot be distinguished from Virgo ellipticals (green symbols), contradicting the naive expectation from the models of [3].

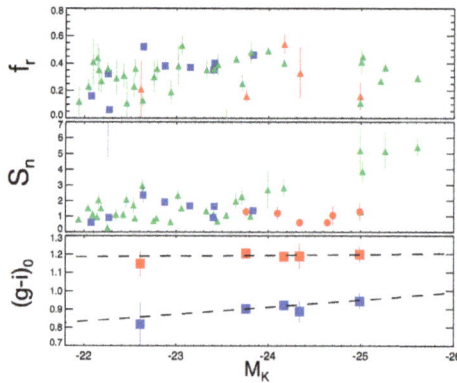

Figure 1. Upper panel: Globular cluster (GC) red fraction as function of 2MASS *K*-band luminosity, as a proxy for stellar mass. Red symbols represent our sample, while blue squares are taken from the [12] sample of 10 "low-density" early-type galaxies. Green triangles are the results from the ACS Virgo ClusterSurvey [1]; **Middle panel**: GC specific frequency. Same symbols, including the isolated ellipticals N7507 [7], N3585 and N5812 [8]; **Lower panel**: position of the red and blue peaks as a function of *K* luminosity for the five isolated ellipticals in [10].

Another useful diagnostic is the specific frequency of GCs, S_N, which measures the efficiency of cluster formation compared to the stellar population. Figure 1 (middle panel) shows another comparison between field, isolated, and Virgo ellipticals. The most relevant feature is the constancy of S_N for field/isolated ellipticals with $S_N \sim 1.5$, regardless of the host galaxy mass. On the other hand, Virgo ellipticals show a departure from field ellipticals starting at $M_K = -24$, which only accentuates at brighter luminosities. Finally, the bottom panel indicates the peak color of the red and blue sub-populations as a function of luminosity. Even though the slope for the blue peak is consistent with cluster ellipticals, we do not yet have an explanation for the flat behavior of the red peak.

3. Deep Imaging of Isolated Ellipticals

The results from GCSs of isolated ellipticals give inconclusive evidence about the assembly of those galaxies. While bimodality and red fractions are similar to the ones in cluster ellipticals, hinting at a GCS formation process independent of environment, S_Ns indicate that some process must be different, at least in the bright end. Was their isolation a common feature throughout their lifetimes implying a poor merger history, or is it only a transient feature, in the sense that their isolation only arises from a rich accretion history where all the environment has been merged to form this single galaxy?

One way to glean insight into the accretion history of galaxies is by measuring their outer stellar profile. Galaxies which have experienced a rich accretion history are expected to have an extended outer profile (formed by accreted material), while the stellar halo of truly isolated systems is expected to be sharply truncated at some radius (e.g., [4]). The low S_N means GCSs of these galaxies will be relatively poor, and testing this prediction relies not on the outer distribution of GCs, which is necessarily deficient when based only on optical colors, but on the diffuse stellar component itself.

3.1. Observations

Imaging for six ellipticals in our isolated sample was conducted with two of the 1 m telescopes within the Las Cumbres Observatory network, with total on-source exposures close to 4 h per target. The Sinistro cameras give a field-of-view of 27×27 arcmin^2, which allows the construction of night-sky frames directly from the dithered science frames. Reduction of the images was conducted using the general imaging pipeline THELI [13].

3.2. Results

Surface brightness profiles of the galaxies were measured with IRAF/ellipse after an iterative masking of the surrounding sources. The surface brightness profiles were then modeled with a Sérsic profile. Figure 2 shows a comparison of the Sérsic index of the galaxies in our sample compared to galaxies in [14]. At the same H absolute magnitude (used as a proxy mass), isolated ellipticals show lower Sérsic indices, (between 2 and 5) compared to Virgo ellipticals from [14] (with n up to 12). This is an indication that isolated ellipticals have experienced less mergers and are probably a product of only a major merger.

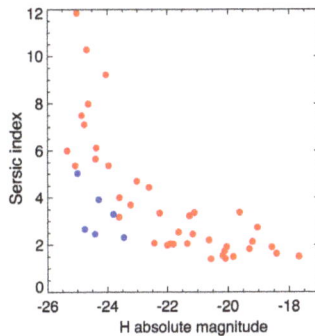

Figure 2. Sérsic index, n, as function of H absolute magnitude derived from 2MASS photometry. Blue circles show Virgo ellipticals from [14], while red symbols are our Las Cumbres sample.

4. Conclusions and Outlook

- Optical color bimodality is a common feature in GCSs, even for the most isolated elliptical galaxies.
- The red fractions of GCs in isolated ellipticals are on average lower than the ones in cluster environments, at odds with the predictions of hierarchical merging (e.g., [3]). Specific frequencies are also lower compared to cluster galaxies, with a remarkably flat behavior around $S_N \sim 1.5$.
- The blue peaks of GCSs in isolated ellipticals present a mild correlation with galaxy luminosity, although this result is driven mostly by the faintest galaxy in the sample (NGC 4240).
- Deep imaging of six isolated ellipticals reaching $\mu_g \sim 29$ show that these galaxies follow a Sérsic profile with low Sérsic index, more akin to dwarf ellipticals than giant ellipticals.
- Spectroscopic follow-up will provide the outer halo dynamics necessary for a systematic study of the dark matter content of isolated ellpticals, which might be rather small, as shown in the case of NGC 7507 [15,16]. Given the large distances to these systems, measuring the ages and metallicities of their GCs spectroscopically is prohibitive; therefore, these quantities will probably only be accessible through near-infrared imaging. A near-infrared imaging campaign for most of the galaxies discussed here is underway.

Acknowledgments: Based on observations obtained at the Gemini Observatory (Gemini programs: GS-2011B-Q-83, GS-2012A-Q-6), which is operated by the Association of Universities for Research in Astronomy, Inc., under a cooperative agreement with the NSF on behalf of the Gemini partnership: the National Science Foundation (United States), the National Research Council (Canada), CONICYT (Chile), Ministerio de Ciencia, Tecnología e Innovación Productiva (Argentina), and Ministério da Ciência, Tecnologia e Inovação (Brazil). T.R. acknowledges support from the BASAL Centro de Astrofísica y Tecnologías Afines (CATA) PFB-06/2007.

Author Contributions: T.R. and R.S. designed the observations and started this collaboration; R.R.L. reduced and analyzed the MOSAIC data; A.A. and R.S. reduced and analyzed the Gemini data, T.R. analyzed the VLT data, N.H., M.S. and R.S. reduced and analyzed the Las Cumbres data. R.S. wrote this contribution.

Conflicts of Interest: The authors declare no conflict of interest.

Abbreviations

The following abbreviations are used in this manuscript:

GC globular cluster
GCS globular cluster system
S_N specific frequency

References

1. Peng, E.W.; Jordán, A.; Côté, P.; Takamiya, M.; West, M.J.; Blakeslee, J.P.; Chen, C.W.; Ferrarese, L.; Mei, S.; Tonry, J.L.; et al. The ACS Virgo Cluster Survey. XV. The Formation Efficiencies of Globular Clusters in Early-Type Galaxies: The Effects of Mass and Environment. *Astrophys. J.* **2008**, *681*, 197–224.

2. Niemi, S.; Heinämäki, P.; Nurmi, P.; Saar, E. Formation, evolution and properties of isolated field elliptical galaxies. *Mon. Not. R. Astron. Soc.* **2010**, *405*, 477–493.

3. Tonini, C. The Metallicity Bimodality of Globular Cluster Systems: A Test of Galaxy Assembly and of the Evolution of the Galaxy Mass-Metallicity Relation. *Astrophys. J.* **2013**, *762*, 39.

4. Abadi, M.G.; Navarro, J.F.; Steinmetz, M. Stars beyond galaxies: The origin of extended luminous haloes around galaxies. *Mon. Not. R. Astron. Soc.* **2006**, *365*, 747–758.

5. Reda, F.M.; Forbes, D.A.; Beasley, M.A.; O'Sullivan, E.J.; Goudfrooij, P. The photometric properties of isolated early-type galaxies. *Mon. Not. R. Astron. Soc.* **2004**, *354*, 851–869.

6. Smith, R.M.; Martínez, V.J.; Graham, M.J. A Sample of Field Ellipticals. *Astrophys. J.* **2004**, *617*, 1017–1021.

7. Caso, J.P.; Richtler, T.; Bassino, L.P.; Salinas, R.; Lane, R.R.; Romanowsky, A. The paucity of globular clusters around the field elliptical NGC 7507. *Astron. Astrophys.* **2013**, *555*, A56.

8. Lane, R.R.; Salinas, R.; Richtler, T. Isolated ellipticals and their globular cluster systems. I. Washington photometry of NGC 3585 and NGC 5812. *Astron. Astrophys.* **2013**, *549*, A148.

9. Richtler, T.; Salinas, R.; Lane, R.R.; Hilker, M.; Schirmer, M. Isolated elliptical galaxies and their globular cluster systems. II. NGC 7796–Globular clusters, dynamics, companion. *Astron. Astrophys.* **2015**, *574*, A21.

10. Salinas, R.; Alabi, A.; Richtler, T.; Lane, R.R. Isolated ellipticals and their globular cluster systems. III. NGC 2271, NGC 2865, NGC 3962, NGC 4240, and IC 4889. *Astron. Astrophys.* **2015**, *577*, A59.

11. Stetson, P.B. DAOPHOT—A computer program for crowded-field stellar photometry. *Publ. Astron. Soc. Pac.* **1987**, *99*, 191–222.

12. Cho, J.; Sharples, R.M.; Blakeslee, J.P.; Zepf, S.E.; Kundu, A.; Kim, H.S.; Yoon, S.J. Globular cluster systems of early-type galaxies in low-density environments. *Mon. Not. R. Astron. Soc.* **2012**, *422*, 3591–3610.

13. Schirmer, M. THELI: Convenient Reduction of Optical, Near-infrared, and Mid-infrared Imaging Data. *Astrophys. J. Suppl. Ser.* **2013**, *209*, 21.

14. Kormendy, J.; Fisher, D.B.; Cornell, M.E.; Bender, R. Structure and Formation of Elliptical and Spheroidal Galaxies. *Astrophys. J. Suppl. Ser.* **2009**, *182*, 216–309.

15. Salinas, R.; Richtler, T.; Bassino, L.P.; Romanowsky, A.J.; Schuberth, Y. Kinematic properties of the field elliptical NGC 7507. *Astron. Astrophys.* **2012**, *538*, A87.

16. Lane, R.R.; Salinas, R.; Richtler, T. Dark matter deprivation in the field elliptical galaxy NGC 7507. *Astron. Astrophys.* **2015**, *574*, A93.

galaxies

MDPI

Letter

The Brazil–Argentina Gemini Group for the Study of Globular Cluster Systems (BAGGs GCs): FLAMINGOS-2 and GMOS Data for NGC 1395

Favio Faifer [1,2,*,†,‡], Carlos Escudero [1,2,‡], Analía Smith Castelli [2,‡], Juan Forte [3,4,‡], Leandro Sesto [1,2,‡], Ana Chies Santos [5,‡], Arianna Cortesi [6,‡] and Claudia Mendes de Oliveira [6,‡]

[1] Facultad de Ciencias Astronómicas y Geofísicas, National University of La Plata, La Plata B1900FWA, Argentina; cgescudero@fcaglp.unlp.edu.ar (C.E.); sesto@fcaglp.unlp.edu.ar (L.S.)
[2] Instituto de Astrofísica de La Plata (CCT La Plata, CONICET-UNLP), La Plata B1904CMC, Argentina; asmith@fcaglp.unlp.edu.ar
[3] The National Scientific and Technical Research Council (CONICET), CABA C1425FQB, Argentina; planeta.jcf@gmail.com
[4] Planetario de la Ciudad de Buenos Aires, CABA C1425FGC, Argentina
[5] Departamento de Astronomia, Instituto de Física, Universidade Federal do Rio Grande do Sul, Porto Alegre 90040-060, Brazil; ana.chies@ufrgs.br
[6] Instituto de Astronomia, Geofísica e Ciências Atmosféricas da U. de São Paulo, Cidade Universitária, 05508-900 São Paulo, Brazil; aricorte@googlemail.com (A.C.); claudia.oliveira@iag.usp.br (M.d.O.)
* Correspondence: favio@fcaglp.unlp.edu.ar; Tel.: +54-221-423-6591
† Current address: Paseo del Bosque s/n, La Plata CP B1900FWA, Argentina.
‡ These authors contributed equally to this work.

Academic Editors: Duncan A. Forbes and Ericson D. Lopez
Received: 1 July 2017; Accepted: 7 August 2017; Published: 14 August 2017

Abstract: In this letter, we present preliminary results of the analysis of Flamingos-2 and GMOS-S photometry of the globular cluster (GC) system of the elliptical galaxy NGC 1395. This is the first step of a long-term Brazilian–Argentinian collaboration for the study of GC systems in early-type galaxies. In the context of this collaboration, we obtained deep NIR photometric data in two different bands (J and Ks), which were later combined with high quality optical Gemini + GMOS photometry previously obtained by the Argentinian team. This allowed us to obtain different color indices, less sensitive to the effect of horizontal branch (HB) stars for several hundreds of GC candidates, and to make an initial assessment of the presence or absence of multiple GC populations in colors in NGC 1395.

Keywords: early-type galaxies; globular cluster; galaxy halos

1. Introduction

Globular clusters (GCs) are powerful probes to study the evolutionary histories of galaxies, as they are good tracers of galactic star forming episodes. They are found around all major galaxies and can be easily observed far beyond the Local Group [1]. As GCs are intrinsically old objects, their integrated properties could give us information about the physical conditions in the interstellar medium at the moment of formation of their host galaxies.

GC systems in massive galaxies are known to present a bimodal optical color distribution ([2,3]). This phenomenon has been identified in different galaxies and it has been shown that this effect is more clearly detectable when metallicity-sensitive color indices (such as $(g' - z')$ or $(C - T_1)$) are used ([4,5]). Therefore, color bimodality is usually interpreted as evidence for the existence of two GC sub-populations: the "blue" subpopulation (metal-poor clusters associated with the halo) and the "red" one (metal-rich clusters linked to bulge/disc). These sub-populations would have been formed in at least two stages (e.g., [6,7]): blue GCs would have been a primordial population formed at high redshifts in protogalactic fragments (e.g., [8,9]), while the red GCs would have formed later, during a gas-rich merging of these fragments (e.g., [10]).

However, it was suggested that the bimodality may be an artifact arising from the non-linearity in the color–metallicity (C–M) relation of GCs ([11,12]). In the [12] scenario, the morphology of the horizontal branch (HB) produces a "wavy" pattern in the C–M relation in such a way that it is possible to obtain a bimodal color distribution from a unimodal metallicity distribution. Despite several empirical C–M calibrations have been published in the last decade, there is still a lot of controversy about the existence of the alleged "wavy" pattern, i.e., how strong it is, and how much it influences our interpretation of multiple GC populations based on broad-band colors ([13–16]).

Although the combination of optical and NIR colors is expected to help to mitigate the "age–metallicity degeneracy" ([17]), it is still not clear how to interpret the lack of multi-modality in some systems observed with optical/NIR filters ([18]). The study of NGC 3115 by [15] (multi-band photometry and spectra centered on the CaT), suggests that if the underlying metallicity bimodality is real, it should be detected in all colors as well as in metal-sensitive indices. This means that, if the bimodality appears only in some colors, it would be due to the "wavy" pattern in the C–M relation.

In the framework of our BAGGs GCs collaboration, we are involved in obtaining deep optical and NIR multi-band photometry, as well as deep MOS spectroscopy, of GC systems belonging to massive early-type galaxies. We expect that this data, combined with different single stellar population models (SSP), helps to mitigate the age–metallicity degeneracy and allows us to measure several color-indices less sensitive to the effect of HB stars. By combining kinematics and color–metallicity measurements of GC systems in a self-consistent manner, we hope to be able to recover the evolutionary histories of the galaxies and shed light on the assembly history of the halos of the galaxies.

As a first step of our collaboration's observing campaign, in semester 2015B, we obtained data of the GC system of the giant elliptical galaxy NGC 1395 (D $\sim$ 21.4 Mpc, $M_B = -21.02$ mag), one of the dominant galaxies of the Eridanus group. This galaxy harbors thousands of GCs and shows a clear bimodal optical color distribution (Escudero+, in prep.). Here, we present preliminary results of an analysis of deep NIR photometric images taken in two different photometric bands (J and Ks), obtained with Gemini+Flamingos-2. The photometry obtained from these images was combined with high quality optical photometry from previous Gemini+GMOS runs.

2. Data

Within the BAGGs GCs collaboration, we obtained NIR images of two fields of the elliptical galaxy NGC 1395 using Gemini + Flamingos-2 in the J and Ks bands (Program GS-2015B-Q-38). The total on-source exposure times were 30×50 s in both fields for the J band, and, in the Ks band, 302×12 s on-source for the field containing the galaxy and 277×12 s on-source for the other field. The results presented here are based on a preliminary reduction performed with THELI ([19]). We used DAOPHOT ([20]) to obtain psf photometry of all the detected sources in the fields. The NIR photometry was calibrated using 2MASS objects present in our fields.

We also worked on a mosaic of four image fields of NGC 1395, previously obtained with Gemini+GMOS by the Argentinian team (Figure 1). These optical images were observed in the g' (4×180.5 s), r' (4×120.5 s), i' (4×150.5 s) and z' (4×150.5 s) bands, using 2×2 binning which gives a scale of $0.146''$/pix (Programs GS-2012B-Q-44 and GS-2014B-Q-28). In this work, we present the results obtained from the two fields that overlap with the NIR fields. The reduction of this data set was performed through the Gemini/GMOS IRAF package in the usual way (see [21] for more details). The detection and classification of the sources was made using a combination of SExtractor ([22]) and different IRAF tasks as is explained in [23]. Finally, we obtained DAOPHOT photometry in each band and we calibrated our final photometric catalogue using Sloan standard stars observed on the same night as our target. The complete optical photometric analysis will be presented in Escudero+ (in prep.).

We used RA and DEC coordinates to match the optical candidates and to build a master catalogue which includes all the unresolved sources detected and measured in the g', r', i', z', J and Ks bands. In order to obtain a clean sample of GC candidates as possible and to have reliable colors in all the bands, we cut the photometric sample in the range $18 < g' < 24$ mag with color cuts similar to those in Escudero+ (2015a). The final catalogue includes 650 candidates detected in all the bands (red circles in Figure 1).

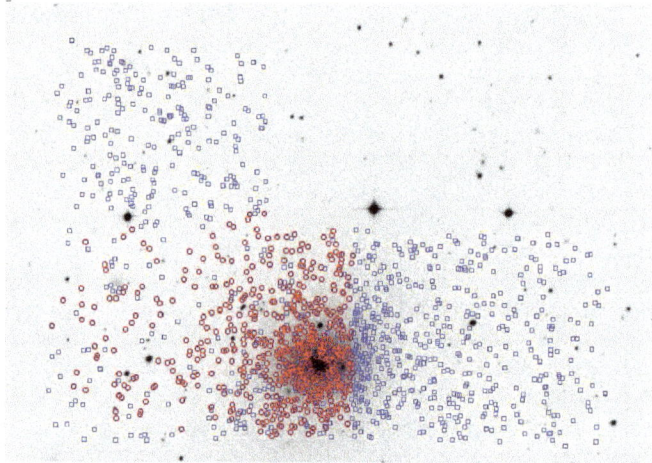

Figure 1. DSS red image of NGC 1395 showing the optical globular cluster (GC) candidates (blue squares). The GC candidates detected in the NIR images ($\sim$650) are shown in red. The size of this image is 12×17 arcmin. North is up and east is to the left.

3. Results

Our first goal is to try to identify the presence of different color GC subpopulations. To this aim, we built several color histograms and color–color diagrams. As an example, in Figure 2, we present the different color histograms obtained by combining g' magnitudes with the rest of our photometric bands. At this stage of our work, we did not apply any statistical test but rather made a preliminary identification of the modal peaks of the different colors. As in other giant early-type galaxies, the $(g' - z')$ color distribution of NGC 1395 looks bimodal ([24]). Two main peaks are clearly identified in modal colors of $(g' - z') = 0.93$ and $(g' - z') = 1.33$ mag. However, the appearance of the color distribution strongly depends on the combined bands. All the optical, optical-NIR and most of the purely NIR color distributions show signs of multiple subpopulations. Some of them, such as $(g' - i')$, $(g' - z')$ and $(g' - K_s)$, look strongly bimodal with possible substructures. As previously noticed by other authors (e.g., [25]), in the particular case of colors involving only NIR bands, the situation is not so clear.

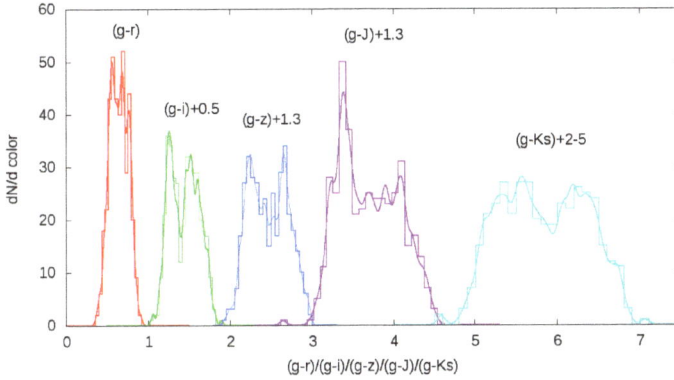

Figure 2. Histograms and density color distributions for different band combinations and for objects with $g' < 24$ mag. The considered bins are in the range 0.04–0.1 mag. The density distributions were built using a Gaussian kernel with $0.02 < \sigma < 0.05$ mag. To avoid superposition, we shifted some histograms by a constant value, as indicated in the figures.

The color–color diagrams show signs of different degrees of non-linearity. As an example, Figure 3 shows a $(g' - K_s)$ vs. $(g' - z')$ diagram with evidence of a slight non-linearity which is in good agreement with the results of previous studies ([26–28]). Interestingly, most of the smoothed diagrams obtained from our photometry suggest that, at least, two main subpopulations in color are present.

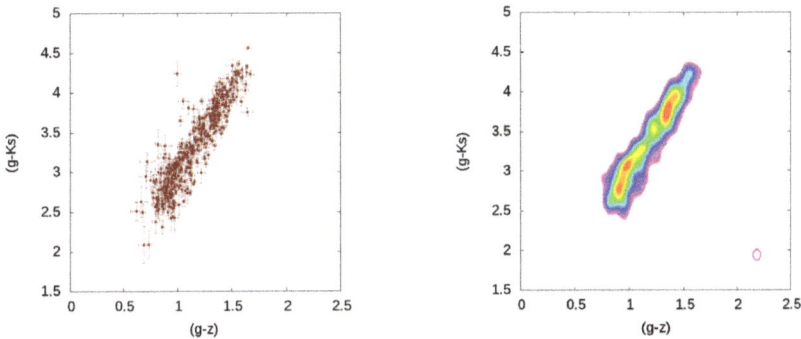

Figure 3. (**Left**): Color–color $(g' - K_s)$ vs. $(g' - z')$ diagram built from our optical and NIR photometric sample (objects with $g' < 24$ mag); (**Right**): Smoothed version of the same diagram as that on the left. The elliptical symbol depicts the kernel used to smooth the original diagram.

4. Discussion

As it has already been noticed by other authors, in some galaxies, the usual description of color histograms on unimodal or bimodal bases by fitting Gaussian distribution seems to be simplistic. For example, [29] has shown that the $(g' - z')$ color distributions of the GCs of several galaxies in Virgo and Fornax seem to display a common and synchronized multi-population pattern. Furthermore, [16] proposed that in some massive galaxies, the metallicity distribution might be thought of as the result of a rapid sequence of individual GC formation events leading to an apparent "continuous" metallicity distribution. In the particular case of NGC 1395, though preliminary, our results seem to fit in these pictures. That is, they seem to present a bimodal color distribution in some photometric band combinations, but show evidence of substructures in these color patterns.

To clarify all these issues, high-quality multi-band photometry and deep spectroscopy for a large sample of GCs belonging to different systems, are clearly needed. In the framework of our BAGGs GCs collaboration, we expect to obtain deep spectroscopy for a sub-sample of GC candidates in order to determine spectroscopic ages and metallicities to test the photometric results.

Acknowledgments: This research was funded with grants from Consejo Nacional de Investigaciones Científicas y Técnicas de la República Argentina, Agencia Nacional de Promoción Científica y Tecnológica, and Universidad Nacional de La Plata, Argentina.

Conflicts of Interest: The authors declare no conflict of interest.

References

1. Brodie, J.; Strader, J. Extragalactic Globular Clusters and Galaxy Formation. *Ann. Rev. Astron. Astrophys.* **2006**, *44*, 193.
2. Ashman, K.; Zeph, S. *Merging and Interacting Galaxies: Sites of Globular Cluster Formation*; Astronomical Society of the Pacific: San Francisco, CA, USA, 1993; Volume 48, p. 776.
3. Elson, R.; Santiago, B. The globular clusters in M87: A bimodal colour distribution. *Mon. Not. R. Astron. Soc.* **1996**, *278*, 617.
4. Barmby, P.; Huchra, J.P.; Brodie, J.P.; Forbes, D.A.; Schroder, L.L.; Grillmair, C.J. M31 Globular Clusters: Colors and Metallicities. *Astron. J.* **2000**, *119*, 727.
5. Sinnott, B.; Hou, A.; Anderson, R.; Harris, W.E.; Woodley, K.A. New g'r'i'z' Photometry of the NGC 5128 Globular Cluster System. *Astron. J.* **2010**, *140*, 2101.
6. Forbes, D.A.; Brodie, J.P.; Grillmair, C.J. On the Origin of Globular Clusters in Elliptical and cD Galaxies. *Astron. J.* **1997**, *113*, 1652.
7. Forte, J.; Vega, E.; Faifer, F.; Smith Castelli, A.; Escudero, C.; González, N.; Sesto, L. Globular clusters: DNA of early-type galaxies? *Mon. Not. R. Astron. Soc.* **2014**, *441*, 1391.
8. Beasley, M.; Baugh, C.; Forbes, D.; Sharples, R.; Frenk, C. On the formation of globular cluster systems in a hierarchical Universe. *Mon. Not. R. Astron. Soc.* **2002**, *333*, 383.
9. Muratov, A.; Gnedin, O. Modeling the Metallicity Distribution of Globular Clusters. *Astrophys. J.* **2010**, *718*, 1266.
10. Li, H.; Gnedin, O.Y. Modeling the Formation of Globular Cluster Systems in the Virgo Cluster. *Astrophys. J.* **2014**, *796*, 10.
11. Richtler, T. Some remarks on extragalactic globular clusters. *Bull. Astron. Soc. India* **2006**, *34*, 83.
12. Yoon, S.; Yi, S.; Lee, Y. Explaining the Color Distributions of Globular Cluster Systems in Elliptical Galaxies. *Science* **2006**, *311*, 1129.
13. Moyano Loyola, G.; Faifer, F.; Forte, J. Globular Clusters: Chemical Abundance—Integrated Colour calibration. *Bol. Asoc. Argent. Astron.* **2010**, *53*, 133.
14. Usher, C.; Forbes, D.; Brodie, J.; Foster, C.; Spitler, L.; Arnold, J.; Romanowsky, A.; Strader, J.; Pota, V. The SLUGGS survey: Calcium triplet-based spectroscopic metallicities for over 900 globular clusters. *Mon. Not. R. Astron. Soc.* **2012**, *426*, 1475.
15. Cantiello, M.; Blakeslee, J.; Raimondo, G.; Chies-Santos, A.; Jennings, Z.; Norris, M.; Kuntschner, H. Globular clusters of NGC 3115 in the near-infrared. Demonstrating the correctness of two opposing scenarios. *Astron. Astrophys.* **2014**, *564*, L3.
16. Harris, W.; Ciccone, S.; Eadie, G.; Gnedin, O.; Geisler, D.; Rothberg, B.; Bailin, J. Globular Cluster Systems in Brightest Cluster Galaxies. III: Beyond Bimodality. *Astrophys. J.* **2017**, *835*, 101.

17. Worthey, G. Comprehensive stellar population models and the disentanglement of age and metallicity effects. *Astrophys. J. Suppl.* **1994**, *95*, 107.

18. Chies-Santos, A.; Larsen, S.; Cantiello, M.; Strader, J.; Kuntschner, H.; Wehner, E.M.; Brodie, J.P. An optical/NIR survey of globular clusters in early-type galaxies. III. On the colour bimodality of globular cluster systems. *Astron. Astrophys.* **2012**, *539*, 54.

19. Schirmer, M. THELI: Convenient Reduction of Optical, Near-infrared, and Mid-infrared Imaging Data. *Astrophys. J. Suppl.* **2013**, *209*, 21.

20. Stetson, P.B. DAOPHOT—A computer program for crowded-field stellar photometry. *Publ. Astron. Soc. Pac.* **1987**, *99*, 191

21. Escudero, C.; Faifer, F.; Bassino, L.; Calderón, J.; Caso, J. The extremely populated globular cluster system of the lenticular galaxy NGC 6861. *Mon. Not. R. Astron. Soc.* **2015**, *449*, 612.

22. Bertin, E.; Arnouts, S. SExtractor: Software for source extraction. *Astron. Astrophys. Suppl.* **1996**, *117*, 393.

23. Faifer, F.; Forte, J.; Norris, M.; Bridges, T.; Forbes, D.; Zepf, S.; Beasley, M.; Gebhardt, K.; Hanes, D.; Sharples, R. Gemini/GMOS imaging of globular cluster systems in five early-type galaxies. *Mon. Not. R. Astron. Soc.* **2011**, *416*, 155.

24. Escudero, C.; Sesto, L.; González, N.; Faifer, F.; Smith Castelli, A.; Forte, J. Comparación fotométrica en imágenes Gemini/GMOS. *Bol. Asoc. Argent. Astron.* **2015**, *57*, 19.

25. Cho, H.; Blakeslee, J.; Chies-Santos, A.; Jee, M.; Jensen, J.; Peng, E.; Lee, Y. The Globular Cluster System of the Coma cD Galaxy NGC 4874 from Hubble Space Telescope ACS and WFC3/IR Imaging. *Astrophys. J.* **2016**, *822*, 95.

26. Blakeslee, J.; Cho, H.; Peng, E.; Ferrarese, L.; Jordán, A.; Martel, A. Optical and Infrared Photometry of Globular Clusters in NGC 1399: Evidence for Color-Metallicity Nonlinearity. *Astrophys. J.* **2012**, *746*, 88.

27. Forte, J.; Faifer, F.; Vega, E.; Bassino, L.; Smith Castelli, A.; Cellone, S.; Geisler, D. Multicolour-metallicity relations from globular clusters in NGC 4486 (M87). *Mon. Not. R. Astron. Soc.* **2013**, *431*, 1405.

28. Powalka, M.; Lançon, A.; Puzia, T.H.; Peng, E.W.; Liu, C.; Muñoz, R.P.; Blakeslee, J.P.; Côté, P.; Ferrarese, L.; Roediger, J.; et al. The Next Generation Virgo Cluster Survey (NGVS). XXV. Fiducial Panchromatic Colors of Virgo Core Globular Clusters and Their Comparison to Model Predictions. *Astrophys. J. Suppl.* **2016**, *227*, 12.

29. Forte, J. Supra-galactic colour patterns in globular cluster systems. *Mon. Not. R. Astron. Soc.* **2017**, *468*, 3917.

MDPI

Conference Report

The "Building Blocks" of Stellar Halos

Kyle A. Oman [1,*], Else Starkenburg [2] and Julio F. Navarro [1,†]

[1] Department of Physics & Astronomy, University of Victoria, Victoria, BC V8P 5C2, Canada; jfn@uvic.ca
[2] Leibniz Institute for Astrophysics Potsdam (AIP), An der Sternwarte 16, 14482 Potsdam, Germany; estarkenburg@aip.de
* Correspondence: koman@uvic.ca; Tel.: +1-250-721-7747
† Senior CIfAR fellow.

Academic Editors: Duncan A. Forbes and Ericson D. Lopez
Received: 30 June 2017; Accepted: 28 July 2017; Published: 2 August 2017

Abstract: The stellar halos of galaxies encode their accretion histories. In particular, the median metallicity of a halo is determined primarily by the mass of the most massive accreted object. We use hydrodynamical cosmological simulations from the APOSTLE project to study the connection between the stellar mass, the metallicity distribution, and the stellar age distribution of a halo and the identity of its most massive progenitor. We find that the stellar populations in an accreted halo typically resemble the old stellar populations in a present-day dwarf galaxy with a stellar mass $\sim$0.2–0.5 dex greater than that of the stellar halo. This suggests that had they not been accreted, the primary progenitors of stellar halos would have evolved to resemble typical nearby dwarf irregulars.

Keywords: galaxy formation and evolution; stellar halo; numerical simulations

1. Introduction

The accretion of smaller systems is an integral part of galaxy formation. The accretion history of a galaxy is perhaps most clearly encoded in its stellar halo, due to the combination of a relative scarcity of stars formed "in-situ", and long dynamical times which allow orbital information to persist. Motivated by the large apparent differences in mass and metallicity between the stellar halos of the Milky Way (MW) and M 31, [1,2] proposed a picture—supported by cosmological simulation work—in which the metallicity of accreted material reflects the assembly history of the galaxy. The median metallicity of a stellar halo is now thought to be a reflection of the mass of the most massive accreted object [3–5], a notion supported by the recent first observation of the stellar halo mass–metallicity relation [6].

Below, we propose a simple method to explore, within a theoretical framework, the possible link between the accreted halo of a galaxy and present-day dwarf galaxies which may resemble those disrupted to form it.

2. Materials and Methods

We use the APOSTLE[1] suite of cosmological hydrodynamical simulations [7,8]. These comprise twelve volumes, each containing two halos with masses, separations, kinematics, and local environment consistent with the MW, M 31, and the Local Group of galaxies. Each volume is simulated at mutliple resolution levels, with gas particle masses varying from $\sim$10^6 M$_\odot$ at the lowest (L3) resolution level to $\sim$10^4 M$_\odot$ at the highest (L1) level. In this study, we use the intermediate (L2) level with gas particles of $\sim$10^5 M$_\odot$, dark matter particle mass $\sim$6 $\times$ 10^5 M$_\odot$, and $\sim$300 pc force softening.

---

[1] A Project Of Simulating The Local Environment

This is the highest resolution at which all twelve volumes have been integrated to the present day. Each volume samples a region extending to radii $\gtrsim 2\,\text{Mpc}$ around the barycentre of the two central objects. We assume the WMAP7 cosmological parameters [9].

APOSTLE uses the same hydrodynamics and galaxy formations prescriptions as the EAGLE project [10,11]—specifically, the model labelled "Ref" by [10]. The hydrodynamics are solved using the pressure–entropy formulation of smoothed particle hydrodynamics [12], and the ANARCHY collection of numerical methods (for a brief description, see [10]) is used. The model includes prescriptions for radiative cooling [13], star formation [14,15], stellar and chemical enrichment [16], stellar feedback [17], and cosmic reionization [16,18]. The model is calibrated to reproduce the galaxy mass–size relation and galaxy stellar mass function of $M_\star > 10^8\,\text{M}_\odot$ objects [11].

Structures are identified in the simulation output using the friend-of-friends (FoF) [19] and SUBFIND [20,21] algorithms. The former iteratively links particles separated by less than $0.2\times$ the mean inter-particle separation; the latter then identifies self-bound substructures, termed "subhalos", separating them along saddle points in the density distribution. The most massive object in each FoF group is labelled the "central" object, and other objects in the same group are "satellites".

In this study we focus on the stellar halo component of the 24 roughly MW- and M 31-like galaxies (two per volume) in the APOSTLE suite. For each simulated galaxy we identify the progenitor system at earlier times using the merger tree procedure described in [22]. We explicitly check that the progenitor "tracks" are smooth in position and mass (i.e., that the primary progenitor is accurately traced through time). For each system we define the "accreted halo" as the collection of star particles in the SUBFIND group (i.e., gravitationally bound to the host) whose FoF group at their time of formation is not the FoF group hosting the progenitor of the MW- or M 31-like galaxy at the same time. Typically, a satellite galaxy will become FoF-associated with its host before any substantial disturbances to the stellar and gas components of the satellite due to the massive host begin. Our sample therefore does not include stars formed in tidal tails after accretion, "in-situ" halo stars, or stars ejected from the central galaxy. We note that our definition of "accreted halo" includes *all accreted stars*. Many of these are located in the central regions of the object they were accreted by and might be observationally characterized as part of a bulge, rather than a halo, component.

3. Results

In Figure 1 we show the metallicity and formation time[2] distributions of the accreted halo stars. The distributions are normalized to the total stellar mass of each system before they are combined, such that each system contributes equal weight to the distribution. The accreted halos as defined here are significantly more metal-rich (median -0.31) than recent estimates for the MW (median -1.78 [23]), M 31 ($\lesssim -0.7$ [24]), and other galaxies with similar stellar masses [6]. The chemical enrichment prescriptions adopted in the EAGLE-Ref model—and therefore used in the APOSTLE simulations—result in a galaxy stellar mass–median metallicity relation offset to higher metallicities than observed [25] for galaxies with $M_\star \lesssim 10^8\,\text{M}_\odot$. The disruption of these unusually metal-rich satellite galaxies unsurprisingly results in unusually metal-rich stellar haloes. This is fundamentally a shortcoming of the model. A direct comparison with the measurements cited above is further hindered by our selection of accreted particles, which inevitably includes many stars which would not usually be present in observed samples of "halo stars", especially toward the centre of each system. In the context of our analysis below, these differences are of limited concern since our argument concerns mainly relative—rather than absolute—metallicities.

---

[2]   the age of the Universe minus the age of the star

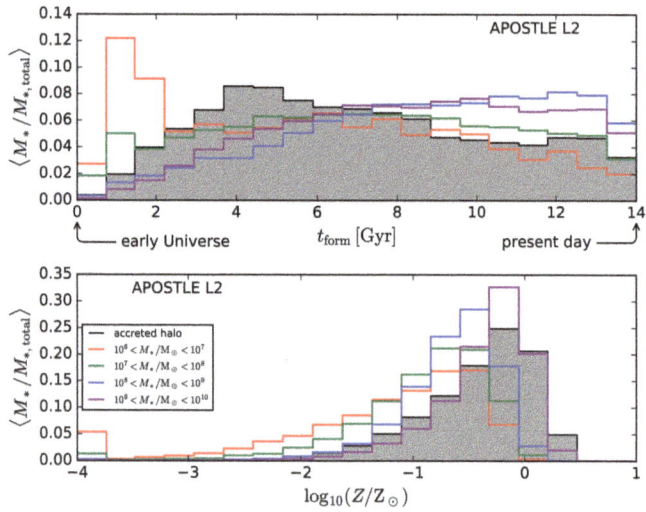

Figure 1. The upper panel shows the formation time distribution for stars in the accreted halos of the 24 Milky Way (MW)- and M 31-like galaxies from the APOSTLE simulation suite (filled histogram) at resolution level L2. The coloured lines show the same distribution for field (central) objects in 4 consecutive 1-dex bins of stellar mass from 10^6 to 10^{10} $M_\odot$ (red, green, blue, purple in order of increasing $M_\star$). The lower panel shows the metallicity distribution for the same classes of objects. The accreted halos have a metallicity distribution similar to that of 10^9–10^{10} $M_\odot$ field objects, but relatively older stellar populations. For all curves, star particles with metallicities < -4 contribute to the lowest metallicity bin shown.

With its tail of recently-formed stars, the formation time distribution may at first glance seem unusual: the MW stellar halo does not have such a population of young stars [26]. However, the M 31 stellar halo has a star formation history which, though it still peaks at ages of 5–13 Gyr, has a clear tail to much younger ages [27]. Furthermore, there is no attempt in the selection of systems for the APOSTLE simulations to choose halos with merger histories similar to the MW and M 31, or even the morphology of the galaxies: a few of the systems have had recent major mergers and still show clear morphological disturbances. These systems make the largest (but not the entire) contribution to the tail of young stars. How the APOSTLE sample of stellar haloes compares to the similarly-sized GHOSTS sample of stellar haloes recently observed in detail [6,28] (see also the compilation in [29]) is a topic we hope to pursue in a future contribution.

In Figure 1 we also show the formation time and metallicity distributions for other isolated (central) galaxies within the same APOSTLE simulation volumes, binned by stellar mass in 1 dex bins from 10^6–10^{10} $M_\odot$. The metallicity distribution of the accreted halos is roughly similar to that of the field objects in the highest mass bin ($M_\star \sim 10^{9.5}$), but their formation time distribution is biased to earlier times (older ages). This is unsurprising: 10^{10} $M_\odot$ galaxies in the field are typically actively star forming up to the present day, whereas recently formed stars are excluded from the accreted halo sample which is dominated by disrupted "quenched" sattelites, as enforced by our selection process.

The metallicity distributions in Figure 1 hint that the stellar populations in the accreted halos must be dominated by relatively massive accreted objects—the high median metallicity simply cannot be reached via the accretion of many low mass objects. We now explore this point further. For illustrative purposes, we use a single APOSTLE galaxy from the 7th volume, which we label[3] AP-L2-V7-1-0. This galaxy was chosen "by eye" to be representative of the sample. The formation time and metallicity distributions of this galaxy are shown in Figure 2. The total stellar mass of accreted halo stars in this system is $10^{9.4}$ $M_\odot$ (for comparison, the MW stellar halo mass is $\sim 5.5 \times 10^8$ $M_\odot$ [30], that of M 31 is $\sim 1.5 \times 10^{10}$ $M_\odot$ [24]; see also [29]). We also show the formation time and metallicity distributions for present-day dwarf galaxies in the field which have stellar masses slightly (0.2–0.5 dex; the choice of this particular interval is explained below) larger than this accreted halo. The stellar populations in these field galaxies are more metal-rich and younger than those in the accreted halo. However, if we re-weight the star particles in the same field objects such that the age distribution of the accreted halo is exactly matched, the resulting metallicity distribution is a close match to the metallicity distribution of the accreted halo, both in terms of the median (offset by less than 0.05 dex, compared to 0.3 dex before re-weighting) and the shape.

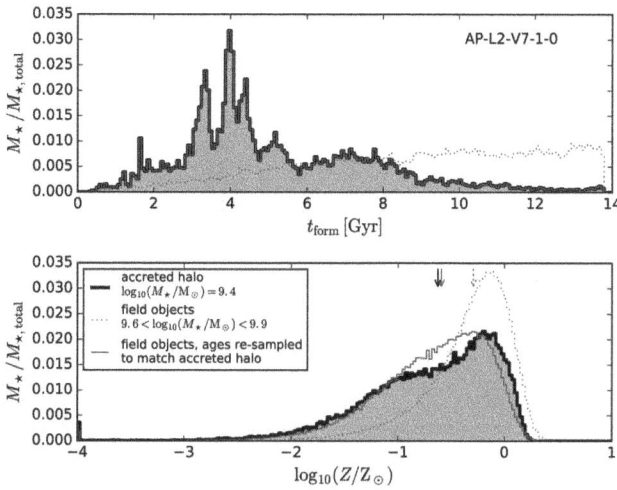

Figure 2. The upper panel shows the formation time distribution for stars in the accreted halo of AP-L2-V7-1-0, one MW or M 31-like galaxy from the APOSTLE suite (filled histogram); the lower panel is similar but for the metallicity distribution. The dotted purple histogram shows the same distributions for field (central) objects which have stellar masses in the range $10^{9.6}$–$10^{9.9}$ $M_\odot$, i.e., 0.2–0.5 dex more massive than the accreted halo of AP-L2-V7-1-0. The result of re-weighting the stellar populations of the field objects, weighted by the formation time distribution of the accreted halo, is shown with the solid purple histogram (by construction, the formation time distribution is then a perfect match). This enforced bias toward older stars in the field objects results in a re-sampled metallicity distribution which resembles much more closely that of the accreted halo. Arrows mark the median of the distribution with corresponding line style.

In Figure 3 we show the result of applying the same process illustrated in Figure 2 to all 24 accreted halos in our sample. We use the same offset of 0.2–0.5 dex in stellar mass for all galaxies and show the median metallicity before and after re-weighting by the formation time distribution of the accreted halo. In most cases, the median of the re-weighted distribution approaches that of the accreted halo, though with significant scatter. The mass offset interval was chosen based on purely empirical considerations,

---

[3] AP-[resolution level]-[volume number]-[FoF group number]-[subgroup number]

by systematically exploring a range of possibilities covering the full mass range of field objects present in the simulations, and various widths for the interval. The 0.2–0.5 dex window is the one which minimizes the scatter in the right panel of Figure 3, without introducing a systematic offset from the line of 1:1 agreement.

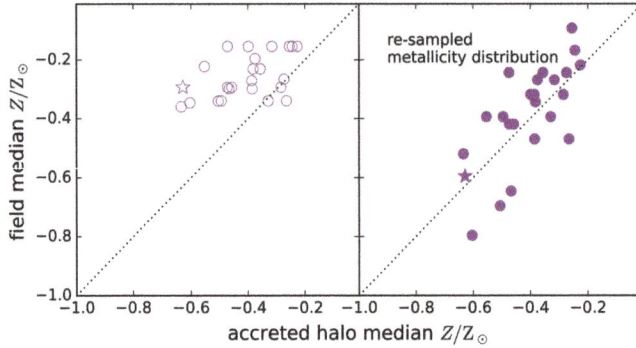

Figure 3. Result of applying the stellar population re-weighting illustrated in Figure 2 to the 24 MW- and M 31-like galaxies in the APOSTLE suite (AP-L2-V7-1-0, the example from Figure 2, is marked with a star). In each case, the field objects in the stellar mass interval between 0.2 and 0.5 dex more massive than the stellar mass of the accreted halo are selected. Before the re-weighting (**left panel**), the field objects typically have a median metallicity greater than that of the corresponding accreted halo (the dotted line indicates 1:1 agreement). After re-weighting (**right panel**), the medians of the metallicity distributions of the field objects and accreted halos agree, albeit with substantial scatter.

4. Discussion

The "accreted halos" from the APOSTLE simulation suite, as we have defined them here, should not be taken as direct detailed models of the MW, M 31, or other galactic stellar halos as defined observationally. They are, however, robust and internally self-consistent models of the assembly of such systems, and simultaneously of the nearby field objects which survive to the present day.

The above results suggest that the mass in the accreted halo of a galaxy is usually dominated by the disrupted content of one or a handful of relatively massive objects. If these had continued to grow and evolve in isolation instead of being accreted and destroyed, we would expect them to resemble present-day dwarf galaxies in the field with masses a factor of $\lesssim 3$ greater than that of the accreted halo. Older stellar populations in relatively massive dwarf galaxies are roughly the surviving analogs of the most massive "building blocks" of stellar halos.

Though it seems that 1–2 disrupted massive systems make up the bulk of most stellar halos, the remains of many lower-mass systems are also expected to be present. These are a nearly insignificant contribution (by mass) to the halo as a whole, but their signature may be detectable as a radial gradient—less massive systems are subject to weaker dynamical friction, and are destroyed at larger radii—and/or as overdense features such as shells or streams. The simulations and method used above offer a means of studying the link between the properties of such features and the types of objects which were destroyed to create them.

Acknowledgments: We thank the other members of the APOSTLE team, especially T. Sawala, A. Fattahi, M. Schaller, C. Frenk, for their efforts in creating the simulations used in this work. We thank the EAGLE simulation collaboration for providing the galaxy formation model, software and calibration. KAO thanks J. Helly for support in using the merger tree software. We thank the anonymous referees for their useful comments and suggestions. This work was supported by the Science and Technology Facilities Council (grant number ST/F001166/1). This work used the DiRAC Data Centric system at Durham University, operated by the Institute for Computation Cosmology on behalf of the STFC DiRAC HPC Facility (www.dirac.ac.uk). This equipment was funded by BIS National E-infrastructure capital grant ST/K00042X/1, STFC capital grant ST/H008519/1, and STFC DiRAC Operations grant ST/K003267/1 and Durham University. DiRAC is part of the National E-Infrastructure. This research has made use of NASA's Astrophysics Data System.

Author Contributions: K.A.O., E.S. and J.F.N. conceived and designed the experiments; K.A.O. performed the experiments; K.A.O. and E.S. and J.F.N. analyzed the data; K.A.O. wrote the paper.

Conflicts of Interest: The authors declare no conflict of interest. The founding sponsors had no role in the design of the study; in the collection, analyses, or interpretation of data; in the writing of the manuscript, and in the decision to publish the results.

References

1. Renda, A.; Gibson, B.K.; Mouhcine, M.; Ibata, R.A.; Kawata, D.; Flynn, C.; Brook, C.B. The stellar halo metallicity-luminosity relationship for spiral galaxies. *Mon. Not. R. Astron. Soc.* **2005**, *363*, L16–L20.

2. Font, A.S.; Johnston, K.V.; Bullock, J.S.; Robertson, B.E. Phase-Space Distributions of Chemical Abundances in Milky Way-Type Galaxy Halos. *Astrophys. J.* **2006**, *646*, 886–898.

3. Robertson, B.; Bullock, J.S.; Font, A.S.; Johnston, K.V.; Hernquist, L. Λ Cold Dark Matter, Stellar Feedback, and the Galactic Halo Abundance Pattern. *Astrophys. J.* **2005**, *632*, 872–881.

4. Deason, A.J.; Mao, Y.Y.; Wechsler, R.H. The Eating Habits of Milky Way-mass Halos: Destroyed Dwarf Satellites and the Metallicity Distribution of Accreted Stars. *Astrophys. J.* **2016**, *821*, 5.

5. D'Souza, R.; Bell, E. Accreted Metallicity-Stellar Mass Relationship. *arXiv* **2017**, arXiv:1705.08442.

6. Harmsen, B.; Monachesi, A.; Bell, E.F.; de Jong, R.S.; Bailin, J.; Radburn-Smith, D.J.; Holwerda, B.W. Diverse stellar haloes in nearby Milky Way mass disc galaxies. *Mon. Not. R. Astron. Soc.* **2017**, *466*, 1491–1512.

7. Sawala, T.; Frenk, C.S.; Fattahi, A.; Navarro, J.F.; Bower, R.G.; Crain, R.A.; Dalla Vecchia, C.; Furlong, M.; Helly, J.C.; Jenkins, A.; et al. The APOSTLE simulations: solutions to the Local Group's cosmic puzzles. *Mon. Not. R. Astron. Soc.* **2016**, *457*, 1931–1943.

8. Fattahi, A.; Navarro, J.F.; Sawala, T.; Frenk, C.S.; Oman, K.A.; Crain, R.A.; Furlong, M.; Schaller, M.; Schaye, J.; Theuns, T.; et al. The APOSTLE project: Local Group kinematic mass constraints and simulation candidate selection. *Mon. Not. R. Astron. Soc.* **2016**, *457*, 844–856.

9. Komatsu, E.; Smith, K.M.; Dunkley, J.; Bennett, C.L.; Gold, B.; Hinshaw, G.; Jarosik, N.; Larson, D.; Nolta, M.R.; Page, L.; et al. Seven-year Wilkinson Microwave Anisotropy Probe (WMAP) Observations: Cosmological Interpretation. *Astrophys. J. Suppl.* **2011**, *192*, 18.

10. Schaye, J.; Crain, R.A.; Bower, R.G.; Furlong, M.; Schaller, M.; Theuns, T.; Dalla Vecchia, C.; Frenk, C.S.; McCarthy, I.G.; Helly, J.C.; et al. The EAGLE project: Simulating the evolution and assembly of galaxies and their environments. *Mon. Not. R. Astron. Soc.* **2015**, *446*, 521–554.

11. Crain, R.A.; Schaye, J.; Bower, R.G.; Furlong, M.; Schaller, M.; Theuns, T.; Dalla Vecchia, C.; Frenk, C.S.; McCarthy, I.G.; Helly, J.C.; et al. The EAGLE simulations of galaxy formation: Calibration of subgrid physics and model variations. *Mon. Not. R. Astron. Soc.* **2015**, *450*, 1937–1961.

12. Hopkins, P.F. A general class of Lagrangian smoothed particle hydrodynamics methods and implications for fluid mixing problems. *Mon. Not. R. Astron. Soc. Lett.* **2013**, *428*, 2840–2856.

13. Wiersma, R.P.C.; Schaye, J.; Smith, B.D. The effect of photoionization on the cooling rates of enriched, astrophysical plasmas. *Mon. Not. R. Astron. Soc.* **2009**, *393*, 99–107.

14. Schaye, J. Star Formation Thresholds and Galaxy Edges: Why and Where. *Astrophys. J.* **2004**, *609*, 667–682.

15. Schaye, J.; Dalla Vecchia, C. On the relation between the Schmidt and Kennicutt-Schmidt star formation laws and its implications for numerical simulations. *Mon. Not. R. Astron. Soc.* **2008**, *383*, 1210–1222.

16. Wiersma, R.P.C.; Schaye, J.; Theuns, T.; Dalla Vecchia, C.; Tornatore, L. Chemical enrichment in cosmological, smoothed particle hydrodynamics simulations. *Mon. Not. R. Astron. Soc.* **2009**, *399*, 574–600.

17. Dalla Vecchia, C.; Schaye, J. Simulating galactic outflows with thermal supernova feedback. *Mon. Not. R. Astron. Soc.* **2012**, *426*, 140–158.

18. Haardt, F.; Madau, P. Modelling the UV/X-ray cosmic background with CUBA. In *Clusters of Galaxies and the High Redshift Universe Observed in X-rays,Proceedings of the XMM-Newton and Chandra, XXXVIth Rencontres de Moriond , XXIst Moriond Astrophysics Meeting, Savoie, France, 10–17 March 2001*; Neumann, D.M., Tran, J.T.V., Eds.; CEA: Gif-sur-Yvette, France, 2001; p. 64.

19. Davis, M.; Efstathiou, G.; Frenk, C.S.; White, S.D.M. The evolution of large-scale structure in a universe dominated by cold dark matter. *Astrophys. J.* **1985**, *292*, 371–394.

20. Springel, V.; White, S.D.M.; Tormen, G.; Kauffmann, G. Populating a cluster of galaxies-I. Results at z = 0. *Mon. Not. R. Astron. Soc.* **2001**, *328*, 726–750.

21. Dolag, K.; Borgani, S.; Murante, G.; Springel, V. Substructures in hydrodynamical cluster simulations. *Mon. Not. R. Astron. Soc.* **2009**, *399*, 497–514.

22. Helly, J.C.; Cole, S.; Frenk, C.S.; Baugh, C.M.; Benson, A.; Lacey, C. Galaxy formation using halo merger histories taken from N-body simulations. *Mon. Not. R. Astron. Soc.* **2003**, *338*, 903–912.

23. An, D.; Beers, T.C.; Johnson, J.A.; Pinsonneault, M.H.; Lee, Y.S.; Bovy, J.; Ivezić, Ž.; Carollo, D.; Newby, M. The Stellar Metallicity Distribution Function of the Galactic Halo from SDSS Photometry. *Astrophys. J.* **2013**, *763*, 65.

24. Ibata, R.A.; Lewis, G.F.; McConnachie, A.W.; Martin, N.F.; Irwin, M.J.; Ferguson, A.M.N.; Babul, A.; Bernard, E.J.; Chapman, S.C.; Collins, M.; et al. The Large-scale Structure of the Halo of the Andromeda Galaxy. I. Global Stellar Density, Morphology and Metallicity Properties. *Astrophys. J.* **2014**, *780*, 128.

25. Kirby, E.N.; Cohen, J.G.; Guhathakurta, P.; Cheng, L.; Bullock, J.S.; Gallazzi, A. The Universal Stellar Mass-Stellar Metallicity Relation for Dwarf Galaxies. *Astrophys. J.* **2013**, *779*, 102.

26. Carollo, D.; Beers, T.C.; Placco, V.M.; Santucci, R.M.; Denissenkov, P.; Tissera, P.B.; Lentner, G.; Rossi, S.; Lee, Y.S.; Tumlinson, J. The age structure of the Milky Way's halo. *Nat. Phys.* **2016**, *12*, 1170–1176.

27. Brown, T.M.; Smith, E.; Ferguson, H.C.; Rich, R.M.; Guhathakurta, P.; Renzini, A.; Sweigart, A.V.; Kimble, R.A. The Detailed Star Formation History in the Spheroid, Outer Disk, and Tidal Stream of the Andromeda Galaxy. *Astrophys. J.* **2006**, *652*, 323–353.

28. Monachesi, A.; Bell, E.F.; Radburn-Smith, D.J.; Bailin, J.; de Jong, R.S.; Holwerda, B.; Streich, D.; Silverstein, G. The GHOSTS survey-II. The diversity of halo colour and metallicity profiles of massive disc galaxies. *Mon. Not. R. Astron. Soc.* **2016**, *457*, 1419–1446.

29. Bell, E.F.; Monachesi, A.; Harmsen, B.; de Jong, R.S.; Bailin, J.; Radburn-Smith, D.J.; D'Souza, R.; Holwerda, B.W. Galaxies Grow Their Bulges and Black Holes in Diverse Ways. *Astrophys. J. Lett.* **2017**, *837*, L8.

30. Bland-Hawthorn, J.; Gerhard, O. The Galaxy in Context: Structural, Kinematic, and Integrated Properties. *Annu. Rev. Astron. Astrophys.* **2016**, *54*, 529–596.

galaxies

MDPI

Conference Report

The Extended Baryonic Halo of NGC 3923

Bryan W. Miller [1,*], Tomás Ahumada [1], Thomas H. Puzia [2], Graeme N. Candlish [3], Stacy S. McGaugh [4], J. Christopher Mihos [4], Robyn E. Sanderson [5], Mischa Schirmer [1], Rory Smith [6] and Matthew A. Taylor [7]

[1] Gemini Observatory, Casilla 603, La Serena 1700000, Chile; tahumada@gemini.edu (T.A.); mschirmer@gemini.edu (M.S.)

[2] Institute of Astrophysics, Pontificia Universidad Católica de Chile, Santiago 7820436, Chile; tpuzia@astro.puc.cl

[3] Physics and Astronomy Institute, Universidad de Valparaíso, Valparaíso 2360102, Chile; graeme.candlish@ifa.uv.cl

[4] Department of Astronomy, Case Western Reserve University, Cleveland, OH 44106, USA; stacy.mcgaugh@case.edu (S.S.M.); mihos@case.edu (J.C.M.)

[5] Department of Astronomy, California Institute of Technology, Pasadena, CA 91125, USA; robyn.sanderson@gmail.com

[6] Optical Astronomy Division, Korea Astronomy and Space Science Institute, Daedeokdae-ro 776, Yuseong-gu, Daejeon 34055, Korea; rorysmith274@gmail.com

[7] Gemini Observatory, Hilo, HI 96720, USA; mtaylor@gemini.edu

* Correspondence: bmiller@gemini.edu; Tel.: +56-51-2205618

Academic Editors: Emilio Elizalde, Duncan A. Forbes and Ericson D. Lopez
Received: 1 July 2017; Accepted: 14 July 2017; Published: 20 July 2017

Abstract: Galaxy halos and their globular cluster systems build up over time by the accretion of small satellites. We can learn about this process in detail by observing systems with ongoing accretion events and comparing the data with simulations. Elliptical shell galaxies are systems that are thought to be due to ongoing or recent minor mergers. We present preliminary results of an investigation of the baryonic halo—light profile, globular clusters, and shells/streams—of the shell galaxy NGC 3923 from deep Dark Energy Camera (DECam) g and i-band imaging. We present the 2D and radial distributions of the globular cluster candidates out to a projected radius of about 185 kpc, or $\sim 37 R_e$, making this one of the most extended cluster systems studied. The total number of clusters implies a halo mass of $M_h \sim 3 \times 10^{13}$ M$_\odot$. Previous studies had identified between 22 and 42 shells, making NGC 3923 the system with the largest number of shells. We identify 23 strong shells and 11 that are uncertain. Future work will measure the halo mass and mass profile from the radial distributions of the shell, N-body models, and line-of-sight velocity distribution (LOSVD) measurements of the shells using the Multi Unit Spectroscopic Explorer (MUSE).

Keywords: galaxies: elliptical and lenticular, cD; galaxies: halos; galaxies: individual (NGC 3923); galaxies: structure; galaxies: star clusters: general

1. Introduction

Galaxies at high redshift are much more compact at a given mass than in the local universe, implying that galaxies and their halos grow with time due to the accretion of mostly lower-mass galaxies [1–3]. Therefore, we can learn about the details of this process by observing nearby systems that are in the process of merging or accreting. Merger remnants and galaxies with tidal streams are good candidates. Elliptical shell galaxies are other promising environments for studying the build-up of halos. Shell galaxies are surrounded by interleaved "umbrellas" or shells of stellar material that match the appearance of structures created during the accretion of low-mass satellites on near-radial orbits in simulations (Figure 1) [4].

In this contribution, we present preliminary results of a study of the globular cluster system and shells of the elliptical shell galaxy NGC 3923. NGC 3923 is one of the most studied shell galaxies because it has the largest number of detected shells. Different studies have identified between 22 and 42 shells [5,6]. Therefore, new, deeper data will help determine the true number of shells and look for additional shells at larger radii. Mergers will also bring in new globular clusters (GCs) so the distribution and total number of GCs can be used to estimate the halo mass. We adopt a distance to NGC 3923 of 21.3 ± 1.4 Mpc ($(m-M)_0 = 31.64 \pm 0.14$) [7]. Using a total apparent magnitude of $V_T^0 = 9.69$ [8] results in an absolute magnitude of $M_V = -21.95$.

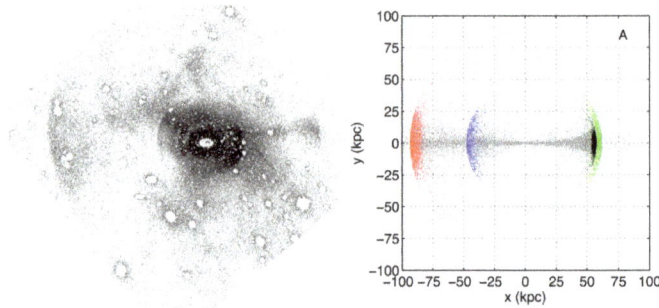

Figure 1. The observed shells around NGC 3923 (left panel) compared with the distribution of particles stripped from an accreting satellite galaxy on a radial orbit in an N-body simulation (right panel; reproduced from Figure 3 of [4]). The underlying light of the galaxy has been removed from the image using the ARCHANGEL ellipse-fitting package [9]. The simulation is of a 2.2×10^8 M$_\odot$ satellite falling into a spherical potential with total mass of 2.7×10^{12} M$_\odot$.

2. Observations and Reduction

NGC 3923 was observed on 22–23 April 2015 using Dark Energy Camera (DECam) on the Blanco 4-m telescope at the Cerro Tololo Inter-American Observatory (CTIO) [10]. The conditions were photometric with seeing of about 1 arcsec. Total integration times of 10,400 sec in g'-band and 10,800 sec in i'-band were obtained. A Fermat spiral dither pattern was used to maximize uniformity across the gaps between the 62 detectors. Sky subtraction is critical since we are searching for low-surface brightness features on scales of a degree on the sky. Therefore, exposures alternated between the NGC 3923 field and five different sky fields (Figure 2).

These observations were done as part of the larger *Neighborhood Watch* survey of the baryonic structures within about 20 Mpc using DECam $u'g'i'$ and VIRCAM J and K_s imaging. The goals are to use globular clusters, dwarf galaxies, and shells to study stellar populations and the formation of structures in the local universe. Targets include the Fornax Cluster and the CenA, Sombrero, NGC 2997, NGC 6744, and NGC 3115 groups.

The sky-subtraction algorithms used by the DECam community pipeline [11] are not designed for extended objects and over-subtract the sky in the vicinity of large, bright galaxies. Therefore, we reduced the data using the THELI reduction software [12] that used the images of the sky fields to create the background model. This eliminated the problem of sky over-subtraction.

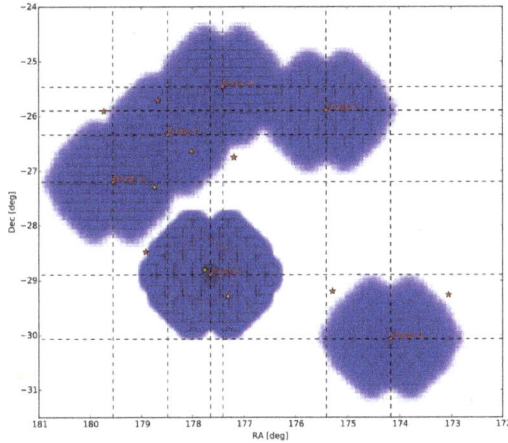

Figure 2. DECam observing strategy for NGC 3923. The yellow star marks the center of NGC 3923 and the red dashed circle indicates the approximate extent of the shell system. Red stars indicate bright foreground stars that we attempted to avoid. Orange diamonds indicate other objects of interest. The sky fields (Probes 1–5) were selected to not have bright galaxies or extended structures.

3. Results

3.1. The Globular Cluster System

At at distance of 21.3 Mpc, GCs are unresolved from the ground in 1 arcsec seeing. Therefore, we identified candidate GCs by selecting objects consistent with being point sources and then doing statistical background subtraction. The candidate GCs from [13]—taken in much better image quality using GMOS-S —were used as a "training" set to guide our selection. Source detection and photometry were done using SExtractor and PSFEx [14,15]. Principal component analysis (PCA) was applied to the photometry parameters in order to select point sources (Figure 3a). The eigenvectors are dominated by the FLUX_RADIUS and SPREAD_MODEL parameters. We then applied cuts in color-magnitude space as shown in Figure 3b. Finally, we fit a plane to the source density of objects more than 30 arcmin from NGC 3923, and statistically (randomly) subtracted this background.

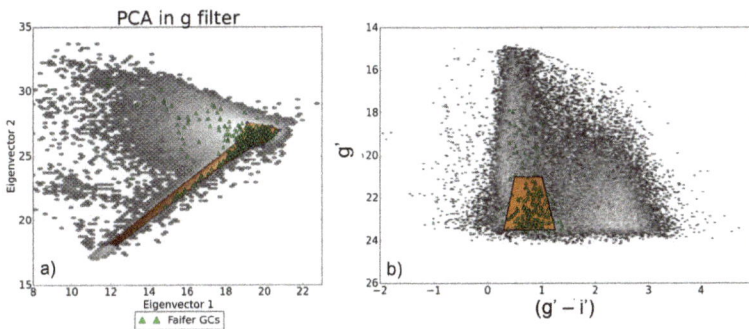

Figure 3. (**a**) Star+globular cluster (GC)/galaxy separation using principal component analysis (PCA). Most of the power is in the FLUX_RADIUS and SPREAD_MODEL parameters. Green triangles represent GCs and ultra-compact dwarfs (UCDs) from [13]. (**b**) Globular cluster candidates were then selected to be in the orange polygon in the color-magnitude diagram (CMD).

This process resulted in the selection of 500 candidate GCs detected by DECam within a projected radius of 0.5 degree, which corresponds to 185 kpc, about $37R_e$ (Figure 4a). This makes the NGC 3923 globular cluster system (GCS) one of the most extended known, comparable to the GCSs of the Milky Way and M31 [16]. This is mainly due to the very wide-field imaging used here. As other galaxies are observed on this scale, we expect that most GCSs of massive galaxies will be found to have similar extents.

In the GCSs of other elliptical galaxies, the "blue" and "red" GCs usually have different radial distributions [13]. We explore the radial distributions of the NGC 3923 GCs by dividing the candidates at $(g' - i') = 0.85$, the minimum between the blue and red color peaks from [13], and plotting the radial distributions in Figure 4b. For this and further analysis, we include 249 GCs from [13] that are detected closer to the center of the galaxy than our current method can find. As found in previous work, the distribution of the blue GCs is more extended than that of the red GCs. The colors of the blue GCs are consistent with the colors of GCs in dwarf galaxies [17,18], and spectroscopic abundances show that the blue clusters are more metal-poor than the red clusters [19]. This is evidence that the two populations are distinct and may have had different formation histories. It is commonly thought that many of the blue GCs may have been accreted from merging dwarf galaxies, but the orbits of outer GCs can be unexpectedly tangential (see [19]), so more kinematic information is needed to test this hypothesis.

The total number of GCs associated with NGC 3923 can be estimated by integrating over the globular cluster luminosity function (GCLF). Following [20], we assume that the GCLF is a Gaussian with a peak at $M_g^0 = -7.2$ or $g^0 = 24.44$. Completeness tests have not been performed yet, so the faint magnitude limit ($g' = 23.5$) was chosen to be well below the magnitude where incompleteness is significant. Fitting the observed luminosity function gives a reasonable Gaussian sigma of 1.4 mag and 3142 total clusters. This gives a globular cluster specific frequency of $S_N = N_{GC}10^{0.4(M_V+15)} = 5.2$, similar to previous results [13].

Harris et al. (2017) [21] has shown that the total number of GCs is proportional to the total halo mass. With this method, the 3142 GCs in NGC 3921 imply a halo mass of $\mathcal{M}_h \sim 3.5 \times 10^{13}$ $M_\odot$. This compares to a dynamical mass of $\mathcal{M}_{dyn} \sim 2 \times 10^{12}$ $M_\odot$ at 30 kpc [22].

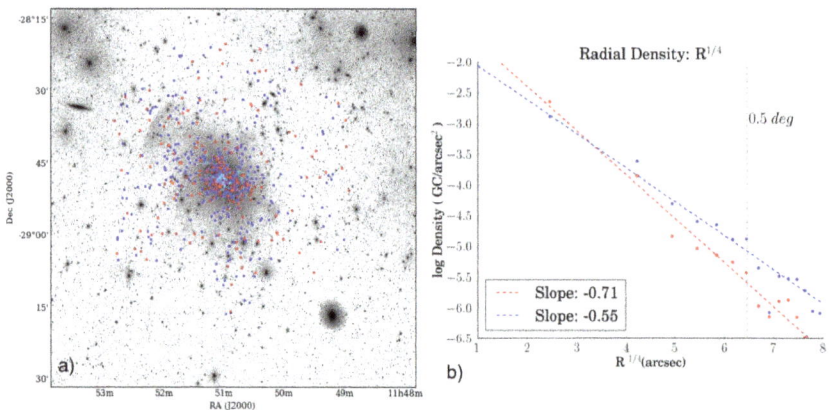

Figure 4. (**a**) Dark Energy Camera (DECam) g'-band image with GC candidates indicated. Magenta and cyan points are GC candidates from [13], while the red and blue points are DECam detections. Magenta and red points indicate indicate the "red" objects with $(g' - i') \geq 0.85$, while cyan and blue points are for "blue" objects with $(g' - i') < 0.85$. (**b**) The radial distributions of the "red" and "blue" GC candidates as separated in panel a). As in other galaxies, the bluer GC candidates have a more extended distribution.

3.2. Shells

The shells around NGC 3923 were identified by eye after removing the underlying galaxy light. The galaxy subtraction was done using a two-Sersic model with *Imfit* image-fitting code [23]. We also enhanced the shell edges in the inner regions by applying an erosion filter edge detection algorithm [5]. The results are given in Figure 5. We have identified 23 strong shells in common with [5]. Eleven shells are classified as uncertain, and eight shells from [5] are not detected. No new shells have been detected yet, but analysis is ongoing.

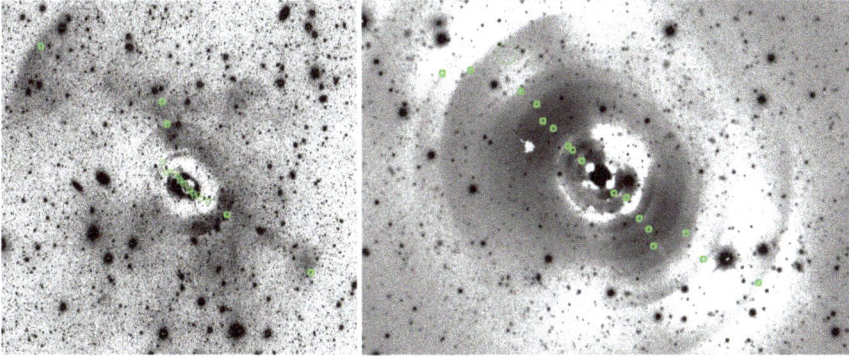

Figure 5. The green squares mark the rims of the clear, strong shells identified in NGC 3923. The underlying background galaxy light has been removed by fitting a two-Sersic model with *Imfit* [23]. The left panel shows the full shell system and the right panel is an expanded view of the inner region.

The shells from a single minor merger should have an alternating, or interleaved, pattern relative to the center of the galaxy. The pattern for the 23 strong shells that we identified in NGC 3923 is shown in Figure 6. The pattern does not alternate as expected. In future work we will look for multiple patterns and additional shells. The spacing of the shells can then be used to constrain the potential and the enclosed mass [24].

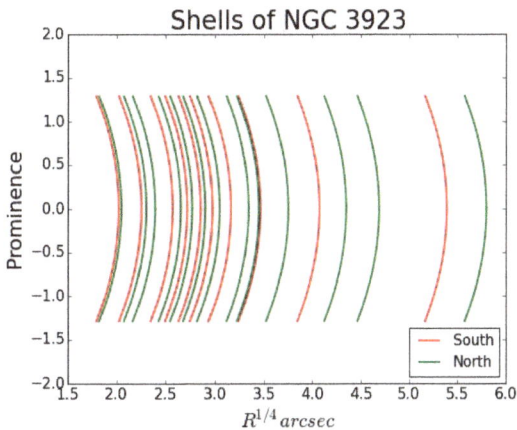

Figure 6. A representation of the interleaving pattern of the shells in NGC 3923. R is the projected radius of the shell rim from the galaxy center. Green arcs indicate shells to the north of the center and red arcs are those to the south. Red and green arcs should alternate, but the outer shells do not follow this pattern.

4. Summary and Future Work

In this work we have identified and characterized the GCs and shells in the halo of NGC 3923. A total of 500 GC candidates are detected in deep DECam imaging out to a projected radius of 185 kpc. This implies a total GC population of about 3100 and a halo mass of $\mathcal{M}_h \sim 3 \times 10^{13}\ M_\odot$. In continuing work, we are trying to detect fainter sources and characterize the photometric completeness. Twenty-three interleaved shells are confirmed, and so far we have not detected previously uncatalogued shells. We are currently applying edge-detection and other algorithms that may be useful for detecting, defining, and characterizing the shells.

Future analysis will focus on deriving the mass profile of NGC 3923 out to a projected radius of about 130 kpc. We will use N-body simulations to reproduce the positions and morphologies of the shells and the velocities of satellite galaxies. This will put constraints on the shape and profile of the potential. Measuring the kinematics of the GCs will also be extremely important. Finally, we will be using the Multi Unit Spectroscopic Explorer (MUSE) instrument on the European Southern Observatory Very Large Telescope (ESO VLT) to measure the line-of-sight velocity distribution (LOSVD) of the stars in the low-surface brightness shells themselves ($B \sim 27$ mag arcsec^2). Fitting the width of the LOSVD with radius from the shell edge can give the gravitational acceleration and the enclosed mass at the shell edge [4].

Elliptical shell galaxies are excellent systems for studying the extent of galaxy halos and the processes that form them. They allow us to use the multiple constraints of globular clusters, shells, and satellite companions to constrain models of their structure and formation.

Acknowledgments: Based on observations at Cerro Tololo Inter-American Observatory, National Optical Astronomy Observatory (NOAO Prop. ID: 2015A-0630; PI: T. Puzia), which is operated by the Association of Universities for Research in Astronomy (AURA) under a cooperative agreement with the National Science Foundation. Funding for the DES Projects has been provided by the DOE and NSF (USA), MISE (Spain), STFC (UK), HEFCE (UK), NCSA (UIUC), KICP (U. Chicago), CCAPP (Ohio State), MIFPA (Texas A&M), CNPQ, FAPERJ, FINEP (Brazil), MINECO (Spain), DFG (Germany) and the collaborating institutions in the Dark Energy Survey, which are Argonne Lab, UC Santa Cruz, University of Cambridge, CIEMAT-Madrid, University of Chicago, University College London, DES-Brazil Consortium, University of Edinburgh, ETH Zürich, Fermilab, University of Illinois, ICE (IEEC-CSIC), IFAE Barcelona, Lawrence Berkeley Lab, LMU München and the associated Excellence Cluster Universe, University of Michigan, NOAO, University of Nottingham, Ohio State University, University of Pennsylvania, University of Portsmouth, SLAC National Lab, Stanford University, University of Sussex, and Texas A&M University. This research has made use of the NASA/IPAC Extragalactic Database (NED) which is operated by the Jet Propulsion Laboratory, California Institute of Technology, under contract with the National Aeronautics and Space Administration. This research made use of Astropy, a community-developed core Python package for Astronomy [25].

Author Contributions: B.W.M. and T.H.P. conceived and designed the experiment. B.W.M. wrote this paper. T.A. reduced the data and performed the detection and photometry of the GCs and shells. S.S.M., J.C.M., M.A.T., and M.S. assisted with the observing strategy, data reduction, and analysis. R.E.S., G.N.C., and R.S. provided simulations and theoretical analysis of the shells.

Conflicts of Interest: The authors declare no conflict of interest.

Abbreviations

The following abbreviations are used in this manuscript:

CMD	Color-Magnitude Diagram
CTIO	Cerro Tololo Inter-American Observatory
DECam	Dark Energy Camera
ESO	European Southern Observatory
GC	Globular Cluster
GCLF	Globular Cluster Luminosity Function
GCS	Globular Cluster System

GMOS-S	Gemini Multi-Object Spectrograph - South
LOSVD	Line-of-Sight Velocity Distribution
MUSE	Multi Unit Spectroscopic Explorer
PCA	Principal Component Analysis
UCD	Ultra-Compact Dwarf
VLT	Very Large Telescope

References

1. Carrasco, E.R.; Conselice, C.J.; Trujillo, I. Gemini K-band NIRI Adaptive Optics Observations of massive galaxies at 1 < z < 2. *Mon. Not. Roy. Astro. Soc.* **2010**, *405*, 2253.
2. Oser, L.; Ostriker, J.P.; Naab, T.; Johansson, P.H.; Burkert, A. The Two Phases of Galaxy Formation. *Astrophys. J.* **2010**, *725*, 2312.
3. Van Dokkum, P.G.; Nelson, E.J.; Franx, M.; Oesch, P.; Momcheva, I.; Brammer, G.; Schreiber, N.M.F.; Skelton, R.E.; Whitaker, K.E.; van der Wel, A.; et al. Forming Compact Massive Galaxies. *Astrophys. J.* **2015**, *813*, 23.
4. Sanderson, R.E.; Helmi, A. An analytical phase-space model for tidal caustics. *Mon. Not. Roy. Astro. Soc.* **2013**, *435*, 378.
5. Bílek, M.; Cuillandre, J.C.; Gwyn, S.; Ebrová, I.; Bartošková, K.; Jungwiert, B.; Jílková, L. Deep imaging of the shell elliptical galaxy NGC 3923 with MegaCam. *Astron. Astrophys.* **2016**, *588*, A77.
6. Prieur, J.L. The shell system around NGC 3923 and its implications for the potential of the galaxy. *Astrophys. J.* **1988**, *326*, 596.
7. Tully, R.B.; Courtois, H.M.; Dolphin, A.E.; Fisher, J.R.; Héraudeau, P.; Jacobs, B.A.; Karachentsev, I.D.; Makarov, D.; Makarova, L.; Mitronova, S.; et al. Cosmicflows-2: The Data. *Astron. J.* **2013**, *146*, 86.
8. De Vaucouleurs, G.; de Vaucouleurs, A.; Corwin, H.G., Jr.; Buta, R.J.; Paturel, G.; Fouque, P. *Third Reference Catalog of Bright Galaxies (RC3)*; Springer: New York, NY, USA, 1991.
9. Schombert, J. ARCHANGEL Galaxy Photometry System. *arXiv* **2007**, arXiv:astro-ph/0703646.
10. Flaugher, B.; Diehl, H.T.; Honscheid, K.; Abbott, T.M.C.; Alvarez, O.; Angstadt, R.; Annis, J.T.; Antonik, M.; Ballester, O.; Beaufore, L.; et al. The Dark Energy Camera. *Astron. J.* **2015**, *150*, 43.
11. Valdes, F.; Gruendl, R. DES Project The DECam Community Pipeline. *ASP Conf. Ser.* **2014**, *485*, 379.
12. Schirmer, M. THELI: Convenient Reduction of Optical, Near-infrared, and Mid-infrared Imaging Data. *Astrophys. J. Suppl.* **2013**, *209*, 21.
13. Faifer, F.R.; Forte, J.C.; Norris, M.A.; Bridges, T.; Forbes, D.A.; Zepf, S.E.; Beasley, M.; Gebhardt, K.; Hanes, D.A.; Sharples, R.M. Gemini/GMOS imaging of globular cluster systems in five early-type galaxies. *Mon. Not. Roy. Astro. Soc.* **2011**, *416*, 155.
14. Bertin, E. Automated Morphometry with SExtractor and PSFEx. *ASP Conf. Ser.* **2011**, *442*, 435–438.
15. Bertin, E.; Arnouts, S. SExtractor: Software for source extraction. *Astron. Astrophys. Suppl.* **1996**, *117*, 393.
16. Huxor, A.P.; Mackey, A.D.; Ferguson, A.M.N.; Irwin, M.J.; Martin, N.F.; Tanvir, N.R.; Veljanoski, J.; McConnachie, A.; Fishlock, C.K.; Ibata, R.; et al. The outer halo globular cluster system of M31 - I. The final PAndAS catalogue. *Mon. Not. Roy. Astro. Soc.* **2014**, *442*, 2165–2187.
17. Lotz, J.M.; Miller, B.W.; Ferguson, H.C. The Colors of Dwarf Elliptical Galaxy Globular Cluster Systems, Nuclei, and Stellar Halos. *Astrophys. J.* **2004**, *613*, 262.
18. Peng, E.W.; Jordán, A.; Côté, P.; Blakeslee, J.P.; Ferrarese, L.; Mei, S.; West, M.J.; Merritt, D.; Milosavljević, M.; Tonry, J.L. The ACS Virgo Cluster Survey. IX. The Color Distributions of Globular Cluster Systems in Early-Type Galaxies. *Astrophys. J.* **2006**, *639*, 95.
19. Brodie, J.B.; Romanowsky, A.J.; Strader, J.; Forbes, D.A.; Foster, C.; Jennings, Z.G.; Pastorello, N.; Pota, V.; Usher, C.; Blom, C.; et al. The SAGES Legacy Unifying Globulars and GalaxieS Survey (SLUGGS): Sample Definition, Methods, and Initial Results. *Astrophys. J.* **2014**, *796*, 52.
20. Jordán, A.; McLaughlin, D.; Cote, P. The ACS Virgo Cluster Survey. XII. The Luminosity Function of Globular Clusters in Early-Type Galaxies. *Astrophys. J. Suppl.* **2007**, *171*, 101–145.
21. Harris, W.E.; Blakeslee, J.P.; Harris, G.L.H. Galactic Dark Matter Halos and Globular Cluster Populations. III. Extension to Extreme Environments. *Astrophys. J.* **2017**, *836*, 67.

22. Norris, M.A.; Gebhardt, K.; Sharples, R.M.; Faifer, F.R.; Bridges, T.; Forbes, D.A.; Forte, J.C.; Zepf, S.E.; Beasley, M.A.; Hanes, D.A.; et al. The globular cluster kinematics and galaxy dark matter content of NGC 3923. *Mon. Not. Roy. Astro. Soc.* **2012**, *421*, 1485–1498.

23. Erwin, P. Imfit: A Fast, Flexible New Program for Astronomical Image Fitting. *Astrophys. J.* **2015**, *799*, 226.

24. Hernquist, L.; Quinn, P.J. Shells and Dark Matter in Elliptical Galaxies. *Astrophys. J.* **1987**, *312*, 1.

25. Astropy Collaboration; Robitaille, T.P.; Tollerud, E.J.; Greenfield, P.; Droettboom, M.; Bray, E.; Aldcroft, T.; Davis, M.; Ginsburg, A.; Price-Whelan, A.M.; et al. Astropy: A community Python package for astronomy. *Astron. Astrophys.* **2013**, *558*, A33.

galaxies

MDPI

Conference Report

The Globular Cluster System of the Galaxy NGC 6876

Ana Inés Ennis [1,2,*], Lilia Patricia Bassino [1,2] and Juan Pablo Caso [1,2]

[1]   Facultad de Ciencias Astronómicas y Geofísicas, Universidad Nacional de La Plata, La Plata 1900, Argentina; lbassino@fcaglp.unlp.edu.ar (L.P.B.); jpceda@fcaglp.unlp.edu.ar (J.P.C.)
[2]   Instituto de Astrofísica de La Plata, CCT La Plata, CONICET-UNLP, La Plata 1900, Argentina
*   Correspondence: anaennis@fcaglp.unlp.edu.ar

Received: 27 May 2017; Accepted: 21 July 2017; Published: 25 July 2017

Abstract: We present preliminary results of the deep photometric study of the elliptical galaxy NGC 6876, located at the center of the Pavo group, and its globular cluster system. We use images obtained with the GMOS camera mounted on the Gemini South telescope, in the g' and i' bands, with the purpose of disentangling the evolutionary history of the galaxy on the basis of its characteristics.

Keywords: elliptical galaxies; globular clusters; evolution

1. Introduction

Globular clusters (GCs) are considered to be among the oldest objects in the Universe, with ages greater than 10 Gyr [1,2]. Because of this, GC systems are widely used as tools to study the first evolutionary stages of the galaxies in which they reside, since they hold a record of both the chemical properties of the environment at the time the galaxy was formed, and of any assembly events that the galaxy has gone through [3]. In bright galaxies, a bimodality in the color distribution of their GC system is usually found, which signals the existence of two sub-populations differentiated by their metallicities [4]. Though all GCs are metal-poor, these sub-populations are refered to as "metal-rich" (red GCs) and "metal-poor" (blue GCs). This bimodality is one of the most studied properties of GC systems, since these sub-populations show distinct intrinsic characteristics such as their spatial distribution and their kinematics.

NGC 6876 is a massive elliptical galaxy situated at the center of the Pavo Group, at an approximate distance of 45 Mpc [5]. This is a moderately massive group, with 13 confirmed members, that appears to be dynamically young considering the several interactions between its members that have been studied. In a recent work by [6], an X-ray trail between NGC 6876 and its spiral neighbor, NGC 6872, was found, hinting at a possible interaction between the two galaxies. This interaction has also been studied in IR and HI [7].

In this work, we present preliminary results for the study of the GC system of NGC 6876, including the color-magnitude diagram, the color distribution, and the spatial, radial and azimuthal distribution for both the entire system and the presumed sub-populations. The long-term aim of this study is to look for evidence of the mentioned interactions with NGC 6872 in the GC population, such as irregularities in the azimuthal distribution, differences between the radial distribution of the sub-populations, etc. [8].

2. Materials and Methods

The images used in this study were obtained in 2013, using the Gemini Multi-Object Spectograph (GMOS) of Gemini South (Cerro Pachón, Chile) and the g' and i' filters (Program GS-2013B-Q-37). They were processed using Gemini routines included in IRAF, and bias and flat-field corrections were applied using the appropriate images obtained from the Gemini Observatory Archive (GOA) (https://archive.gemini.edu). The *i'* filter showed night-sky fringing, which was corrected using *i'* blank sky images, also obtained from the GOA.

The field of standard stars 195940-595000 [9] was observed within the same program, obtaining images of both long and short exposures to allow us to perform photometry on faint and bright stars, respectively. These images were reduced using the same procedure as the science images, and the photometry of the standard stars present in this field was used to calibrate our photometry.

A catalogue of point-like objects was built using the software SExtractor [10] on the images, which detected sources using both a Gaussian filter, and a mexhat one which works as a complement since it is better for detecting sources near the galactic center. Our first selection was then made of detected sources with a "stellarity index" larger than 0.5. The "stellarity index" ranges from 0 to 1, with 0 being the value used for extended objects such as galaxies, and 1 being the value that identifies point-like sources, such as stars. Using DaoPhot tasks, PSF (Point Spread Function) photometry was performed on all objects. The statistic tests provided by these tasks that determine the goodness of the fit (χ^2 and sharpness parameters from ALLSTAR) were then used to make a second selection, ending up with 1631 point-like sources.

3. Results

Figure 1 shows the color-magnitude diagram of all point-like sources detected by SExtractor. The limits in color were taken from previous works in the same photometric system [11]. The upper limit in magnitude was also obtained from the literature [12], to separate possible ultra-compact dwarfs (UCDs) which fall in the same range of color as GCs, but are brighter than them. After estimating the completeness, the lower limit was selected so as to ensure a completeness level greater than 80 per cent. With these constraints, our GC system is finally made of $\approx$917 candidates.

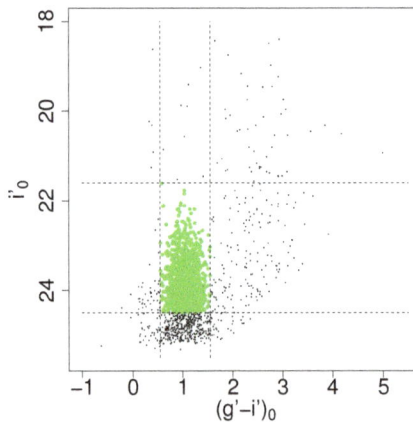

Figure 1. Color-magnitude diagram for all point-like objects. Green dots represent the globular cluster (GC) candidates.

In Figure 2, we present the same four histograms in both panels, each corresponding to a different range of galactocentric distance except for the first one, which shows the entire population. In these histograms, the color distribution of the system is shown with Gaussian functions fitted using GMM [13] for a bimodal distribution (left panel) and a unimodal distribution (right panel). Though it can be seen that in the exterior region and the total population there is a strong tendency towards bimodality, the statistic parameters of these fits (D and the kurtosis) were not conclusive. It is expected that applying the background correction, as we intend to do in the near future, will help produce more definitive results.

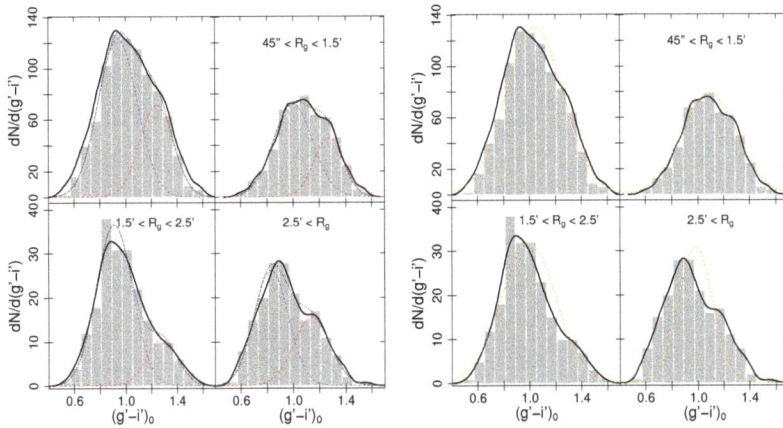

Figure 2. Color distributions for the entire population of GC candidates, and for different ranges of galactocentric distance. The left panel shows a bimodal fit in dashed lines, while the right panel shows a unimodal fit in dashed lines. Solid lines represent smoothed density in both.

The spatial distribution is shown in Figures 3 and 4 as discrete and smooth distributions, respectively. In both figures, it can be seen that there are GCs up to the borders of the image, indicating that the field does not cover the entirety of the system. From the color distribution, a limit of $(g' - i')_0 = 1.0$ is taken to separate the richer (redder) and the poorer (bluer) GCs. The sub-populations show different behaviors, since the red sub-population is more concentrated towards the center whereas the blue sub-population is more disperse.

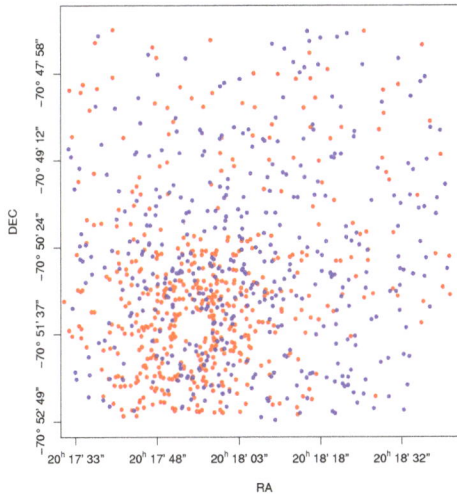

Figure 3. Spatial distribution for all GCs. Red and blue dots separate the metal-rich and metal-poor sub-populations, with the limit between them (obtained from the color distribution) at $(g' - i') = 1.0$.

The previous result is confirmed by the radial distribution, shown in Figure 5a. Here, a power law was fitted to the distributions, and the slope obtained for the red sub-population is considerably larger than the one obtained for the blue sub-population and the one for the entire system, indicating that it is more concentrated towards the center. The plot for the totality of the GCs reaches the inner regions of the galaxy, showing that the distribution in this region turns flat. This is both because of the saturation of the images in this area, and because of the disruption of GCs being more frequent closer to the galactic center due to tidal forces being stronger.

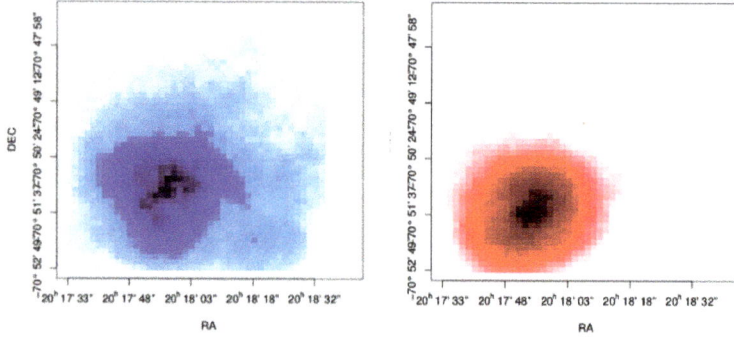

Figure 4. Smoothed spatial distribution for both sub-populations, blue on the left, red on the right.

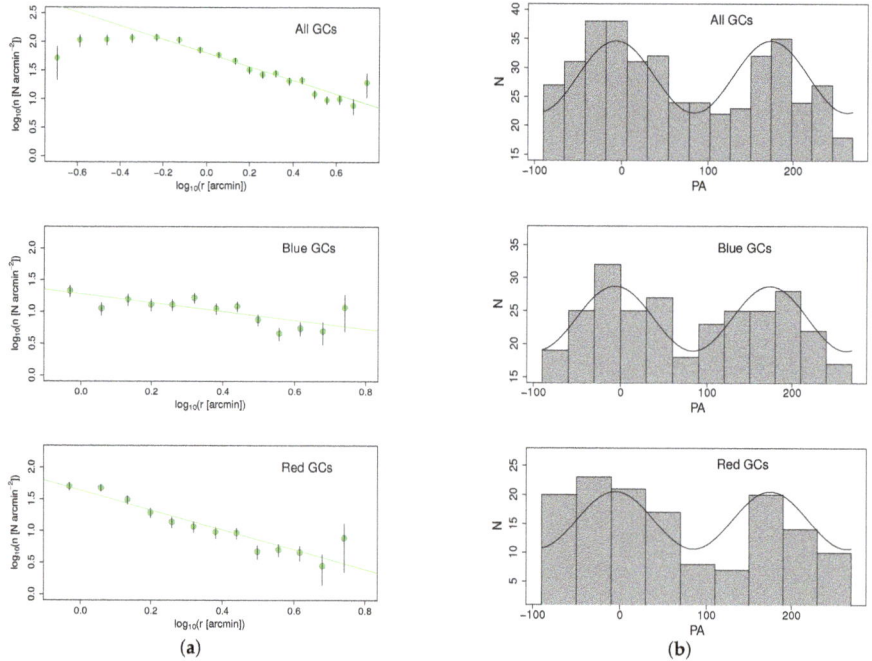

Figure 5. (a) In the left panel, we have the radial distribution for the entire system and both sub-populations. A power law (solid line) was fitted to the three distributions; (b) In the right panel, the azimuthal distribution for the entire system and both sub-populations is presented. The black line represents the sinusoidal fit.

In Figure 5b, we show the azimuthal distribution for the entire system and both sub-populations, with sinusoidal functions fitted in order to obtain the position angle of the semi-major axis of the ellipse described by each of them. In all cases, we obtained a position angle of $\approx 100°$, indicating that both sub-populations are oriented in the same direction as the total system.

Acknowledgments: This work was funded with grants from Consejo Nacional de Investigaciones Científicas y Técnicas de la República Argentina (PIP 112-201101-00393), Agencia Nacional de Promoción Científica y Tecnológica (PICT-2013-0317), and Universidad Nacional de La Plata (UNLP 11-G124), Argentina. Based on observations obtained at the Gemini Observatory, which is operated by the Association of Universities for Research in Astronomy, Inc., under a cooperative agreement with the NSF on behalf of the Gemini partnership: the National Science Foundation (United States), the National Research Council (Canada), CONICYT (Chile), Ministerio de Ciencia, Tecnología e Innovación Productiva (Argentina), and Ministério da Ciência, Tecnologia e Inovação (Brazil). The Gemini programme ID is GS-2013B-Q-37. This research has made use of the NED, which is operated by the Jet Propulsion Laboratory, Caltech, under contract with the National Aeronautics and Space Administration.

Author Contributions: L.P.B. and J.P.C. conceived and designed the observing proposal. A.I.E. performed the data reduction as a part of her "Tesis de Licenciatura" at the UNLP and wrote the paper. All authors contributed to the data analysis.

Conflicts of Interest: The authors declare no conflict of interest.

Abbreviations

The following abbreviations are used in this manuscript:

GC   Globular Cluster
PSF   Point Spread Function
GOA   Gemini Observatory Archive

References

1. Mendel, J.T.; Proctor, R.N.; Forbes, D.A. The age, metallicity and α-element abundance of Galactic globular clusters from single stellar population models. *Mon. Notic. Roy. Astron. Soc.* **2007**, *379*, 1618–1636.
2. Tonini, C. The Metallicity Bimodality of Globular Cluster Systems: A Test of Galaxy Assembly and of the Evolution of the Galaxy Mass-Metallicity Relation. *Astrophys. J.* **2013**, *762*, 39.
3. Bassino, L.P.; Caso, J.P. The merger remnant NGC 3610 and its globular cluster system: A large-scale study. *Mon. Notic. Roy. Astron. Soc.* **2017**, *466*, 4259–4271.
4. Brodie, J.P.; Strader, J. Extragalactic Globular Clusters and Galaxy Formation. *Annu. Rev. Astron. Astrophys.* **2006**, *44*, 193–267.
5. Blakeslee, J.P.; Lucey, J.R.; Barris, B.J.; Hudson, M.J.; Tonry, J.L. A synthesis of data from fundamental plane and surface brightness fluctuation surveys. *Mon. Notic. Roy. Astron. Soc.* **2001**, *327*, 1004–1020.
6. Machacek, M.E.; Nulsen, P.; Stirbat, L.; Jones, C.; Forman, W.R. XMM-Newton Observation of an X-ray Trail between the Spiral Galaxy NGC 6872 and the Central Elliptical Galaxy NGC 6876 in the Pavo Group. *Astrophys. J.* **2005**, *630*, 280–297.
7. Machacek, M.; Ashby, M.L.N.; Jones, C.; Forman, W.R.; Bastian, N. A Multiwavelength View of Star Formation in Interacting Galaxies in the Pavo Group. *Astrophys. J.* **2009**, *691*, 1921–1935.
8. Bassino, L.P.; Faifer, F.R.; Forte, J.C.; Dirsch, B.; Richtler, T.; Geisler, D.; Schuberth, Y. Large-scale study of the NGC 1399 globular cluster system in Fornax. *Astron. Astrophys.* **2006**, *451*, 789–796.
9. Smith, J.A.; Tucker, D.L.; Kent, S.; Richmond, M.W.; Fukugita, M.; Ichikawa, T.; Hamabe, M. The $u'g'r'i'z'$ Standard-Star System. *Astron. J.* **2002**, *123*, 2121–2144.
10. Bertin, E.; Arnouts, S. SExtractor: Software for source extraction. *Astron. Astrophys. Suppl. Ser.* **1996**, *117*, 393–404.
11. Faifer, F.R.; Forte, J.C.; Norris, M.A.; Bridges, T.; Forbes, D.A.; Zepf, S.E.; Sharples, R.M. Gemini/GMOS imaging of globular cluster systems in five early-type galaxies. *Mon. Notic. Roy. Astron. Soc.* **2011**, *416*, 155–177.
12. Mieske, S.; Hilker, M.; Infante, L.; Jordán, A. Spectroscopic Metallicities for Fornax Ultracompact Dwarf Galaxies, Globular Clusters, and Nucleated Dwarf Elliptical Galaxies. *Astron. J.* **2006**, *131*, 2442–2451.
13. Muratov, A.L.; Gnedin, O.Y. Modeling the Metallicity Distribution of Globular Clusters. *Astrophys. J.* **2010**, *718*, 1266–1288.

Conference Report

The HI Distribution Observed toward a Halo Region of the Milky Way

Ericson López *, Jairo Armijos , Mario Llerena and Franklin Aldás

Observatorio Astronómico de Quito, Escuela Politécnica Nacional, Av. Gran Colombia S/N, Quito 170403, Ecuador; jairo.armijos@epn.edu.ec (J.A.); mario.llerena01@epn.edu.ec (M.L.); franklin.aldas@epn.edu.ec (F.A.)
* Correspondence: ericsson.lopez@epn.edu.ec; Tel.: +593-22570765

Academic Editor: Emilio Elizalde
Received: 18 July 2017; Accepted: 24 August 2017; Published: 28 August 2017

Abstract: We use observations of the neutral atomic hydrogen (HI) 21-cm emission line to study the spatial distribution of the HI gas in a $80° \times 90°$ region of the Galaxy halo. The HI column densities in the range of $3–11 \times 10^{20}$ cm^{-2} have been estimated for some of the studied regions. In our map—obtained with a spectral sensitivity of $\sim$2 K—we do not detect any HI 21-cm emission line above 2σ at Galactic latitudes higher than $\sim$46°. This report summarizes our contribution presented at the conference on the origin and evolution of barionic Galaxy halos.

Keywords: 21-cm emission line; Galaxy halo; interstellar medium

1. Introduction

Neutral atomic hydrogen (HI) is the most abundant element in the interstellar medium, and its 21-cm emission line is a powerful tool to trace the structure and dynamics of the Milky Way Galaxy (Kalverla and Jürgen [1]). The HI gas in our own galaxy has a two-component structure; one is composed of cold neutral gas with temperatures $\lesssim$300 K and the other by warm neutral gas with temperatures $\gtrsim$5000 K (Kalverla and Jürgen [1]). The HI disk of the Milky Way together with its spiral arms extend up to a radius of $\sim$35 kpc (Kalverla and Jürgen [1]). There is also a galaxy halo composed of HI gas with densities of 10^{-3} cm^{-3} that extends in vertical height up to a distance of $\sim$4 kpc, and radially HI gas is detected in the outskirts ($\gtrsim$35 kpc) of the Milky Way (Kalverla and Jürgen [1]). All these important features have been discovered mainly thanks to studies carried out using the HI 21-cm emission line.

The HI column density (N_{HI}) has been estimated across the galaxy by Dickey and Lockman [2], who estimated the highest and lowest N_{HI} values of 2.6×10^{22} at $(l,b) = (339°,0°)$ and of 4.4×10^{19} cm^{-2} at $(l,b) = (152°,62°)$, respectively. A recent HI 4π survey has been used to obtain all-sky column density maps of HI for the Milky Way (Ben Bekhti et al. [3]). The data of this survey have a spectral sensitivity of 43 mK and a spatial resolution of 16.2 arc-min.

In this work, we aim to map a defined region (chosen randomly) of the Milky Way ($87° \leq l \leq 180°$ and $13° \leq b \leq 180°$) at the HI 21-cm emission line, which can help us to identify a disk-halo transition region. The data employed in our study have been obtained using the 2.3 m SALSA radio telescopes[1] of the Onsala Space Observatory located in Sweden, which are operating at a wavelength of 21 cm. We also aim to check whether our N_{HI} estimates derived for our selected regions are in good agreement with previous values derived by other authors (Dickey and Lockman [2],

---

[1] http://vale.oso.chalmers.se/salsa/

Ben Bekhti et al. [3]). Even working with a small telescope mainly used for student experiences, we expect to achieve acceptable results in comparison with those observations obtained from larger telescopes like Parkes and Dwingeloo.

2. Observations and Data Reduction

As mentioned previously, we used two small radio telescopes (see Figure 1) of the Onsala Space Observatory to carry out our observations. The observations were performed in September 2016. The telescopes provide a spatial resolution of 6° at 21 cm, and their receivers have a spectral resolution of 7.8 KHz per channel and 2 MHz bandwidth [4]. Using both telescopes, we mapped a region between galactic longitudes 87° and 180° and galactic latitudes 13° and 85°, in steps of 5.5°.

The data processing and analysis was done using the SalsaSpectrum software[2]. Unfortunately, the data obtained with the 2.3 m radio telescopes are not flux calibrated. To carry out this process, we observed several positions toward the galactic plane. Then, the obtained HI 21-cm spectra were compared with calibrated spectra of the same galactic positions previously observed (Higgs and Tapping [5]), obtaining in this way a flux calibration factor. The calibration observations were carried out toward four positions: $l = 74°$, $b = 1°$; $l = 119°$, $b = -1°$; $l = 131.2°$, $b = 1°$; and $l = 140°$, $b = -3°$. The intensity of our calibrated spectra is given in brightness temperature (T_B^*). We noted that the HI 21-cm line intensities—observed toward the calibration positions—did not change by more than 20% as the integration time varied from 20 s to 160 s; therefore, our spectra are affected by 20% uncertainties in the intensity. We used an integration time of 20 s to observe the calibration positions and the target regions of the Milky Way.

Figure 1. The 2.3 m radio telescopes of the Onsala Space Observatory.

3. Results

As we expected, the strongest HI 21-cm lines were detected at the lowest galactic latitudes ($\sim$8°) covered in our study. In Figure 2 we show a sample of HI 21-cm emission profiles observed toward six selected positions in the Milky Way. As seen in this figure, the 21-cm lines show central line velocities of $\sim$0 km s^{-1}, that correspond at low latitudes to our own galactic arm and at high latitudes to the galactic halo. The spectrum observed at $l = 125°$ and $b = 13°$ shows the HI emission line clearly tracing two gas components with velocities of 0 and -20 km s^{-1}. The gas with negative velocities is likely associated with another spiral arm of the galaxy. We also show a peak-intensity map of the HI lines in Figure 3. We did not detect a HI 21-cm emission line above 2σ ($\sigma \approx 2$ K) at galactic latitudes higher than $\sim$46°. For these regions, in our map we used 2σ limits in the peak-intensity. This fact defines the lower boundary of the galactic halo, so we use this criterion based on the lack of a HI 21-cm emission line above 2σ to define the start of the galactic halo. Nevertheless, another criteria may be used to

---

[2]   http://vale.oso.chalmers.se/salsa/software

define the start of the galactic halo; for example, a threshold in the decreasing number of Milky Way HI clouds as a function of the galactocentric radius (Ford et al. [6]).

On the other hand, we used the SalsaSpectrum software to subtract the underlying continuum emission and fit Gaussians to the H 21-cm lines. Then, we were able to estimate the integrated line intensities ($\int T_b dV$), which are listed in Table 1 for the galactic positions indicated in Figure 2. In this figure, the last two HI 21-cm line profiles (GLAT: 35 ° and 40.5 °) show the lowest signal-to-noise ratio, so they were not considered. The remaining profiles were used to derive the hydrogen column density (N_{HI}), which is proportional to $\int T_b dV$ (assuming an optically thin medium), and it can be estimated following the expression given by Dickey and Lockman [2]:

$$\left(\frac{N_H}{cm^{-2}} \right) = 1.82 \times 10^{18} \int T_b dV \tag{1}$$

Using this expression, we estimated the hydrogen column density for our halo sub-regions confined between 13° and 30° of galactic latitude. The obtained values are given in Table 1.

Table 1. Parameters derived for four positions of the Milky Way.

l Degrees	*b* Degrees	$\int T_b dV$ (K km s^{-1})	N_H ($\times 10^{20}$ cm^{-2})
125	13.0	623.5	11.4
125	18.5	514.7	9.4
125	24.0	391.7	7.1
125	29.5	193.4	3.5

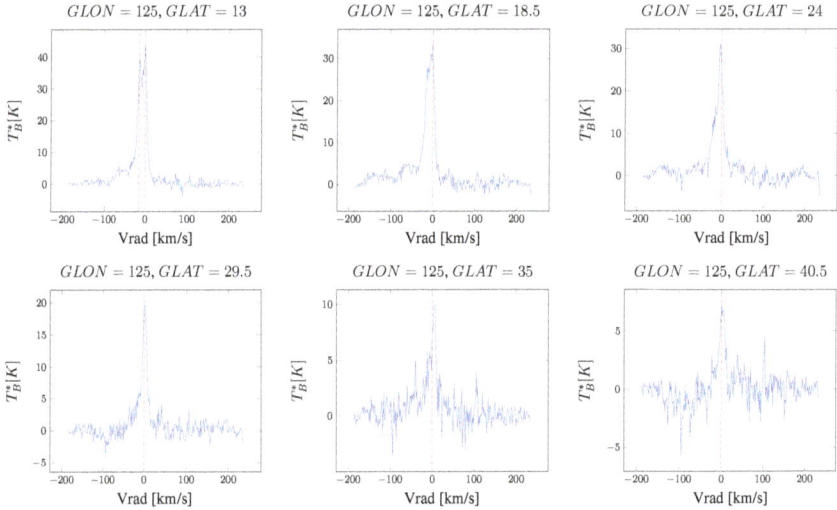

Figure 2. Neutral atomic hydrogen (HI) 21-cm emission line observed toward six positions in the Milky Way. The dashed red line indicates the 0 km s^{-1} velocity in all of the panels. In the upper-left panel, the velocity of −20 km s^{-1}—corresponding to another gas component with a radial velocity different than 0 km s^{-1}—is also indicated.

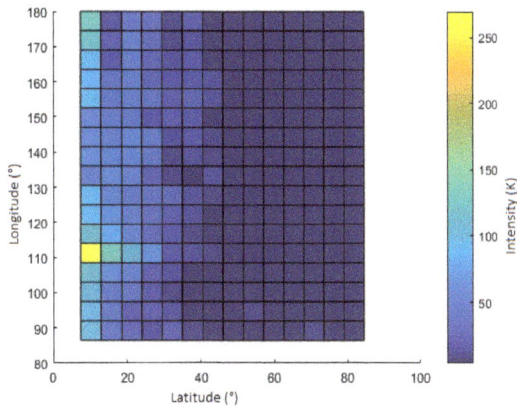

Figure 3. Peak-intensity map of HI 21-cm lines covering an $80° \times 90°$ area of the Milky Way.

4. Conclusions and Future Work

Using data obtained from the small SALSA telescope, we were able to obtain a peak-intensity map of the HI 21-cm emission line for a $80° \times 90°$ region of the Milky Way. We did not detect any emission line above 2σ at Galactic latitudes higher than $\sim 46°$, so we have considered this latitude to define the galactic halo lower boundary.

We have also estimated atomic hydrogen column density N_{HI} values, which fall in the range of 3–11×10^{20} cm^{-2}, for our four randomly-chosen positions of the Milky Way, under the assumption of an optically thin medium.

Despite the low spatial resolution of our observations, the derived N_{HI} values are consistent with the values of $\sim 10^{20}$–10^{21} cm^{-3}, provided for the same regions studied in our paper with a much higher spatial resolution instrument (Ben Bekhti et al. [3]). In the same context, our results are also in good agreement with the N_{HI} range of $4.4 \times 10^{19} - 2.6 \times 10^{22}$ cm^{-2} obtained by Dickey and Lockman [2] in their study of the HI gas across the galaxy.

Regarding our future work, instead of continuing the study of the large-scale distribution of HI gas in the Milky Way, we are interested in the study of the physical and dynamical properties of galactic high-velocity clouds. For that, we plan to apply for observing time in large radio telescopes such IRAM, Parkes, among others—instruments with much better spatial resolutions and data quality.

Author Contributions: All authors carried out the observations and performed the data reduction and analysis. E. López and J. Armijos wrote the manuscript.

Conflicts of Interest: The authors declare no conflict of interest.

References

1. Kalverla, P.M.W.; Jürgen, K. The HI Distribution of the Milky Way. *Annu. Rev. Astron. Astrophys.* **2009**, *47*, 27–61.
2. Dickey, J.M.; Lockman, F.J. HI in the Galaxy. *Annu. Rev. Astron. Astrophys.* **1990**, *28*, 215–259.
3. Ben Bekhti, N.; Flöer, L.; Keller, R.; Kerp, J.; Lenz, D.; Winkel, B.; Bailin, J.; Calabretta, M.R.; Dedes, L.; Ford, H.A.; et al. HI4PI: A full-sky HI survey based on EBHIS and GASS. *Astron. Astrophys.* **2016**, *594*, A116.
4. SALSA User Manual: Telescope Control and Data Analysis. Available online: http://vale.oso.chalmers.se/salsa/support (accessed on 17 June 2017).

5. Higgs, L.A.; Tapping, K.F. The low-resolution DRAO survey of HI emission from the Galactic plane. *Astron. J.* **2000**, *120*, 2471–2487.

6. Ford, H.A.; Lockman, F.J.; McClure-Griffiths, N.M. Milky Way disk-halo transition in HI: Properties of the cloud population. *Astrophys. J.* **2010**, *722*, 367–379.

galaxies

Article

The Outer Halos of Very Massive Galaxies: BCGs and their DSC in the Magneticum Simulations

Rhea-Silvia Remus [1,*], Klaus Dolag [1,2] and Tadziu L. Hoffmann [1]

[1] Faculty of Physics, Universitäts-Sternwarte München, Ludwig-Maximilians-Universität, Scheinerstr. 1, D-81679 München, Germany; dolag@usm.lmu.de (K.D.); hoffmann@usm.lmu.de (T.L.H.)
[2] Max Planck Institut for Astrophysics, D-85748 Garching, Germany
* Correspondence: rhea@usm.lmu.de

Academic Editors: Duncan A. Forbes and Ericson D. Lopez
Received: 1 August 2017; Accepted: 28 August 2017; Published: 1 September 2017

Abstract: Recent hydrodynamic cosmological simulations cover volumes up to Gpc^3 and resolve halos across a wide range of masses and environments, from massive galaxy clusters down to normal galaxies, while following a large variety of physical processes (star-formation, chemical enrichment, AGN feedback) to allow a self-consistent comparison to observations at multiple wavelengths. Using the Magneticum simulations, we investigate the buildup of the diffuse stellar component (DSC) around massive galaxies within group and cluster environments. The DSC in our simulations reproduces the spatial distribution of the observed intracluster light (ICL) as well as its kinematic properties remarkably well. For galaxy clusters and groups we find that, although the DSC in almost all cases shows a clear separation from the brightest cluster galaxy (BCG) with regard to its *dynamic* state, the radial stellar *density* distribution in many halos is often characterized by a single Sérsic profile, representing both the BCG component and the DSC, very much in agreement with current observational results. Interestingly, even in those halos that clearly show two components in both the dynamics and the spatial distribution of the stellar component, no correlation between them is evident.

Keywords: galaxy clusters; intracluster light; numerical simulation

---

1. Introduction

Brightest cluster galaxies (BCGs), residing in the centers of galaxy clusters, are the most massive and luminous galaxies in the universe. During their lifetime, they experience frequent interactions with satellite galaxies, and their growth is dominated by merger events. These merger events also lead to the buildup of a diffuse stellar component (DSC), which very likely contains a significant fraction of the total stellar mass of the galaxy clusters (see Murante et al., 2007 [1] and references therein). The velocities of the stars in the BCG and the DSC have distinct kinematic distributions, which can be characterized by two superposed Maxwellian distributions, as demonstrated by Dolag et al., 2010 [2]. While the velocity dispersion of the stars in the BCG represents the central mass of the stars, the velocity dispersion of the DSC is much larger and is comparable to that of the dark matter halo (see, for example, Dolag et al., 2010 [2], Bender et al., 2015 [3], and Longobardi et al., 2015 [4]). More details on this matter can also be found in a recent review by Mihos et al., 2016 [5].

Similarly, early simulations of galaxy clusters found that the density distributions of BCGs in clusters can be described by a superposition of two extended components as well (e.g., Puchwein et al., 2010 [6]). However, more recent simulations find the opposite, namely that in many cases the radial density profiles can be described by a single profile, which is in good agreement with observations. These simulations also indicate that a double-component fit to the radial density profiles is only needed in rare cases. Interestingly, the three-dimensional distribution of these outer stellar halos seems to be described universally by a so-called Einasto profile over a wide range of halo masses, as shown by Remus et al., 2016 [7], where the curvature of the radial profiles appears to be more closely linked to the cluster's assembly history than the separation of the radial profiles into distinct components.

In this study we analyse the velocity distributions as well as the projected radial surface density profiles of the stellar component in galaxy clusters selected from a state-of-the-art cosmological simulation, and test for possible correlations between these distributions.

2. Simulations

We use galaxy clusters selected from the Magneticum Pathfinder (www.magneticum.org) simulation set. This suite of fully hydrodynamic cosmological simulations comprises a broad range of simulated volumes, with box lengths of 2688 Mpc/h to 18 Mpc/h, covering different resolution levels of stellar particle masses from $m_{Star} = 6.5 \times 10^8\ M_\odot/h$ at the lowest resolution level down to particle masses of $m_{Star} = 1.9 \times 10^6\ M_\odot/h$ at the highest resolution level. For this work, we use two different simulations, Box2b and Box4, with the smaller one (Box4) having a higher resolution. The details of these two simulations are summarized in Table 1.

Table 1. Magneticum simulations used in this work.

	Box Size	N_{part}	m_{Star}	ϵ_{Star}
Box2b hr	910 Mpc	2×2880^3	$3.5 \times 10^7\ M_\odot/h$	2 kpc/h
Box4 uhr	68 Mpc	2×576^3	$1.9 \times 10^6\ M_\odot/h$	0.7 kpc/h

All simulations of the Magneticum Pathfinder simulation suite are performed with an advanced version of the tree-SPH code P-Gadget3 (Springel, 2005 [8]). They include metal-dependent radiative cooling, heating from a uniform time-dependent ultraviolet background, star formation according to Springel & Hernquist, 2003 [9], and the chemo-energetic evolution of the stellar population as traced by SN Ia, SN II, and AGB stars, including the associated feedback from these stars (Tornatore et al., 2007 [10]). Additionally, they follow the formation and evolution of supermassive black holes, including their associated quasar and radio-mode feedback. For a detailed description, see Dolag et al. (in prep), Hirschmann et al., 2014 [11], and Teklu et al., 2015 [12].

Galaxy clusters are chosen according to the total mass of a structure as found by the baryonic SUBFIND algorithm (see Dolag et al., 2009 [13]). For the larger, less resolved volume (Box2b), we classify all structures with masses of $M_{tot} > 2 \times 10^{14}\ M_\odot$ as clusters, independent of their dynamical state, and find 890 objects. For the smaller volume (Box4), there are no massive galaxy clusters, but the increased resolution enables us to utilize halos with masses down to $1 \times 10^{13}\ M_\odot < M_{tot} < 1 \times 10^{14}\ M_\odot$, and therefore allows us to add galaxy groups down to the limit of massive field galaxies to this study. Including the three clusters and 35 groups from the smaller volume simulation, we end up with a total sample of 928 objects, which is an unprecedentedly large sample of simulated galaxy clusters and groups for which we here, for the first time, provide a statistically representative analysis of the decomposition of the stellar components into the BCG and the DSC, providing predictions for future observational studies of the ICL and the BCGs.

3. Velocity Distributions and Radial Surface Density Profiles

In their detailed study, Dolag et al., 2010 [2] demonstrated that the two dynamical components found in the velocity distribution of the stellar component of galaxy clusters very well represent the stellar component of the BCGs and the DSC, the latter of which is itself a good approximation of the observed ICL in galaxy clusters. Following their approach, we subtract all substructures (identified with SUBFIND) from the stellar component of each cluster and use the remaining stars for this analysis. First, we calculate the velocities of all stellar particles in a cluster and bin them in small equal-width bins of $\Delta v = 10$ km/s, thereby obtaining the intrinsic 3D velocity distribution of the stars in each cluster. Similarly, we choose a random viewing angle and calculate the projected radius of each stellar particle. Subsequently, we radially bin these particles using equal-particle bins, thus obtaining radial surface density distributions, effectively mimicking the radial surface brightness profiles that are commonly observed for galaxies and galaxy clusters, assuming a constant mass-to-light ratio. Examples of the velocity distributions and surface density profiles obtained by this methods are shown in the lower panels of Figures 1–4.

To obtain the different components of BCG and DSC in the simulations, we again follow Dolag et al., 2010 [2]. First, we fit a superposition of two Maxwellian distributions

$$N(v) = k_1 v^2 \exp\left(-\frac{v^2}{\sigma_1^2}\right) + k_2 v^2 \exp\left(-\frac{v^2}{\sigma_2^2}\right) \tag{1}$$

to the velocity distribution of each cluster. Additionally, we fit a single Maxwellian distribution to the velocity distributions for comparison purposes. In most cases, a double-Maxwellian fit is needed to properly represent the underlying velocity distributions, as shown, for example, in the lower left panels of Figures 1 and 2. For comparison, we also show the stellar particle surface density map of the clusters including all substructures in the large image at the top of these figures. The white contours show equal-density lines of the stellar distribution without the substructures. For both clusters shown in Figures 1 and 2, the BCG is clearly visible, but while the velocity distributions can be well described by double-Maxwellian fits in both cases, the morphological appearance of the two clusters is very different: while the cluster shown in Figure 1 is clearly elongated with a massive colliding structure clearly visible even in the dark matter component (upper small image), the other cluster shown in Figure 2 shows no signs of ongoing substantial accretion, and is only slightly elongated. This is true even in the X-ray map (middle small image), where the elongation is clearly visible for the first cluster while the second cluster shows a more compact shape. We also do not find a similarity between these clusters with regard to their shock properties: whereas the cluster shown in Figure 1 has a clearly visible shock front in the upper right area of the cluster, indicating a recent merging event, the cluster shown in Figure 2 shows no clear signs of such a recent merger event in the shock map (bottom small image). This clearly indicates that the velocity distribution of the cluster remembers the merger history of the cluster over a much larger timescale than other tracers like shocks or satellite distributions, which provide information only about the more recent mass assembly history of a cluster.

In some cases, there is no improvement to the description of the velocity distribution of an individual cluster by using a double-Maxwellian distribution for the fit, as the velocity distributions of that particular cluster is already well described by a single Maxwellian distribution (see, for example, the lower left panel of Figure 3). While the single-Maxwellian fit is a good approximation to the velocity distribution of the cluster stellar light, the stellar light map in the large image in the same figure clearly shows that the cluster is currently accreting another, relatively massive, substructure. This can be seen not only in the stellar component but also in the X-ray emission (middle small image) and the shock map (bottom small image). Thus, this clearly shows that the contribution from this merger to the cluster's DSC is not very large yet and does not show up in the velocity distribution of the cluster as most stars that are brought in through the merger event have been subtracted by SUBFIND. Only in the very-high velocity end of the velocity distribution, the newly accreted component starts to be visible. This also indicates that the DSC of this cluster is very rich, which is caused by a rather diverse accretion history. However, these cases are very rare as we will show later on in this work.

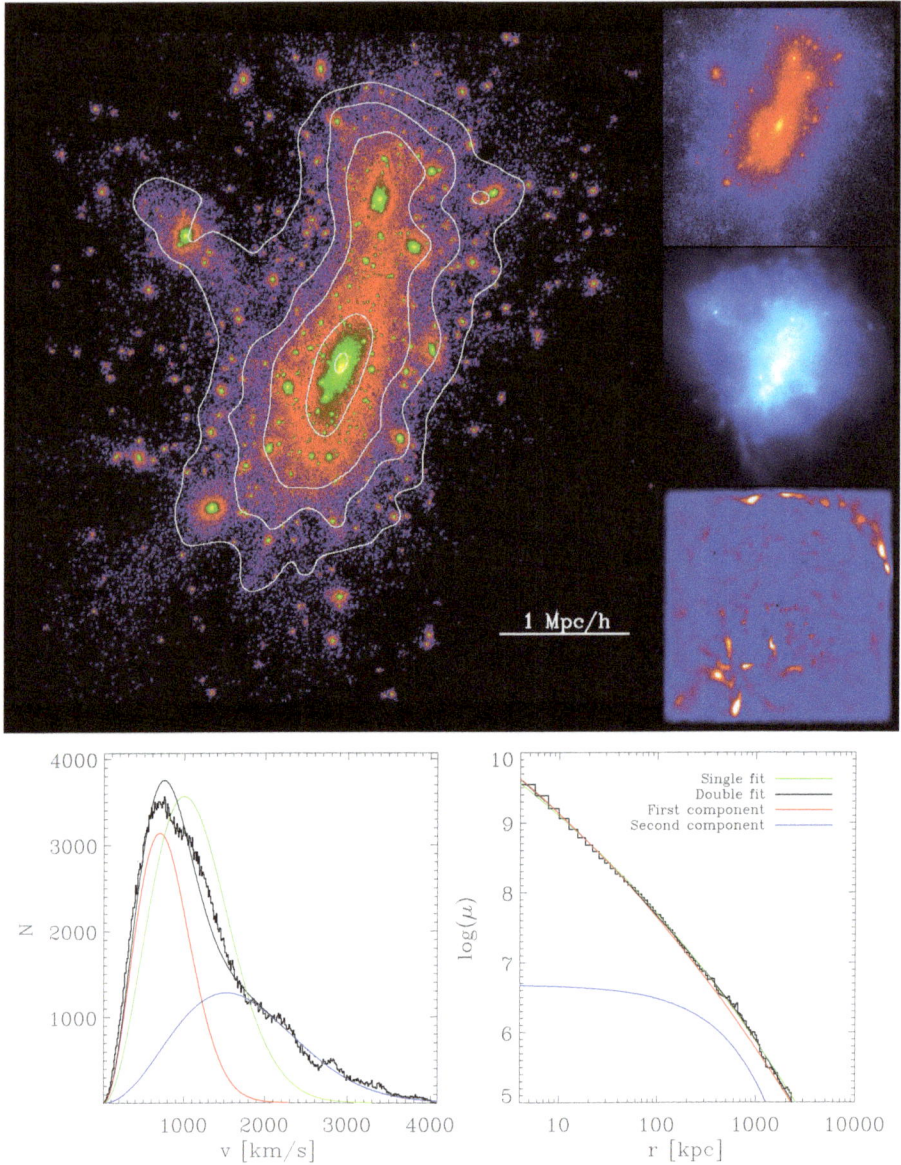

Figure 1. Example of a galaxy cluster where the velocity distribution is best described by a double Maxwellian fit, while the radial surface density profile can be well described by a single Sérsic profile (class d/s). **Upper left panel:** Stellar particle density map of the cluster, with the densest areas in yellow/green and the least dense areas in blue/black. White contours mark the iso-brightness lines of the DSC with the galaxies subtracted. **Upper right panels, from top to bottom:** Total matter density map; X-ray surface brightness map; unsharp-masked image of the pressure map to visualize shock fronts, as indicated by the large, arc-like feature in the upper right corner. **Lower left panel:** Velocity histogram for the stellar particles within the cluster, excluding those from substructures.

(Continued in Figure 2.)

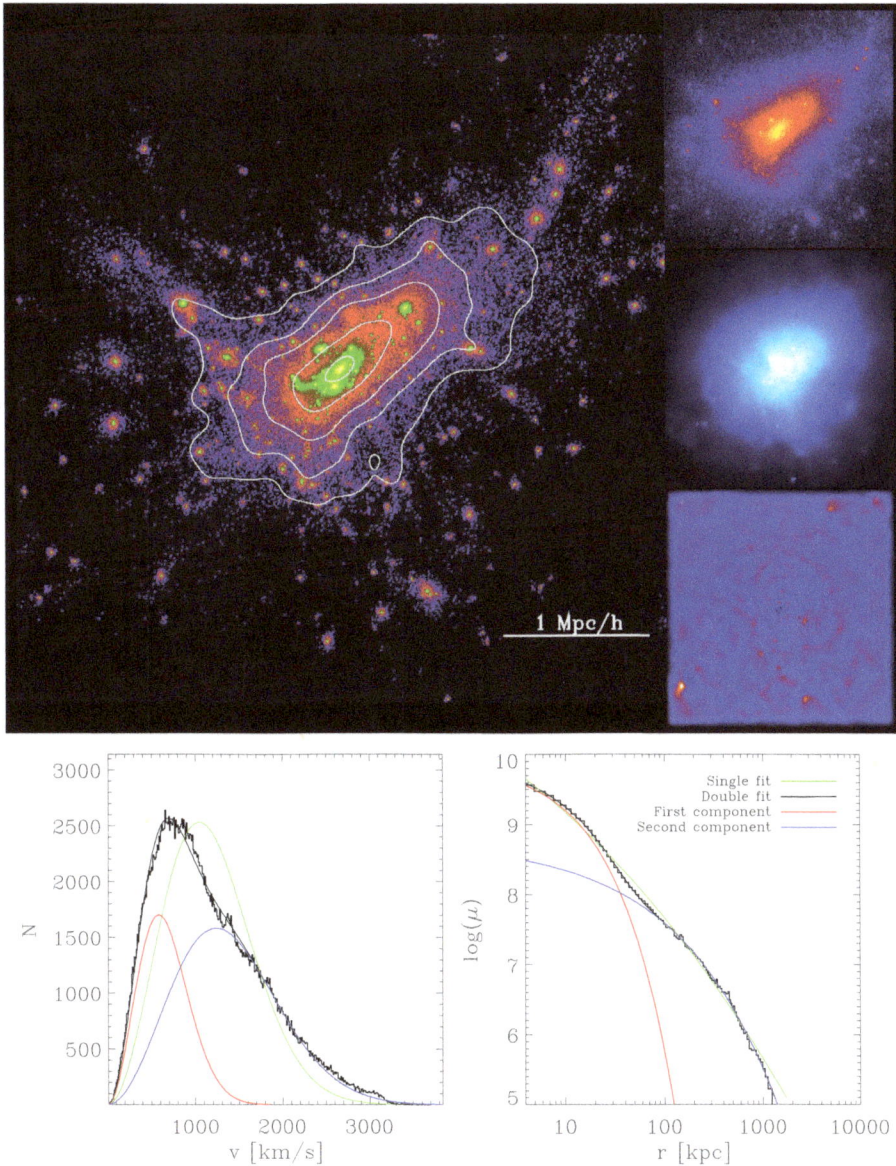

Figure 2. Same as Figure 1 but for a cluster with a double Maxwellian distribution in the velocity and a double Sérsic radial surface density profile as best representation (class d/d). Although the cluster is quite extended, there are no shocks visible, indicating a very late state of the merger.

(Continued from Figure 1.) The green line shows the best single-Maxwell fit to the histogram, while the black line shows the best double-Maxwell fit to the histogram with the red and blue lines indicating the individual Maxwellians of the BCG (first) component and the DSC (second) component. **Lower right panel:** Projected radial stellar surface density profile (with substructures already subtracted) of the cluster centered around the BCG—colours as in the left panel but for the Sérsic fits.

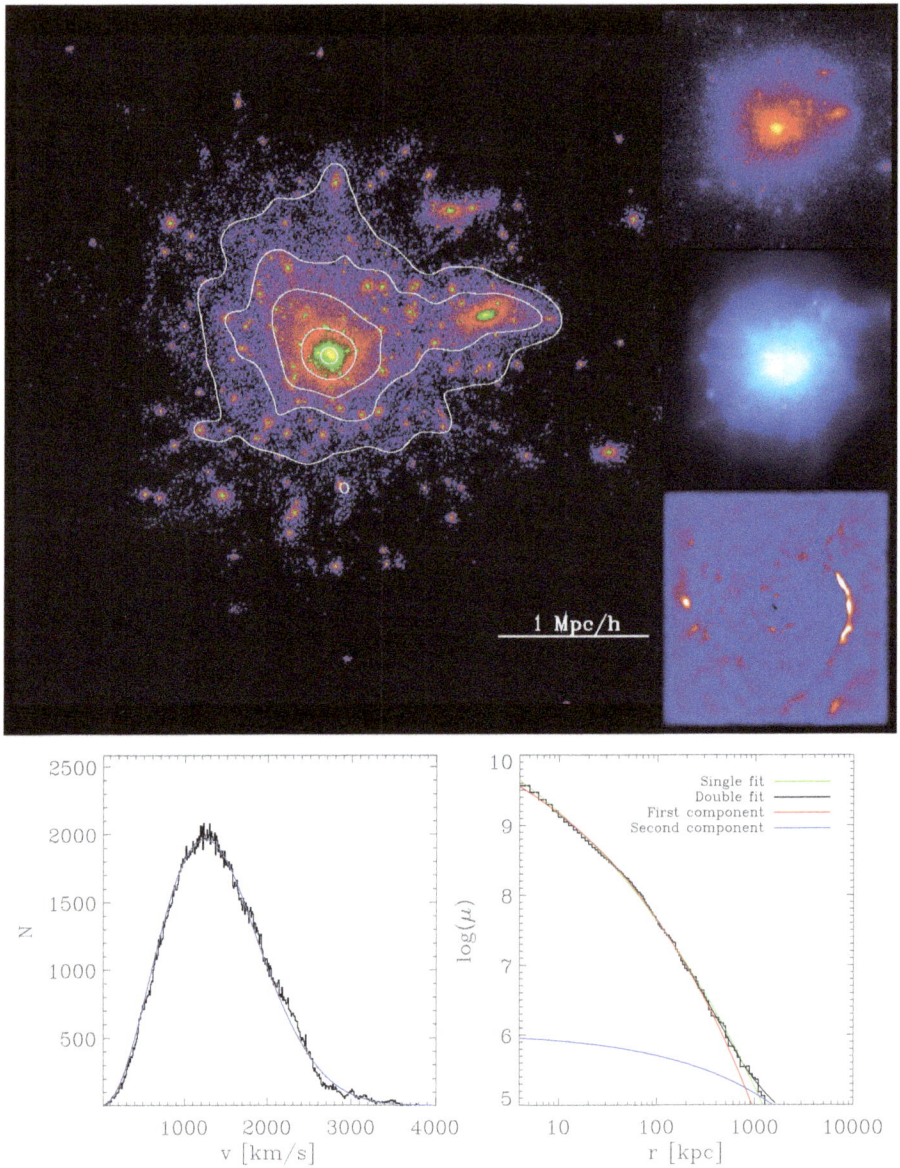

Figure 3. Same as Figure 1 but for a cluster where the velocity distribution is well described by a single Maxwellian and the radial surface density profile is also well described by a single Sérsic profile (class s/s). Although the outwards-moving substructure to the right has lost all its gas component, the injected shock within the ICM is still clearly visible as a large arc.

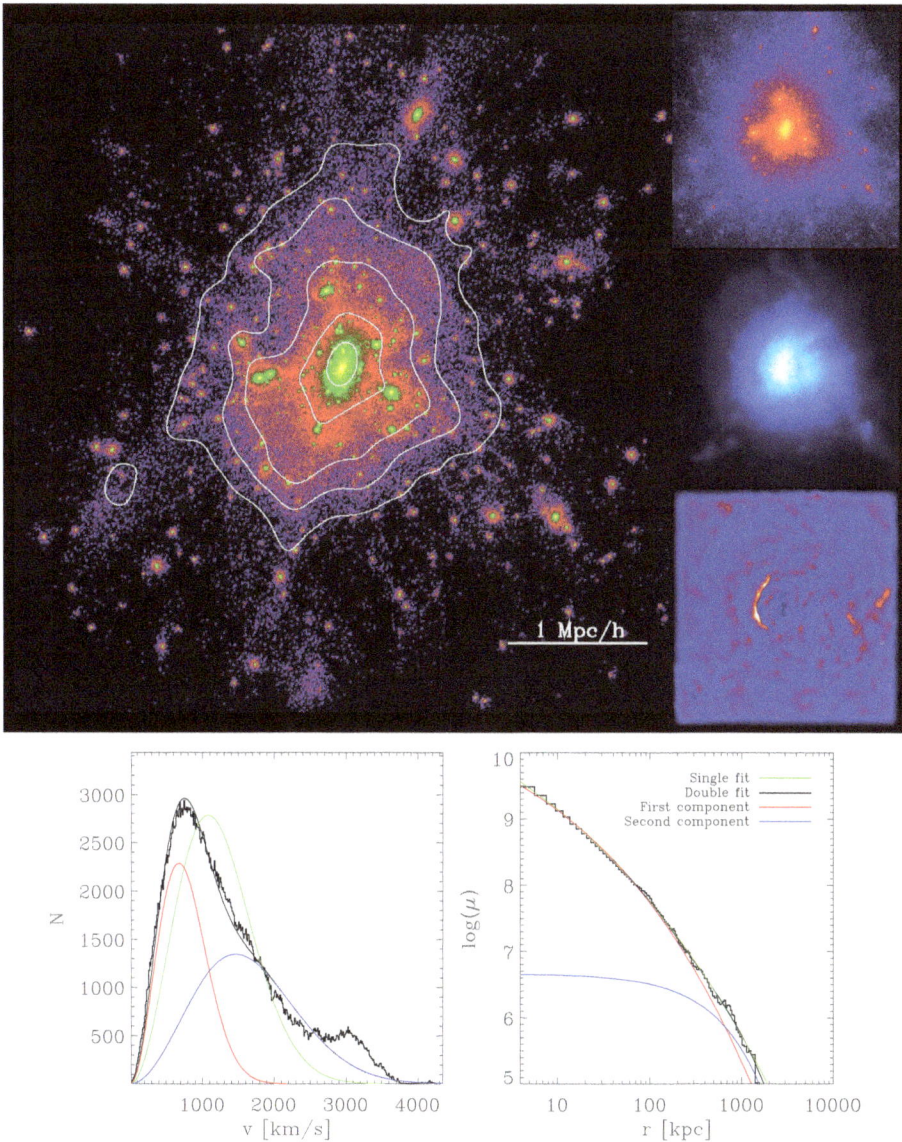

Figure 4. Same as Figure 1 but for a special case where a third component would be needed to describe the velocity distribution, while the radial surface density profile does not show signs of a third component. Since we do not explicitly study these special cases in this work, this cluster is classified as belonging belong to the d/d class, as both the velocity and the radial density clearly are multi-component systems.

The lower left panel of Figure 4 shows a very interesting albeit rare case for the velocity distribution of a galaxy cluster: for this cluster, a double-Maxwellian fit is still not sufficient and a third superposed component would be needed to actually capture all features visible in the velocity distribution. Interestingly, this new accretion is not strongly visible in the maps in the upper panels of this figure. The stellar component of the merger ongoing in the central part of the cluster—clearly visible through the displacement of the central X-ray emission from the center of mass towards the left as well as through the clearly visible shock moving from the center leftwards—is most likely the third, high velocity component in the velocity distribution. As these cases are rare and need a visual inspection of all 928 clusters to identify them, we will not introduce these objects as a separate class of velocity distributions in this work, and will here simply classify them as double-Maxwellian velocity distributions.

While in this three-dimensional analysis the two different velocity components of the BCG and the DSC are nicely visible, this information cannot be drawn directly from observations. However, observationally, the line-of-sight velocity can be measured and used as a substitute to distinguish the two components: In projection, the measured velocity is not represented by a Maxwellian distribution but by a Gaussian distribution, an example of which is shown in Figure 5, where we plot the line-of-sight velocity distribution of the cluster shown in Figure 1. As for the three-dimensional case, also in the projected case a superposition of two Gaussian fits is needed to represent the velocity distribution profile, clearly indicating the two-component structure of the BCG and the DSC. In case of an ideal spherically symmetric relaxed system, the projected Gaussian fits predict the same mass fractions and velocity dispersions for the BCG component and the DSC as the intrinsic Maxwellian distribution fits, but projection effects, asymmetries as well as distortion effects through accretion events can lead to (slightly) different values. However, we do not investigate these issues further in this work.

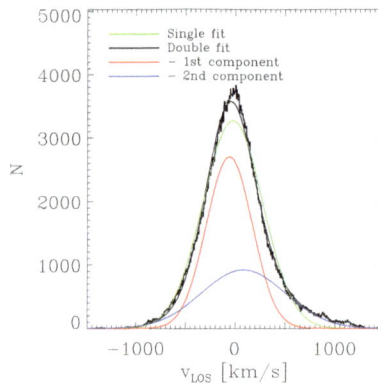

Figure 5. Line-of-sight velocity distributions of the stellar particles (excluding those bound to substructures) of the cluster shown in Figure 1, from a random projection. The green line is the best single Gaussian fit to the distribution, while the black line is the best double-Gaussian fit with its components shown in red and blue.

For the radial surface density profiles, we use a similar approach as for the velocity distributions. We fit the superposition of two Sérsic profiles

$$\mu(r) = \mu_1 \exp\left(-\left(\frac{r}{r_1}\right)^{1/n_1}\right) + \mu_2 \exp\left(-\left(\frac{r}{r_2}\right)^{1/n_2}\right) \tag{2}$$

to the radial surface density profiles of each cluster, and, for comparison, we also fit a single Sérsic profile as well.

Examples for the resulting Sérsic fits to the surface density distributions are shown in the lower right panels of Figures 1–4. As for the velocity distributions, we also see that there is no clear correlation between the visual appearance of the clusters, neither in the stellar nor the X-ray or the shock appearance, and the necessity of a double-Sérsic fit. Interestingly, we can already see from these four examples that there is also no clear correlation between the presence of a second component in the velocity distribution and the presence of a second component in the radial density profile: as shown in Figure 1, there are clusters that display two components in the velocity distribution but only a single component in the radial surface density profile, while Figure 2 shows a cluster where both distributions have a double structure. In the following, we will study this behaviour in more detail.

Statistical Properties

As we cannot check the properties of all 928 galaxy clusters from our sample individually, we now try to quantify their behaviour in a more statistical way. The biggest issue here is that a double fit with twice as many free parameters as a single fit will always yield a better fit, or one at least as good, according to simple statistical tests like χ^2 or Komolgorov–Smirnov, if the number of degrees of freedom is much larger than the number of fit parameters (as is the case here). Thus, we need to find a better way to decide which fit adequately characterizes the properties in velocity and surface brightness of a cluster.

One way to do this is to use the double-Maxwell and double-Sérsic fits of a cluster and integrate over each of the two components. This way, assuming that the two components always represent an inner, slower component that describes the BCG and an outer, faster component that describes the DSC, we can obtain the fraction of mass associated with each component, relative to the total stellar mass given by the full velocity distribution and the full radial surface density profile.

The left panel of Figure 6 shows a histogram of the mass fractions of the BCG, f_{BCG}, obtained with both methods for all 928 halos. The blue line shows the distribution found from the double-Maxwell fits, while the red line shows the fractions obtained from the double-Sérsic fits. The right panel of the same figure shows a scatter plot of the BCG mass fractions obtained with both methods. As can clearly be seen, there is no correlation at all between the mass fractions resulting from the two methods: while there is only a small amount of clusters that have BCG mass fractions below 10% and none with BCG mass fractions above 90% according to the double-Maxwell method, the double-Sérsic method results in about half of the clusters having BCG mass fractions of about 0 or 100%, clearly indicating that in those cases a double-Sérsic fit is not necessary and their radial surface density profiles can be well described by a single Sérsic profile.

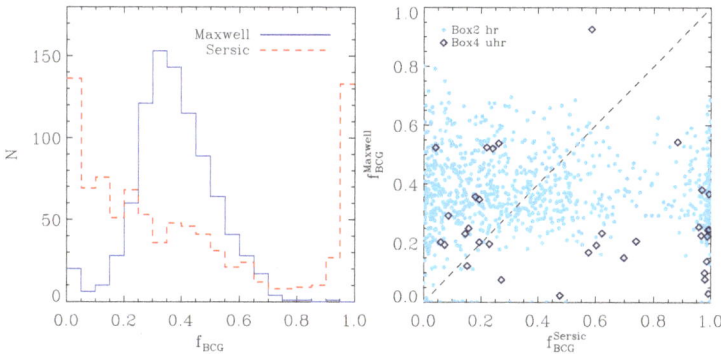

Figure 6. Left panel: Histogram of the mass fraction ascribed to the BCG according to the double-Maxwell fits (blue) and the double-Sérsic fits (red). The mass ascribed to the BCG is always the mass of the slower (Maxwellian fits) or the innermost (Sérsic fits) component. **Right panel:** BCG mass fractions obtained from the Maxwell fits versus those obtained with the Sérsic fits. Light blue symbols show the clusters from Box2b, while dark blue open diamonds mark the galaxy groups selected from Box4. There is no evident correlation between the mass partitioning obtained with the two different methods, and there is also no difference between galaxy groups and clusters.

The latter is in good agreement with observations of radial surface brightness profiles for massive elliptical galaxies, where both single- and double-Sérsic-profiles are observed without a clear correlation to the global dynamical state of the cluster. From the double-Maxwell method, we find BCG mass fractions generally ranging between $20\% < f_{BCG} < 70\%$, but the large mass fractions are rare and most of the BCGs have mass fractions between 30 and 40%, which is in agreement with observational fractions obtained for the BCGs in very massive clusters (e.g., Presotto et al., 2014 [14], Burke et al., 2015 [15]).

We use the BCG mass fractions obtained with both methods to decide whether a double-component fit is needed for the velocity distributions and the surface density profiles or if a single-component fit is sufficient: if the BCG mass fraction obtained through a double-Maxwell fit is below $f_{BCG} = 10\%$ or above $f_{BCG} = 90\%$, we judge that there is no clear signal of a second component in this fit and we thus classify these clusters as single-Maxwell clusters. If the BCG mass fraction is between these values, i.e., $10\% < f_{BCG} < 90\%$, we gauge the double-Maxwell-fit to be necessary and thus classify the cluster as double-Maxwell cluster. As shown in the upper part of Table 2, the fraction of single-Maxwell clusters is below 6%, and nearly all clusters show velocity distributions that reflect two-component systems. Therefore, we conclude that the typical galaxy cluster shows a two-component behaviour in its velocity distribution, in agreement with recent observations—for example, by Longobardi et al., 2015 [4] and Bender et al., 2015 [3].

Similarly, we classify a galaxy cluster as a single-Sérsic cluster if the BCG mass fraction obtained from the double-Sérsic fit is below $f_{BCG} = 10\%$ or above $f_{BCG} = 90\%$, while we classify a cluster as double-Sérsic cluster if the BCG mass fraction is between $10\% < f_{BCG} < 90\%$. Here, we clearly see the same split-up that we already saw from Figure 6, i.e., that about half of the clusters are single-Sérsic clusters while the other half are double-Sérsic clusters, with a slight trend towards the latter (see upper part of Table 2).

Table 2. Relative fractions of the 928 clusters and groups with regard to their Maxwell- and Sérsic-fit properties.

	$N_{cluster}$	$f_{cluster}$ (%)
Single Maxwell sufficient	53	5.7
Double Maxwell needed	875	94.3
Single Sérsic sufficient	386	41.6
Double Sérsic needed	542	58.4
Single Maxwell, Single Sérsic (s/s)	29	3.1
Double Maxwell, Single Sérsic (d/s)	357	38.5
Single Maxwell, Double Sérsic (s/d)	24	2.6
Double Maxwell, Double Sérsic (d/d)	518	55.8

Using both classifications, we can now test how many clusters show a double-fit-behaviour in both the velocity distribution and the surface density profiles. We find that this is the case for more than half of the galaxy clusters in our sample, as shown in the lower part of Table 2 and Figure 7, while about 40% of the clusters are double-Maxwell but single-Sérsic clusters. The single-Maxwell clusters represent less than 6% of all our clusters; they are roughly evenly distributed between single-Sérsic and double-Sérsic cases.

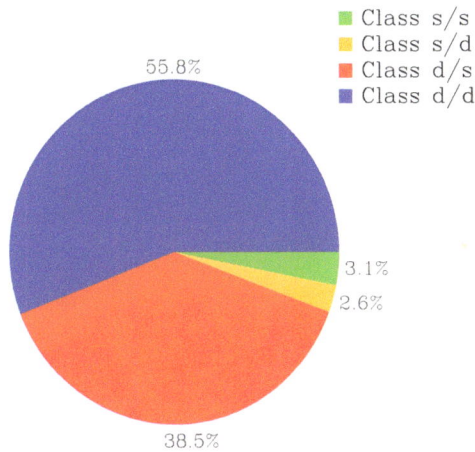

Figure 7. Fraction of galaxy clusters and groups that can be described best by a single Maxwell distribution and a single Sérsic profile (class s/s; green), a double Maxwell distribution and a single Sérsic profile (class d/s; red), a single Maxwell distribution and a double Sérsic profile (class s/d; yellow), and a double Maxwell distribution and a double Sérsic profile (class d/d; blue).

From these results, we conclude that the velocity distribution of a cluster can still distinguish between the component that belongs to the direct potential of the BCG and the outer component that was accreted onto the cluster and stored in the outer regions of the BCG through stripping and flyby events, building up the DSC component that still retains this memory of the assembly history. On the other hand, the imprint of this assembly history is not always visible in the radial surface density profiles of the cluster BCGs as a separate component, where only in some cases the BCG can be separated from the DSC through the surface density profiles, while, in other cases, this is not possible. Whether the assembly history can be traced from the shape of the outer stellar halo radial density profiles of BCGs and galaxies in general will be part of a forthcoming study (see Remus et al., 2016 [7] for a preview on these results).

4. Mass–Velocity-Dispersion Relation

Finally, we want to see if a correlation exists between the velocity dispersion obtained from the Maxwell fits for the BCG and the DSC and the virial mass of the host cluster, as presented by Dolag et al., 2010 [2]. For this purpose, the left panel of Figure 8 shows the velocity dispersion of the BCG component versus the virial mass of the cluster in red, and the velocity dispersion of the DSC versus the virial mass of the cluster in blue. As can clearly be seen, we find a strong correlation for both components with the virial mass of the cluster, and these correlations hold even for the galaxy groups in the lower mass regime, indicating that the split-up between the brightest group galaxies (BGGs) and the Intra-group light (IGL) behaves similarly to that of clusters, clearly hinting at a similar growth mechanism for the IGL through stripping.

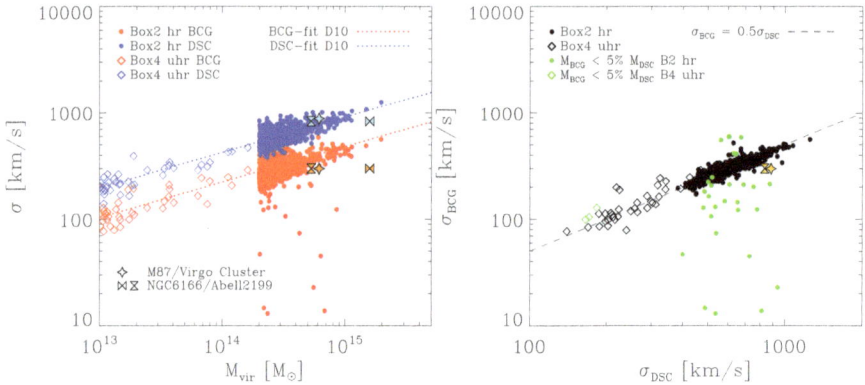

Figure 8. Left panel: Velocity dispersion σ obtained from the double-Maxwellian fit versus virial mass M_{vir} for the BCG-component (red) and the DSC-component (blue) for all galaxy clusters in Box2b (filled circles) and all galaxy groups in Box4 (open diamonds). The red and blue lines are not fitted to the data presented here but are those from Dolag et al., 2010 [2]. The orange and light blue symbols show velocity dispersions from observations of NGC 6166 in the cluster Abell 2199 (bowties/hourglasses) from Bender et al., 2015 [3] and of M 87 in the Virgo cluster (stars) from Longobardi et al., 2015 [4]. The virial mass for Virgo is taken from Tully, 2015 [16] using the average of the masses based on the total K-band luminosity and the virial mass inferred from the zero velocity surface. For Virgo, this is in agreement with measurements by PLANCK [17]; for Abell 2199, we included measurements of the virial mass from the PLANCK mission [17] (hourglasses) as they give significantly smaller values than the one inferred by Tully, 2015 [16] (bowties). **Right panel:** BCG velocity dispersion σ_{BCG} versus DSC velocity dispersion σ_{DSC} obtained from the double-Maxwellian fits for the same clusters (filled circles) and groups (open diamonds) as in the left panel. The grey dashed line shows the $\sigma_{BCG} = 0.5\,\sigma_{DSC}$ relation. Green symbols mark all clusters and groups for which the BCG-component has less than 5% of the stellar mass of the total stellar mass of the system, according to the mass partitioning obtained from the double-Maxwellian fit. The yellow symbols show the same observations as in the left panel.

The left panel of Figure 8 also shows the fits to the velocity-dispersion–virial-mass relation presented by Dolag et al., 2010 [2] as red and blue dotted lines for the BCGs and the DSC, respectively. Although not fitted to the current simulation set but obtained from a less advanced simulation set of the local universe, these relations perfectly describe the behaviour found for the Magneticum simulation sample of galaxy clusters and galaxy groups, even at the low mass end. This additionally proves that this behaviour is independent of the details of the subgrid models included in the simulations.

As can also be seen from Figure 8, the relation between the velocity dispersion and the virial mass for the BCGs and the DSC has the same slope, with the DSC simply having overall larger velocity dispersions than the BCGs. More precisely, as shown in the right panel of Figure 8, the relation between the velocity dispersions of the BCG and the DSC is very tight and can be described as

$$\sigma_{BCG} = 0.5\,\sigma_{DSC}. \tag{3}$$

Again, this behaviour holds even at the galaxy group mass scale, as indicated by the open diamonds marking the groups selected from the smaller volume Box4 with the higher resolution. In addition, this not only demonstrates that galaxy groups and clusters show a similar behaviour, but it also proves that the correlations presented here are independent of the resolution of the simulation and thus only driven by physical processes like accretion and star formation.

Interestingly, we can also explain the few outliers that can be seen in both the velocity-dispersion–virial-mass relation and the velocity-dispersion relation between the BCGs and their DSC: if we mark all clusters (and groups) where the BCG mass fraction obtained from the double-Maxwellian fit is below 5%

(green circles and diamonds in the right panel of Figure 8), all outliers are captured. This clearly indicates that for all galaxy clusters where a double-Maxwellian fit is the better representation of the velocity distribution, the discussed relation between both components and the virial mass of the cluster is present and very tight, and driven by the assembly history of the clusters.

In addition, we also included the observations for NGC 6166 in the cluster Abell 2199 from Bender et al., 2015 [3] and M 87 in the Virgo cluster from Longobardi et al., 2015 [4] in both panels of Figure 8 (for details on the virial mass estimated for these clusters, see the figure caption). Both observations are in excellent agreement with the correlations found in this study, especially with respect to the BCG–DSC velocity dispersion correlation.

5. Discussion and Conclusion

In this work, we presented a detailed and statistically sound analysis of the stellar velocity distributions and the projected stellar radial surface density profiles of galaxy clusters and galaxy groups selected from the Magneticum pathfinder simulation sample. Using two volumes of different sizes and resolutions, we showed that for more than 90% of all 928 clusters and groups in our sample the velocity distributions are represented best by a superposition of two Maxwellian distributions, with the slower component representing the BCG of the cluster and the faster component representing the DSC. We demonstrated that the relative mass fractions of the BCGs found through these fits is in agreement with recent observations. This behaviour in the velocity distribution strongly supports the idea that the DSC is built up from stripping of smaller satellites within the cluster potential close to its center, where the BCG resides.

Furthermore, we found that there is a clear and tight correlation between the velocity dispersions of the two components obtained by these fits and the virial mass of the host clusters, and that this correlation holds down to group-mass scales. We also demonstrated that the velocity dispersions of both components are correlated tightly, with the BCG having about half the velocity dispersion of the DSC, and that the few available observations that distinguish between both components are in excellent agreement with our results.

Additionally, we tested if the same separation into two distinct components is reflected in the projected radial surface density profiles of the cluster. Interestingly, we could only find a separation of the radial profiles into two components in about half of the clusters that exhibit a double-Maxwell imprint in the velocity distribution, clearly showing that the radial profile is not always suitable for distinguishing the two components and that further indicators are needed in the radial stellar profiles to obtain information about the assembly history of galaxy clusters. This issue will be addressed in a forthcoming study.

Lastly, we also tested if we can find a correlation between the velocity distribution behaviour of the DSC and the X-ray or shock properties in the four objects we have examined in detail. No such correlations were evident. On the contrary, we find indications that the X-ray and shock properties describe the very recent assembly history of the cluster, where the presence of shocks and X-ray offsets indicate an ongoing merger event, while the velocity distribution of the stellar component of the galaxy clusters and the BCGs appears to be an indicator for the earlier assembly history of the cluster.

Acknowledgments: We thank the anonymous referees for their helpful comments. The Magneticum Pathfinder simulations were partly performed at the Leibniz-Rechenzentrum with CPU time assigned to the Project "pr86re". This work was supported by the DFG Cluster of Excellence "Origin and Structure of the Universe". We are especially grateful for the support by M. Petkova through the Computational Center for Particle and Astrophysics (C2PAP).

Author Contributions: K.D. performed the simulations; R.-S.R. and T.L.H. analysed the data; R.-S.R. wrote the paper.

Conflicts of Interest: The authors declare no conflict of interest.

References

1. Murante, G.; Giovalli, M.; Gerhard, O.; Arnaboldi, M.; Borgani, S.; Dolag, K. The importance of mergers for the origin of intracluster stars in cosmological simulations of galaxy clusters. *Mon. Not. R. Astron. Soc.* **2007**, *377*, 2–16.
2. Dolag, K.; Murante, G.; Borgani, S. Dynamical difference between the cD galaxy and the diffuse, stellar component in simulated galaxy clusters. *Mon. Not. R. Astron. Soc.* **2010**, *405*, 1544–1559.
3. Bender, R.; Kormendy, J.; Cornell, M.E.; Fisher, D.B. Structure and Formation of cD Galaxies: NGC 6166 in ABELL 2199. *Astrophys. J.* **2015**, *807*, 56.
4. Longobardi, A.; Arnaboldi, M.; Gerhard, O.; Hanuschik, R. The outer regions of the giant Virgo galaxy M 87: Kinematic separation of stellar halo and intracluster light. *Astron. Astrophys.* **2015**, *579*, A135.
5. Mihos, J.C. Intragroup and Intracluster Light. In Proceedings of the IAU Symposium 317: The General Assembly of Galaxy Halos: Structure, Origin and Evolution, Honolulu, HI, USA, 3–7 August 2015; Volume 317, pp. 27–34.
6. Puchwein, E.; Springel, V.; Sijacki, D.; Dolag, K. Intracluster stars in simulations with active galactic nucleus feedback. *Mon. Not. R. Astron. Soc.* **2010**, *406*, 936–951.
7. Remus, R.S.; Burkert, A.; Dolag, K. A "Universal" Density Profile for the Outer Stellar Halos of Galaxies. *ArXiv* **2016**, arXiv:1605.06511.
8. Springel, V. The cosmological simulation code GADGET-2. *Mon. Not. R. Astron. Soc.* **2005**, *364*, 1105–1134.
9. Springel, V.; Hernquist, L. Cosmological smoothed particle hydrodynamics simulations: A hybrid multiphase model for star formation. *Mon. Not. R. Astron. Soc.* **2003**, *339*, 289–311.
10. Tornatore, L.; Borgani, S.; Dolag, K.; Matteucci, F. Chemical enrichment of galaxy clusters from hydrodynamical simulations. *Mon. Not. R. Astron. Soc.* **2007**, *382*, 1050–1072.
11. Hirschmann, M.; Dolag, K.; Saro, A.; Bachmann, L.; Borgani, S.; Burkert, A. Cosmological simulations of black hole growth: AGN luminosities and downsizing. *Mon. Not. R. Astron. Soc.* **2014**, *442*, 2304–2324.
12. Teklu, A.F.; Remus, R.S.; Dolag, K.; Beck, A.M.; Burkert, A.; Schmidt, A.S.; Schulze, F.; Steinborn, L.K. Connecting Angular Momentum and Galactic Dynamics: The Complex Interplay between Spin, Mass, and Morphology. *Astrophys. J.* **2015**, *812*, 29.
13. Dolag, K.; Borgani, S.; Murante, G.; Springel, V. Substructures in hydrodynamical cluster simulations. *Mon. Not. R. Astron. Soc.* **2009**, *399*, 497–514.
14. Presotto, V.; Girardi, M.; Nonino, M.; Mercurio, A.; Grillo, C.; Rosati, P.; Biviano, A.; Annunziatella, M.; Balestra, I.; Cui, W.; et al. Intracluster light properties in the CLASH-VLT cluster MACS J1206.2-0847. *Astron. Astrophys.* **2014**, *565*, A126.
15. Burke, C.; Hilton, M.; Collins, C. Coevolution of brightest cluster galaxies and intracluster light using CLASH. *Mon. Not. R. Astron. Soc.* **2015**, *449*, 2353–2367.
16. Tully, R.B. Galaxy Groups: A 2MASS Catalog. *Astron. J.* **2015**, *149*, 171.
17. Collaboration, P.; Ade, P.A.R.; Aghanim, N.; Arnaud, M.; Ashdown, M.; Aumont, J.; Baccigalupi, C.; Banday, A.J.; Barreiro, R.B.; Barrena, R.; et al. Planck 2015 results. XXVII. The second Planck catalogue of Sunyaev-Zeldovich sources. *Astron. Astrophys.* **2016**, *594*, A27.

galaxies

MDPI

Conference Report

The SLUGGS Survey: Understanding Lenticular Galaxy Formation via Extended Stellar Kinematics

Sabine Bellstedt

Centre for Astrophysics and Supercomputing, Swinburne University of Technology, Hawthorn 3122, Melbourne, Australia; sbellstedt@swin.edu.au

Academic Editors: Duncan A. Forbes and Ericson D. Lopez
Received: 4 May 2017; Accepted: 26 May 2017; Published: 30 May 2017

Abstract: We present the latest published and preliminary results from the SLUGGS Survey discussing the formation of lenticular galaxies through analysis of their kinematics. These include a comparison of the measured stellar spin of low-mass lenticular galaxies to the spin of remnant galaxies formed by binary merger simulations to assess whether a merger is a likely formation mechanism for these galaxies. We determine that while a portion of lenticular galaxies have properties consistent with these remnants, others are not, indicating that they are likely "faded spirals". We also present a modified version of the spin–ellipticity diagram, which utilises radial tracks to be able to identify galaxies with intermediate-scale discs. Such galaxies often have conflicting morphological classifications, depending on whether photometric or kinematic measurements are used. Finally, we present preliminary results on the total mass density profile slopes of lenticular galaxies to assess trends as lower stellar masses are probed.

Keywords: galaxies: formation; galaxies: evolution; galaxies: elliptical and lenticular

1. Introduction

Lenticular galaxies share similarities with both elliptical galaxies (in that they are quenched systems) and spiral galaxies (in that many lenticular galaxies have large-scale discs). The manner in which these systems form is still not entirely understood, and it is likely that multiple processes contribute to their formation. Simulations have shown that it is possible to form a lenticular galaxy through binary mergers of disc galaxies. In addition, environmental processes such as ram pressure stripping, galaxy harassment, and starvation have been proposed as mechanisms by which to transform spiral galaxies into lenticulars as they fall into galaxy clusters.

We analyse the kinematics of lenticular galaxies studied within the SLUGGS survey that extend to $\sim$2–3 effective radii (R_e) with the aim of identifying whether merger or environmental processes are more dominant in shaping lenticular (S0) galaxies. We focus in particular on low-mass (i.e., $M_* < 10^{11}\,M_\odot$) lenticular galaxies.

2. Results

2.1. Comparison of Observations with Binary Merger Simulations

For each of the galaxies in the SLUGGS survey (both elliptical and lenticular), we measured the stellar spin λ_R within $1\,R_e$ within [1], given by

$$\lambda_R = \frac{\sum_{i=1}^{n} F_i R_i V_i}{\sum_{i=1}^{n} F_i R_i \sqrt{V_i^2 + \sigma_i^2}},$$ (1)

where i represents each pixel within the selected aperture. This aperture was selected because leading binary merger simulations make such measurements for simulated galaxies. We then compared these values, as plotted within Figure 1. The left panel compares our results with the simulations of [2], while the right panel compares our results with [3]. The merger progenitors in the simulations of the left panel are shown as the grey region, and the coloured regions in the lower region of the plot indicate the merger remnants. While these remnants coincide well with the SLUGGS elliptical galaxies, they do not coincide with the region occupied by SLUGGS lenticular galaxies. Instead, it seems that the progenitor galaxies themselves have more in common with these lenticular galaxies. In the right-hand panel, the progenitor galaxies are shown as cyan stars. These progenitors have a greater stellar spin than those of the [2] progenitor galaxies, and as a result, the remnant galaxies also have a slightly increased stellar spin. The result of this increase is that a portion of the observed lenticular galaxies are consistent with the merger remnants.

However, around half of the S0 galaxies still have a stellar spin greater than those of the merger remnants. We interpret this to indicate that these galaxies are unlikely to have been formed by a merger of two galaxies, and rather due to environmental processes that transformed a spiral into a lenticular. The CALIFA sample of spiral galaxies [3] has been included in Figure 1, which highlights that spiral galaxies themselves occupy a large part of this parameter space. It is notable that the SLUGGS S0 sample and the CALIFA spiral galaxies have a very similar distribution—another indicator that S0 galaxies may have formed through a "fading" of spiral galaxies.

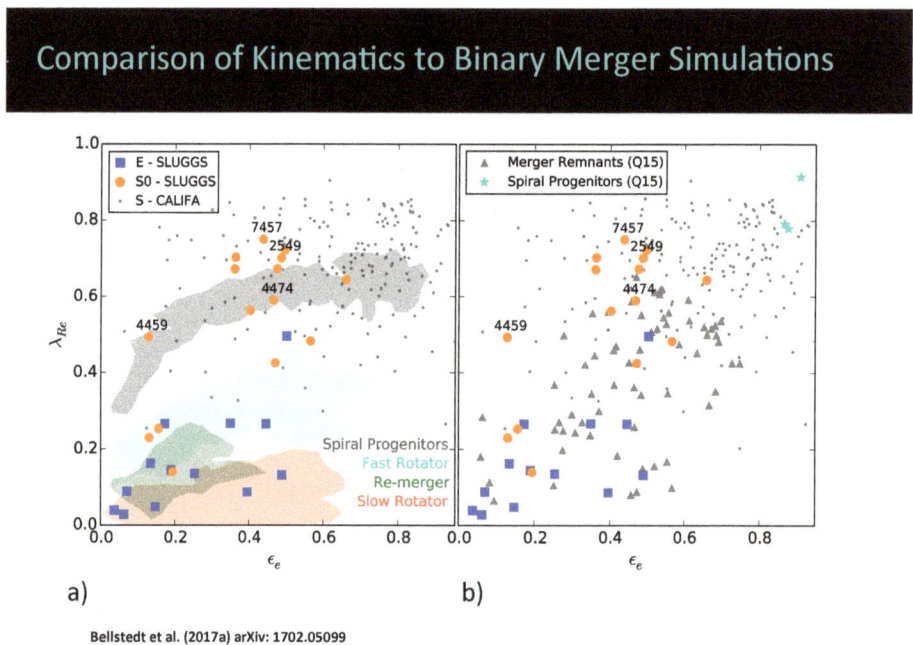

Bellstedt et al. (2017a) arXiv: 1702.05099

Figure 1. Stellar spin against ellipticity for SLUGGS galaxies, compared with results from two different binary merger simulations. (**a**) Comparing with the simulations of [2]. Grey shaded region shows the distribution of merger progenitors, while the other shaded regions indicate the different types of remnants produced by the simulations. (**b**) Comparison with the simulations of [3]. Cyan stars depict the merger progenitors, while grey triangles represent merger remnants.

2.2. Annular Measurements of Stellar Spin

In addition to measuring the total stellar spin of galaxies, it is useful to measure the change in stellar spin with radius, as has been done in numerous studies. This is often done simply by measuring the total stellar spin within a specified aperture, using Equation (1). Plotting these measurements against radius gives an aperture-based stellar spin profile. However, since each measurement is flux-weighted, the stellar spin measured in the outer regions of the galaxy will be more highly weighted towards the spin in the central regions, making it difficult to distinguish changes in spin with greater radii. As described in [1], we rather make these measurements over isophotal annuli.

Using this technique, a number of galaxies in the SLUGGS survey were noted [1] to have strongly downturning stellar spin profiles at larger radii. For such galaxies, the traditional $\lambda_R - \epsilon$ diagram only conveys the behaviour in the central regions. Hence, we have presented a modified version of this diagram in the form of a spin–ellipticity track [4]. Figure 2 shows what such tracks look like for both a typical lenticular galaxy (NGC 1023) and a typical elliptical galaxy (NGC 4365). While ellipticals generally tend to have a low spin and ellipticity at all radii, resulting in stagnant tracks, lenticular galaxies have large-scale discs that result in both spin and ellipticity increasing with radii, resulting in tracks that tend to go from the bottom-left to the top-right regions of the parameter space. For galaxies with downturning spin profiles, these tracks look rather different, as shown in Figure 3.

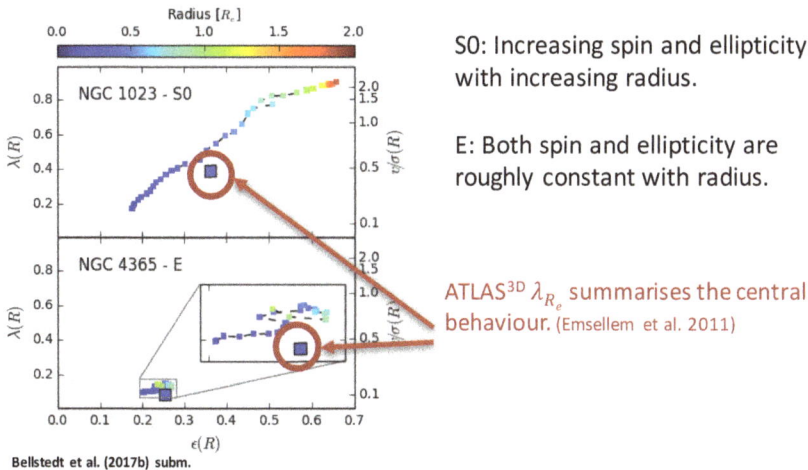

Figure 2. Example spin–ellipticity tracks for "typical" lenticular and elliptical galaxies. In addition to the SLUGGS tracks, we include the central aperture-based value from the ATLAS^{3D} Survey [5].

The four galaxies from the SLUGGS survey identified as having downturning stellar spin profiles are NGC 821, NGC 3377, NGC 4278, and NGC 4473 [1]. What the spin–ellipticity tracks have in common for each of these galaxies is that they rise in the central regions, and then turn back on themselves at larger radii. For NGC 821, NGC 3377, and NGC 4278, the tracks turn back on themselves in the anticlockwise direction; however, due to the double-sigma nature of NGC 4473, the downturn in ellipticity occurs at a larger radius than the downturn in spin, resulting in a clockwise track. Figure 3 also features a velocity map for galaxy NGC 3377, which highlights that the highly rotating nature of the central region of this galaxy does not extend into its outskirts.

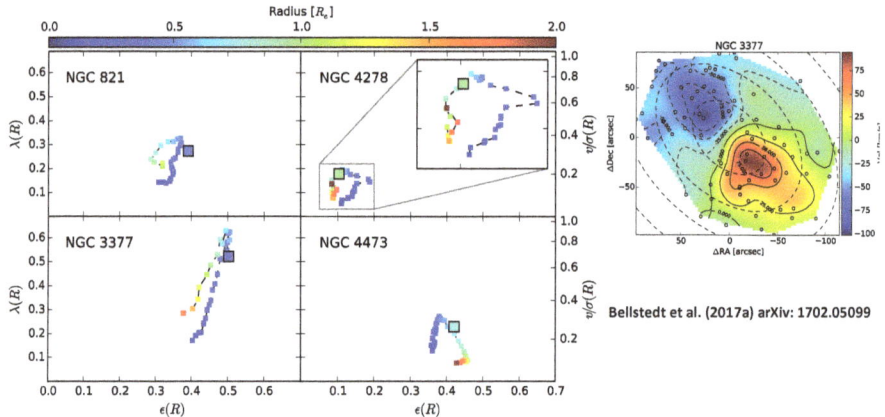

Figure 3. Spin–ellipticity diagrams for the four SLUGGS galaxies with downturning $\lambda(R)$ profiles at larger radii. Note how the inner behaviour is akin to that of S0 galaxies, whereas the tracks at higher radii turn back to the region generally occupied by E galaxies. The right-hand figure is the velocity map for NGC 3377, which shows how the rotating nature of the galaxy does not extend to its outer regions.

2.3. Preliminary Results Measuring Mass Density Profile Slopes of Low-Mass S0 Galaxies

A property of galaxies thought to provide information on their formation histories is the total mass density profile slope, referred to as γ_{tot}. Since this is a value that depends on the total mass and not only the stellar component of the galaxy, dynamical modelling is essential. We apply JAM (Jeans Anisotropic MGE[1]) modelling [6], which utilises the Jeans equations in an MCMC (Markov Chain Monte Carlo) fashion. The input to the JAM code is the observed $V_{\rm rms}$ map of each galaxy. The model matches to these $V_{\rm rms}$ maps are shown in Figure 4.

Preliminary values of γ_{tot} for these galaxies are plotted in blue against M_* in Figure 5. In addition, the values for the higher-mass sample of the SLUGGS survey are plotted in red [7], and simulated results by [8] are plotted in open circles. The dashed line includes early theoretical predictions by [9], which are in disagreement with the observational results. We attribute this to the lack of AGN feedback in these simulations. It can be seen that there is a clear agreement between the observational measurements and the simulated data of [8]—an indication of the improvement in predictions of newer theoretical work. While more galaxies are required in order to make conclusions about any trends in γ_{tot} below $10^{11} M_\odot$, it seems as though γ_{tot} values tend to increase (become more shallow) at lower stellar masses.

A shallower slope is often attributed to a history of accretion, where minor mergers added mass to the outskirts of galaxies over time to make the total mass density slope less steep. This is not a likely

---

[1]   Multi Gaussian Expansion.

explanation for why lower-M_* galaxies have shallower slopes, as it is generally the most massive galaxies in the universe that have experienced the most accretion.

JAM Models for Low-Mass S0 Galaxies !

JAM (Jeans Anisotropic MGE) modelling (Cappellari 2008). Combination of ATLAS3D in the inner regions, and SLUGGS data in the outer regions.

A power law has been assumed as the structure of the gravitational potential of the galaxy.

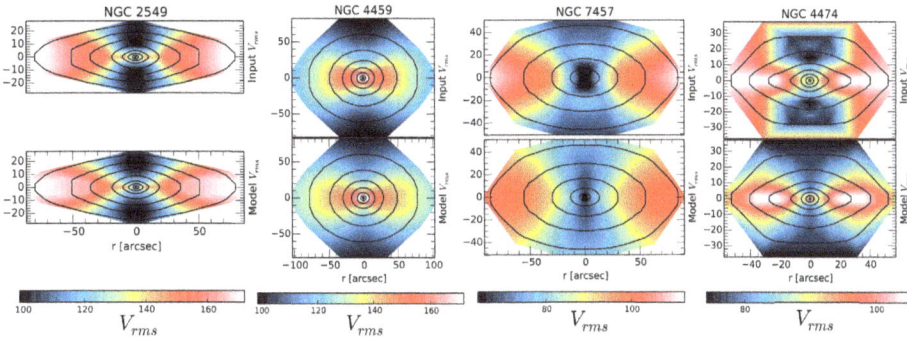

Figure 4. Preliminary JAM V_{rms} maps of four lenticular galaxies of the SLUGGS survey.

Comparison to Simulations !

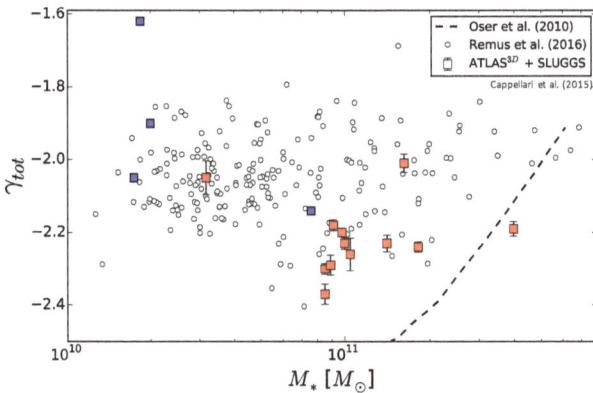

Figure 5. Preliminary plot of γ_{tot} against stellar mass M_* for SLUGGS galaxies. Blue squares represent new measurements. Red squares indicate previously published JAM results from SLUGGS [7], and simulated results [8] are plotted in addition to an older prediction [9].

Galaxies **2017**, *5*, 26

3. Summary

Comparison of extended stellar kinematics of low-mass lenticular galaxies indicates that while a portion of them are consistent with being formed by the merger of two disc galaxies, this is not the case for the full sample. This comparison highlights that—for low-mass S0s in particular—these galaxies are likely "faded" spirals that have transformed from spirals to S0s through environmental processes.

We have additionally identified a modified version of the spin–ellipticity diagram which presents radial tracks of individual galaxies, rather than single aperture values, to describe the behaviour of the galaxy. This representation of the kinematic and photometric data makes it very straightforward to identify galaxies with intermediate-scale discs at a glance.

Finally, we presented preliminary results using JAM modelling to measure the total mass density profile slope γ_{tot} of a selection of low-mass lenticular galaxies. These preliminary results seem to hint that at lower stellar masses, the γ_{tot} values become larger, indicating a shallower density slope. While a shallower slope is often explained by the presence of accretion events in the galaxy's past, this is unlikely to be the cause for increasingly shallow density slopes at lower stellar masses.

Acknowledgments: S.B. gratefully thanks the entire SLUGGS team for their contributions and assistance provided for the work presented in these proceedings. S.B. would also like to acknowledge the Astronomical Society of Australia for providing funding towards attendence at the conference *On the Origin and Evolution of Baryonic Galaxy Halos*.

Conflicts of Interest: The authors declare no conflict of interest.

References

1. Bellstedt, S.; Forbes, D.A.; Foster, C.; Romanowsky, A.J.; Brodie, J.P.; Pastorello, N.; Alabi, A.; Villaume, A. The SLUGGS survey: Using extended stellar kinematics to disentangle the formation histories of low-mass S0 galaxies. *Mon. Not. Roy. Astron. Soc.* **2017**, *467*, 4540–4557.

2. Bois, M.; Emsellem, E.; Bournaud, F.; Alatalo, K.; Blitz, L.; Bureau, M.; Cappellari, M.; Davies, R.L.; Davis, T.A.; de Zeeuw, P.T.; et al. The ATLAS^{3D} project—VI. Simulations of binary galaxy mergers and the link with fast rotators, slow rotators and kinematically distinct cores. *Mon. Not. Roy. Astron. Soc.* **2011**, *416*, 1654–1679.

3. Querejeta, M.; Eliche-Moral, M.C.; Tapia, T.; Borlaff, A.; Rodríguez-Pérez, C.; Zamorano, J.; Gallego, J. Formation of S0 galaxies through mergers. Bulge-disc structural coupling resulting from major mergers. *Astron. Astrophys.* **2015**, *573*, A78.

4. Bellstedt, S.; Graham. A.W.; Forbes, D.A.; Romanowsky, A.J.; Brodie, J.P.; Strader, J. The SLUGGS Survey: Trails of SLUGGS galaxies in a modified spin-ellipticity diagram. *Mon. Not. Roy. Astron. Soc.* **2017**, accepted.

5. Emsellem, E.; Cappellari, M.; Krajnovic, D.; Alatalo, K.; Blitz, L.; Bois, M.; Bournaud, F.; Bureau, M.; Davies, R.L.; Davis, T.A.; et al. The ATLAS^{3D} project—III. A census of the stellar angular momentum within the effective radius of early-type galaxies: Unveiling the distribution of fast and slow rotators. *Mon. Not. Roy. Astron. Soc.* **2011**, *414*, 888.

6. Cappellari, M. Measuring the inclination and mass-to-light ratio of axisymmetric galaxies via anisotropic Jeans models of stellar kinematics. *Mon. Not. Roy. Astron. Soc.* **2008**, *390*, 71–86.

7. Cappellari, M.; Romanowsky, A.J.; Brodie, J.P.; Forbes, D.A.; Strader, J.; Foster, C.; Kartha, S.S.; Pastorello, N.; Pota, V.; Spitler, L.R.; et al. Small Scatter and Nearly Isothermal Mass Profiles to Four Half-light Radii from Two-dimensional Stellar Dynamics of Early-type Galaxies. *Astrophys. J. Lett.* **2015**, *804*, L21.

8. Remus, R.-S.; Burkert, A.; Dolag, K.; Johansson, P.H.; Naab, T.; Oser, L.; Thomas, J. The Dark Halo–Spheroid Conspiracy and the Origin of Elliptical Galaxies. *Astrophys. J.* **2013**, *766*, 71.

9. Oser, L.; Ostriker, J.P.; Naab, T.; Johansson, P.H.; Burkert, A. The Two Phases of Galaxy Formation. *Astrophys. J.* **2010**, *725*, 2312–2323.

MDPI AG

St. Alban-Anlage 66

4052 Basel, Switzerland

Tel. +41 61 683 77 34

Fax +41 61 302 89 18

http://www.mdpi.com

Galaxies Editorial Office

E-mail: galaxies@mdpi.com

http://www.mdpi.com/journal/galaxies